Attractors Under Autonomous and Non-autonomous Perturbations

Mathematical Surveys and Monographs
Volume 246

Attractors Under Autonomous and Non-autonomous Perturbations

Matheus C. Bortolan
Alexandre N. Carvalho
José A. Langa

AMS
American Mathematical Society
Providence, Rhode Island

2010 *Mathematics Subject Classification.* Primary 34D45, 35B41; Secondary 37C70, 35B20, 37D15.

For additional information and updates on this book, visit
www.ams.org/bookpages/surv-246

Library of Congress Cataloging-in-Publication Data

Names: Bortolan, Matheus C. (Matheus Cheque), 1985- author. | Carvalho, Alexandre Nolasco de, author. | Langa, Joseé A., author. Title: Attractors under autonomous and nonautonomous perturbations / Matheus C. Bortolan, Alexandre N. Carvalho, Joseé A. Langa.

Description: Providence, Rhode Island : American Mathematical Society, [2020] — Series: Mathematical surveys and monographs, 0076-5376 ; volume 246 | Includes bibliographical references and index.

Identifiers: LCCN 2019059810 | ISBN 9781470453084 (hardback) | ISBN 9781470456863 (ebook)

Subjects: LCSH: Attractors (Mathematics) | Perturbation (Mathematics) | AMS: Ordinary differential equations – Stability theory [See also 37C75, 93Dxx] – Attractors [See also 37C70, 37D45]. | Partial differential equations – Qualitative properties of solutions – Attractors. | Dynamical systems and ergodic theory [See also 26A18, 28Dxx, 34Cxx, 34Dxx, 35Bxx, 46Lxx, 58Jxx, 70-XX] – Smooth dynamical systems: general theory [See also 34Cxx, 34Dxx] – Attractors and repellers, | Partial differential equations – Qualitative properties of solutions – Perturbations.

Classification: LCC QA614.813 B67 2020 | DDC 515/.395–dc23

LC record available at https://lccn.loc.gov/2019059810

∞ The paper used in this book is acid-free and falls within the guidelines
established to ensure permanence and durability.
Visit the AMS home page at `https://www.ams.org/`

10 9 8 7 6 5 4 3 2 1 25 24 23 22 21 20

To my loving wife Giovana and my family, for they are my rock.
M. C. Bortolan

To my beloved sons, Felipe and Pedro, for the so much they taught me.
A. N. Carvalho

To Rosario, for her lovely closeness, that sustains me.
J. A. Langa

Contents

Preface

In this book, the reader can find the theory of autonomous and non-autonomous perturbations for attractors of dynamical systems, from the coarsest notion of proximity - upper semicontinuity - to the finest notion of proximity - geometric structural stability. We assume that the reader only has some basic knowledge of metric and Banach spaces in what concerns the theoretical part of the book. For the applications the reader will be required to master the theory of semilinear evolution equations in general spaces and some partial differential equations.

The book is divided into two parts: Part I - Autonomous Theory and Part II - Non-autonomous Theory. Each part has five chapters and they are correlated between Part I and Part II, that is, Chapter 1 has its analogue for the non-autonomous case in Chapter 6, and so on. For the autonomous theory of Part I, Chapter 1 contains the basic definitions and results regarding semigroups and their global attractors. Chapter 2 deals with upper and lower semicontinuity. Chapter 3 deals with dynamically gradient semigroups and the topological structural stability of their attractors. Chapter 4 studies neighborhoods of critical elements, that is, equilibrium points and periodic orbits, and presents the proof of the λ-lemma of J. Palis but in the infinite-dimensional case. This proof was found in D. Henry's handwritten notes "Invariant Manifolds near a Fixed Point", and many other results of this chapter were taken from it as well. Chapter 5 presents the concept of Morse-Smale semigroups and the geometrical structural stability of their attractors, that is, an isomorphism between their phase diagrams. In Part II, as we said, the chapters are the analogues of their counterparts in the autonomous theory. Several results from Part II are new developments of the theory that, we hope, will contribute to further understanding the asymptotics of non-autonomous problems.

We want to express our gratitude to Eriko Hironaka at AMS who, throughout the writing of this book, has been very supportive and patient. Also, we thank the anonymous reviewers for the support and suggestions that led to many improvements. We also wish to thank Yanan Li, Estefani Moraes Moreira and Alexandre N. Oliveira-Sousa who read the entire manuscript and made several suggestions during the preparation of it.

In May 2019 we lost our collaborator and warm-hearted friend Geneviève Raugel. With Genèvieve we developed the results on stability of the phase diagram isomorphisms for Morse-Smale semigroups under non-autonomous perturbations, presented in the second part of this book. We wish to honour her at this opportunity.

Matheus C. Bortolan

Alexandre N. Carvalho

José A. Langa

Introduction

Over the last few decades, the study of qualitative properties of semigroups and evolution processes in Banach spaces has received great attention (see for instance [**22, 78, 89, 98, 132, 133, 153, 159, 189, 200**]). Particularly the study of *attractors*, which has created a deep area of research, greatly improving the understanding of qualitative properties of solutions for both finite and infinite-dimensional dynamical systems.

When dealing with dynamical systems, one important aspect to be taken into account is the following: a mathematical model that describes a real world phenomenon (from Biology, Physics, Economy, etc.) is merely an approximation of the real rule of evolution. To obtain a mathematical model one has to make observations, measurements, to use empirical laws, to make simplifications, all of which, intrinsically, carry imprecisions. Therefore our mathematical study begins with an approximation of the reality. Within this context, the study of robustness of *attractors* for dynamical systems is crucial to the modeling of a real world phenomenon by a dynamical system. This book deals with the study of the behavior of 'attractors' for autonomous or non-autonomous dynamical systems under small perturbations.

To study the behavior of attractors under perturbations, we will need to specify what kind of perturbations will be allowed. The perturbations are expressed through some *continuity properties* of the dynamical systems, which together with suitable natural conditions will result in a *continuity* property of the attractors.

The non-autonomous theory generalizes the autonomous one, and thus, we could present only the non-autonomous framework, and leave the autonomous theory as a particular case. But we chose to present the autonomous framework first, and only then the non-autonomous. Such a choice reflects our experience of several years teaching the subject, and it is a reasonable one, since the non-autonomous theory is a lot more complicated than the autonomous one, but becomes somewhat simpler once the needed intuition has been developed within the autonomous case.

To begin the explanation of the different types of *continuity of attractors* that we will consider, let us start fixing a set X of all possible states of a modeled phenomenon. The variable x may describe a large number of (physical, biological, economical, etc.) quantities. As examples we mention the position of a body in space, the temperature distribution in a body, the concentration of substances in a recipient, the magnitude of electrical impulses, the displacement of a string from its rest position, etc. Hence, the space X (called the *state space*) may be either finite or infinite dimensional and, in general, it will be a metric space.

An *evolution process* on X is a two-parameter family of operators

$$\mathcal{S} = \{S(t,s) \colon t \geqslant s\}$$

from X into itself such that, given that the value of the variable at the instant of time s is x, $S(t,s)x$ is the value of the variable at a later time t (it may be that $S(t,s)$ is not injective nor surjective). If $t, s \in \mathbb{Z}$, we say that $\mathcal{S}$ is a *discrete evolution process* and if $t, s \in \mathbb{R}$ we will simply say that $\mathcal{S}$ is an *evolution process*. For simplicity we will only consider $t, s \in \mathbb{R}$.

This family of operators must obey certain compatibility conditions, namely, if X is a metric space with a metric d, let $\mathcal{C}(X)$ be the set of continuous maps from X into itself. Then, to call $\mathcal{S}$ an **evolution process**, we require that $\mathcal{S} \subset \mathcal{C}(X)$ and also that

(i) $S(t,t)x = x$ for all $t \in \mathbb{R}$, $x \in X$,
(ii) $S(t,r)S(r,s)x = S(t,s)x$ for all $t \geqslant r \geqslant s$, $x \in X$,
(iii) the map $\{(t,s,x) \in \mathbb{R} \times \mathbb{R} \times X : t \geqslant s\} \ni (t,s,x) \mapsto S(t,s)x \in X$ is continuous,

where in condition (ii) above, $S(t,r)S(r,s)$ stands for the composition of $S(t,r)$ with $S(r,s)$.

We will distinguish two main types of evolution processes: autonomous and non-autonomous. An evolution process will be called **autonomous** if it satisfies $S(t,s) = S(t-s,0)$ for all $t \geqslant s$. The remaining evolution processes will be called **non-autonomous**.

Note that for an autonomous evolution process, the evolution of the state x from the initial time s to the state $S(t+s,s)x$ at time $t+s$ is independent of s and depends only on t. For autonomous evolution processes we define $T(t) = S(t,0) \in \mathcal{C}(X)$ and the family $\mathcal{T} = \{T(t) : t \geqslant 0\}$ satisfies

(i') $T(0)x = x$, for all $x \in X$,
(ii') $T(t)T(s) = T(t+s)$ for all $t, s \geqslant 0$ and
(iii') the map $[0,\infty) \times X \ni (t,x) \mapsto T(t)x \in X$ is continuous.

A family $\mathcal{T} = \{T(t) : t \geqslant 0\}$ with the above properties is called a **semigroup**. Clearly, from a semigroup $\mathcal{T}$ we can define an autonomous evolution process $\mathcal{S} = \{S(t,s) : t \geqslant s\}$ setting $S(t,s) = T(t-s)$, $t \geqslant s$. We will use both terms indifferently.

Evolution processes frequently appear associated to ordinary, partial or functional (evolving) differential equations. To fix our ideas, consider the Cauchy problem:

$$\begin{cases} \dfrac{dx}{dt} + Ax = f(t,x), \ t > s, \\ x(s) = x_0 \in X, \end{cases} \tag{0.1}$$

where X is a Banach space, $A\colon D(A) \subset X \to X$ is a linear (possibly unbounded) operator and f is a nonlinear forcing term. Here all are ODE's, the semilinear parabolic equations, the Navier-Stokes equations, the wave equations, the Cahn-Hilliard equation, the Schrödinger equation, and many others.

Assume that for each $s \in \mathbb{R}$ and $x_0 \in X$ there is a unique continuous function $x(\cdot,s,x_0)\colon [s,\infty) \to X$ which is the *weak solution* (related to the way it satisfies the equation) for (0.1). Assume also that the map $\mathcal{P} \times X \ni (t,s,u_0) \mapsto x(t,s,x_0) \in X$ is continuous, where $\mathcal{P} = \{(t,s) \in \mathbb{R}^2 : t \geqslant s\}$.

In this case if $(t,s) \in \mathcal{P}$, we define $S(t,s) \in \mathcal{C}(X)$ by

$$S(t,s)x_0 = x(t,s,x_0),$$

and the operator $S(t,s)$ *evolves* the *phase space* X from the initial time s to the later time t and the properties of an evolution process are satisfied.

REMARK 0.1. In order to define the evolution process for all (t,s) in $\mathcal{P}$ we must have that our forcing f be defined for all $t \in \mathbb{R}$. We will, in general, be interested in the cases where f is defined only for $t \in \mathbb{R}^+$. Even so, the problems for which f is defined for all $t \in \mathbb{R}$ appear naturally as we will see later in the book.

If f is independent of t, the evolution depends on a single vector field (whereas if it depends on t, there are infinitely many of them - one for each instant of time). Hence, the evolution of an initial state x depends only on the elapsed time $t-s$ and not particularly on the initial and final times, s and t. Hence $S(t,s)x = S(t-s,0)x$, $(t,s) \in \mathcal{P}$ and $x \in X$, and the equation generates a semigroup.

REMARK 0.2. In this context, when X is infinite dimensional, there is a large number of important mathematical problems (interesting, difficult, and several of them open) related to the local and global well posedness, continuation, continuation after blow up and regularity of solutions for critical nonlinearities. Some references about local well posedness, continuation and continuation after blow up for critical nonlinearities are [**7, 16, 61, 62, 67**]. We will not pursue this direction, instead we will assume the global well posedness and refer to the literature on it, when needed.

Our goal is to study the 'asymptotic behavior' of evolution processes. To that end, we define (when they exist) the objects that contain all asymptotic states of an evolution process, the *attractors*. To define attractors the notions of *invariance* and *attraction* will be essential. Let us first define *invariance*.

DEFINITION 0.3. A subset A of X is called **invariant** for a semigroup $\mathcal{T}$, or equivalently **$\mathcal{T}$-invariant**, if $T(t)A = A$ for all $t \geqslant 0$. A family $\{A(t)\}_{t\in\mathbb{R}}$ of subsets of X is called **invariant** for an evolution process $\mathcal{S}$, or equivalently **$\mathcal{S}$-invariant**, if $S(t,s)A(s) = A(t)$ for all $t \geqslant s$.

REMARK 0.4. If A is $\mathcal{T}$-invariant for a semigroup $\mathcal{T}$, then it is fixed by $T(t)$ for each $t \geqslant 0$. Notice that we are *not* saying that each element $a \in A$ is fixed. Invariant sets for semigroups or a family of invariant sets for an evolution process have distinguished dynamics. So, if we wish to understand the dynamics of a semigroup or an evolution process we must understand its dynamics restricted to its invariants.

Next we define *global solutions*, which will be important to characterize the attractors.

DEFINITION 0.5. A continuous function $\xi\colon \mathbb{R} \to X$ is called a **global solution** for a semigroup $\mathcal{T}$ if $T(t)\xi(s) = \xi(t+s)$ for all $t \geqslant 0$ and $s \in \mathbb{R}$. A continuous function $\xi\colon \mathbb{R} \to X$ is called a **global solution** for an evolution process $\mathcal{S}$ if $S(t,s)\xi(s) = \xi(t)$ for all $t \geqslant s$. In both cases, if $\xi(0) = x$ we say that ξ is a **global solution through** x.

REMARK 0.6. The concept of invariance of a set (or a family) is fundamentally related to the possibility of constructing global solutions through points of this set (or family), that is, if $\{A(t)\}_{t\in\mathbb{R}}$ is $\mathcal{S}$-invariant for an evolution process $\mathcal{S}$, then for each $s \in \mathbb{R}$ and $x \in A(s)$ there exists a global solution $\xi\colon \mathbb{R} \to X$ of $\mathcal{S}$ such that $\xi(s) = x$ and $\xi(t) \in A(t)$ for all $t \in \mathbb{R}$.

It is important to notice that the notions of invariance for semigroups and for evolution processes do not coincide, not even when evolution process $\mathcal{S}$ is autonomous; that is, when $S(t,s)=T(t-s)$, $(t,s)\in\mathcal{P}$ for a semigroup $\mathcal{T}$. In fact, if ξ is a global solution for an evolution process $\mathcal{S}$, it is clear that the family $\{\xi(t)\}_{t\in\mathbb{R}}$ is $\mathcal{S}$-invariant. If $\mathcal{S}$ is autonomous, the notion of invariance for the semigroup $\mathcal{T}$, where $T(t)=S(t,0)$ for each $t\geqslant 0$, makes no sense for the family $\{\xi(t)\}_{t\in\mathbb{R}}$ (unless ξ is a constant function). However, the set $A=\{\xi(t)\colon t\in\mathbb{R}\}$ is $\mathcal{T}$-invariant.

Recall that the Hausdorff semidistance between two subsets A and B of X is defined by

$$\operatorname{dist}_H(A,B)=\sup_{a\in A}\inf_{b\in B} d(a,b).$$

Note that $\operatorname{dist}_H(A,B)$ is a way of saying *how far A is from being inside B*. Clearly, $\operatorname{dist}_H(A,B)=0$ if and only if $A\subset\overline{B}$.

The last notion that we will need so that we are able to define attractors for semigroups and evolution processes is the notion of *attraction*, which can be defined in several different ways.

Definition 0.7. Attraction for semigroups. Given $B,B_0\subset X$, we say that B_0 **attracts** B under the action of the semigroup $\mathcal{T}$, or equivalently B_0 **$\mathcal{T}$-attracts** B, if

$$\lim_{t\to\infty}\operatorname{dist}_H(T(t)B,B_0)=0.$$

Pullback Attraction 1. Given $B,B_0\subset X$, we say that B_0 **pullback attracts** B at time t under the action of the evolution process $\mathcal{S}$, or equivalently, B_0 **pullback $\mathcal{S}$-attracts** B, if

$$\lim_{s\to-\infty}\operatorname{dist}_H(S(t,s)B,B_0)=0.$$

Pullback Attraction 2. We say that the family $\{B_0(t)\}_{t\in\mathbb{R}}$ **pullback attracts** a set $B\subset X$ under the action of the evolution process $\mathcal{S}$, or equivalently, $\{B_0(t)\}_{t\in\mathbb{R}}$ **pullback $\mathcal{S}$-attracts** B, if

$$\lim_{s\to-\infty}\operatorname{dist}_H(S(t,s)B,B_0(t))=0,$$

for each $t\in\mathbb{R}$.

Uniform attraction. Given $B,B_0\subset X$, we say that B_0 **attracts** B **uniformly** under the action of the evolution process $\mathcal{S}$, or equivalently B_0 **uniformly $\mathcal{S}$-attracts** B, if

$$\lim_{\tau\to\infty}\sup_{s\in\mathbb{R}}\operatorname{dist}_H(S(\tau+s,s)B,B_0)=0.$$

We note that, if B_0 *attracts* (pullback attracts) B, the *evolution* (pullback evolution) $T(t)B$ $(S(t,s)B)$ of B *asymptotically* enters $\overline{B_0}$ *as t tends to infinity (as s tends to minus infinity).*

Remark 0.8. It is easy to see that *uniform attraction* implies *pullback attraction 1* (change $\tau+s$ to t and s to $t-\tau$) and that the notions of *attraction, pullback attraction 1* and *uniform attraction* are equivalent when the evolution process is autonomous.

Now we are ready to define global attractors.

DEFINITION 0.9. We say that $\mathcal{A}$ is a **global attractor** for the semigroup $\mathcal{T}$ if it is compact, $\mathcal{T}$-invariant and $\mathcal{T}$-attracts all bounded subsets of X. A family $\{\mathcal{A}(t)\}_{t\in\mathbb{R}}$ is a **pullback attractor** for the evolution process $\mathcal{S}$ if it is $\mathcal{S}$-invariant, pullback $\mathcal{S}$-attracts bounded subsets of X, $\mathcal{A}(t)$ is compact for each $t \in \mathbb{R}$ and $\bigcup_{t\in\mathbb{R}} \mathcal{A}(t)$ is bounded. A set $\mathcal{A}$ is a **uniform attractor** for the evolution process $\mathcal{S}$ if it is the minimal compact subset of X that uniformly $\mathcal{S}$-attracts all bounded subsets of X.

The global attractors for semigroups and the pullback attractors for evolution processes, when they exist, are unique, and can be characterized in the following manner: if $\mathcal{A}$ is the global attractor for a semigroup $\mathcal{T}$ we have

$$\mathcal{A} = \{\xi(0)\colon \xi \text{ is a bounded global solution of } \mathcal{T}\}.$$

If $\{\mathcal{A}(t)\}_{t\in\mathbb{R}}$ is the pullback attractor for an evolution process $\mathcal{S}$ we have

$$\mathcal{A}(t) = \{\xi(t)\colon \xi \text{ is a global bounded solution of } \mathcal{S}\}.$$

The characterization of the uniform attractor is not that simple, and requires additional refinement of the concepts introduced so far.

REMARK 0.10. After introducing the concepts of *attractors* for semigroups and evolution processes, an initial and important task is to give conditions (that are satisfied for a large class of meaningful applications) under which semigroups and evolution processes have attractors (see [**22, 78, 96, 132, 200**]). We develop some aspects of the theory of existence of global attractors for semigroups in Chapter 1, and of attractors for evolution processes (both pullback and uniform attractors) in Chapter 6. The aim of these chapters is to prepare the reader for the future chapters on robustness of these attractors.

After defining the main object of our studies, the attractors, we begin to try to understand its robustness under perturbation. As observed before, an important aspect to be taken into account when dealing with asymptotic behavior of dynamical systems is that, as a model of a real phenomenon, it is merely an approximation of the real rule of evolution. Hence, the study of robustness of attractors (relative to all kinds of perturbations) has to be considered an important part of the theory and this is the purpose of this book.

In order to describe the contents of Chapters 2–6, we will introduce some more terminology. Firstly, consider a family of semigroups $\{\mathcal{T}_\eta\}_{\eta\in[0,1]}$ (we fix, for simplicity, the parameter in the interval $[0,1]$, but it may represent a very singular perturbation, for instance, a domain perturbation, a homogenization, a change of type in the equation, etc.). Assume that the family $\{\mathcal{T}_\eta\}_{\eta\in[0,1]}$ is continuous at 0, that is, $[0,1]\times\mathbb{R}^+\times X \ni (\eta,t,x)\mapsto T_\eta(t)x\in X$ is continuous at $\eta=0$, uniformly for (t,x) in compact subsets of $\mathbb{R}^+\times X$ (this represents a continuity with respect to both initial data and parameters).

If, for each $\eta\in[0,1]$, $\mathcal{T}_\eta$ has a global attractor $\mathcal{A}_\eta$, we ask whether the family of attractors $\{\mathcal{A}_\eta\}_{\eta\in[0,1]}$ behaves *continuously* at 0. This is what we will call *robustness of attractors*. It just remains to clearly establish the 'metric' we will use for the *continuity of attractors*.

The coarsest continuity property considered is the *upper semicontinuity* of attractors, which is expected for a general class of dynamical systems and is fairly simple to obtain. We sometimes refer to the upper semicontinuity of attractors as the *non-explosion property* of attractors, to express that this property means that the perturbed attractor may not become suddenly much larger than the limiting attractors. We say that the family of attractors $\{\mathcal{A}_\eta\}_{\eta\in[0,1]}$ is **upper semicontinuous** at 0 if

$$\lim_{\eta\to 0} \operatorname{dist}_H(\mathcal{A}_\eta, \mathcal{A}_0) = 0.$$

This is a very common property, that is, only a few conditions on the family $\{\mathcal{T}_\eta\}_{\eta\in[0,1]}$ will be required to obtain it, as we will see in the next theorem.

THEOREM 0.11 (Upper semicontinuity). *Let $\{\mathcal{T}_\eta\}_{\eta\in[0,1]}$ be a family of semigroups which is continuous at 0. If, for each $\eta \in [0,1]$, $\mathcal{T}_\eta$ has a global attractor $\mathcal{A}_\eta$ and $\bigcup_{\eta\in[0,1]} \mathcal{A}_\eta$ is compact in X, then the family $\{\mathcal{A}_\eta\}_{\eta\in[0,1]}$ is upper semicontinuous at 0.*

This result is proved in Theorem 2.5 under even slightly more general conditions. Upper semicontinuity has been studied in a variety of problems and under very singular types of perturbations (see [**8, 14, 15, 18, 37, 57–59, 70–72, 78, 82–85, 114**]). It is the only robustness property of attractors that is independent of the structure of the attractors. When using finer 'metrics' to study the robustness of the attractors, we will necessarily need to know some aspects of the inner structure of the attractors.

The next natural step in the study of robustness of attractors is the notion of *lower semicontinuity*, which is much harder to obtain in applications and not as common as the upper semicontinuity. We sometimes refer to the lower semicontinuity of attractors as the *non-implosion property* of attractors, and express that this property means that the perturbed attractor may not become suddenly much smaller than the limiting attractors. A family of sets $\{\mathcal{A}_\eta\}_{\eta\in[0,1]}$ such that

$$\lim_{\eta\to 0} \operatorname{dist}_H(\mathcal{A}_0, \mathcal{A}_\eta) = 0$$

is called **lower semicontinuous** at 0.

A family of sets $\{\mathcal{A}_\eta\}_{\eta\in[0,1]}$ which is upper and lower semicontinuous at 0 is called **continuous** at 0. This property means that the sets fill 'approximately' the same region in the phase space.

The known proofs of lower semicontinuity allow us the ability to reproduce in $\mathcal{A}_\eta$ the same dynamical structures contained in $\mathcal{A}_0$. That involves the permanence of the elementary dynamical structures (critical elements and their unstable manifolds) of $\mathcal{A}_0$ that are, in general, a consequence of the robustness of the dichotomy (hyperbolicity) under perturbation, and of a certain knowledge of the inner structure of $\mathcal{A}_0$ in terms of elementary dynamical structure (see [**22, 135**]).

Both the upper semicontinuity and the lower semicontinuity of attractors will be considered in Chapter 2

We can see with a simple example that the lower semicontinuity property of attractors under perturbations may fail to hold. Consider a simple ODE in the real line, given by

$$\dot{u} = f(u), \tag{0.2}$$

where $f\colon \mathbb{R} \to \mathbb{R}$ is given by $f(u) = -c(u+1)^2(u-1)$, $c > 0$ is fixed (see Figure 0.1).

It is fairly easy to see that the global attractor for the semigroup associated to (0.2) consists of the closed interval $\mathcal{A}_0 = [-1, 1]$. Consider a small perturbation of

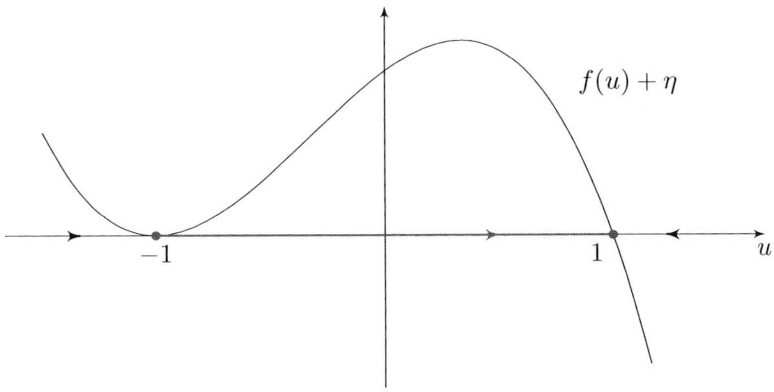

FIGURE 0.1. Graph of the function f.

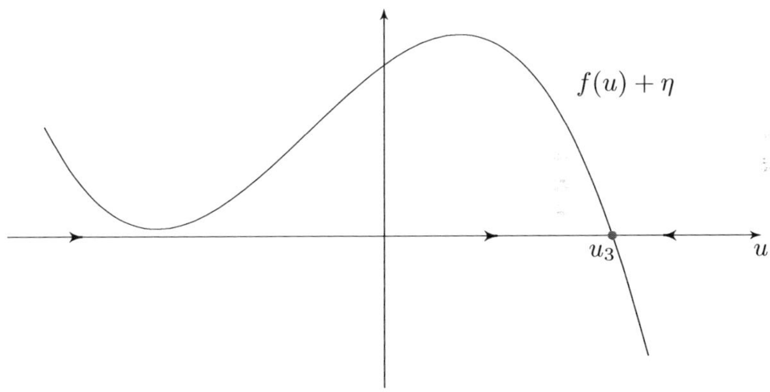

FIGURE 0.2. Graph of the function $f + \eta$.

equation (0.2) of the form

$$\dot{u} = f(u) + \eta, \tag{0.3}$$

for some $\eta > 0$ small (see Figure 0.2).

It is clear that the global attractor for the semigroup associated to (0.3) is given by $\mathcal{A}_\eta = \{u_3\}$, where $u_3 = u_3(\eta)$, and it is easy to see that $u_3 \in (1, 1 + \eta/4)$. The reader may easily check that the family $\{\mathcal{A}_\eta\}_{\eta\in[0,1]}$ is upper semicontinuous at 0 but *not* lower semicontinuous at 0.

Let us give general conditions under which the lower semicontinuity property of global attractors holds. In what follows, we consider a semigroup $\mathcal{T}$ with a global attractor $\mathcal{A}$. Also, for a set $B \subset X$ and $r > 0$, we define the **r-neighborhood of** B by

$$\mathcal{O}_r(B) = \{y \in X \colon d(y, x) < r \text{ for some } x \in B\}.$$

DEFINITION 0.12 (Disjoint collection of isolated invariant sets). We say that $\Xi \subset \mathcal{A}$ is an **isolated invariant set** (with respect to $\mathcal{T}$), when it is the maximal invariant set in a neighborhood of itself, that is, there exists $\delta > 0$ such that if $C \subset \mathcal{O}_\delta(\Xi)$ is $\mathcal{T}$-invariant, then $C \subset \Xi$.

We say that $\Xi = \{\Xi_1, \cdots, \Xi_n\}$ is a **disjoint collection of isolated invariant sets** (with respect to $\mathcal{T}$) if each Ξ_i is an isolated invariant set and there exists $\delta > 0$ such that $\mathcal{O}_\delta(\Xi_i) \cap \mathcal{O}_\delta(\Xi_j) = \varnothing$ for $1 \leqslant i < j \leqslant n$.

We remark that, for instance, a *center* is an isolated equilibrium which is not (as a unitary set) an isolated invariant set. An isolated invariant set enjoys some topological hyperbolicity. Typically, an isolated invariant set will be a hyperbolic equilibrium point or a normally hyperbolic periodic orbit (which will be discussed in Chapter 4).

DEFINITION 0.13. Given an isolated invariant set Ξ, we define the **unstable set of** Ξ by

$$W^u(\Xi) = \{\xi(0)\colon \xi\colon \mathbb{R} \to X \text{ is a global solution of } \mathcal{T} \text{ such that } \lim_{t\to-\infty} \mathrm{dist}_H(\xi(t), \Xi) = 0\},$$

and given a $\delta > 0$, the **(δ-) local unstable set of** Ξ by

$$W^u_\delta(\Xi) = \{\xi(0)\colon \xi\colon \mathbb{R} \to X \text{ is a global solution of } \mathcal{T} \text{ such that } \xi(t) \in \mathcal{O}_\delta(\Xi) \text{ for all } t \leqslant 0 \text{ and } \lim_{t\to-\infty} \mathrm{dist}_H(\xi(t), \Xi) = 0\}.$$

We also use the notation $W^u_{loc}(\Xi)$ to represent a local unstable set of Ξ.

THEOREM 0.14. *Let $\{\mathcal{T}_\eta\}_{\eta\in[0,1]}$ be a family of semigroups which is continuous at 0. Suppose that $\mathcal{T}_\eta$ has a global attractor $\mathcal{A}_\eta$ for each $\eta \in [0,1]$ and that there is an $\eta_0 \in (0,1]$ such that $\bigcup_{\eta\in[0,\eta_0]} \mathscr{A}_\eta$ is compact. Assume also that $\Xi_\eta = \{\Xi_{1,\eta}, \cdots, \Xi_{n,\eta}\}$ is a disjoint collection of isolated invariant sets for $\mathcal{T}_\eta$ for each $\eta \in [0,\eta_0]$ and such that*

$$\mathcal{A}_0 = \bigcup_{j=1}^{n} W^u(\Xi_{j,0}). \tag{0.4}$$

Lastly assume that $\mathrm{dist}_H(W^u_\delta(\Xi_{j,0}), W^u_\delta(\Xi_{j,\eta})) \overset{\eta\to 0}{\longrightarrow} 0$. *Then*

$$\mathrm{dist}_H(\mathcal{A}_\eta, \mathcal{A}_0) + \mathrm{dist}_H(\mathcal{A}_0, \mathcal{A}_\eta) \overset{\eta\to 0}{\longrightarrow} 0,$$

that is, the family of attractors $\{\mathcal{A}_\eta\}_{\eta\in[0,1]}$ is continuous at 0.

This result is proved in Chapter 2 under even more general conditions. Note that, to obtain lower semicontinuity, given the isolated invariant sets $\{\Xi_{1,0}, \cdots, \Xi_{n,0}\}$, we need to ensure their permanence under perturbations (as isolated invariant sets), their continuity with respect to perturbations, the continuity of the local unstable sets of isolated invariant sets as well as to characterize the limiting attractor as (0.4). The permanence of critical elements and the continuity of the local unstable sets of critical elements are studied in Chapter 4.

The applications of the results on lower semicontinuity of global attractors include several singular perturbation problems and a more general class of limiting problems, with respect to the seminal works of [**22, 135**], which include also problems with normally hyperbolic periodic orbits (see [**8, 10, 11, 14, 29, 38, 55, 78**]).

It is important to note that the conditions imposed in Theorem 0.14 could be satisfied in the presence of *homoclinic orbits*. Also note that we do not conclude that $\mathcal{A}_\eta = \bigcup_{j=1}^n W^u(\Xi_{j,\eta})$. See the pictorial example given in Figure 0.3.

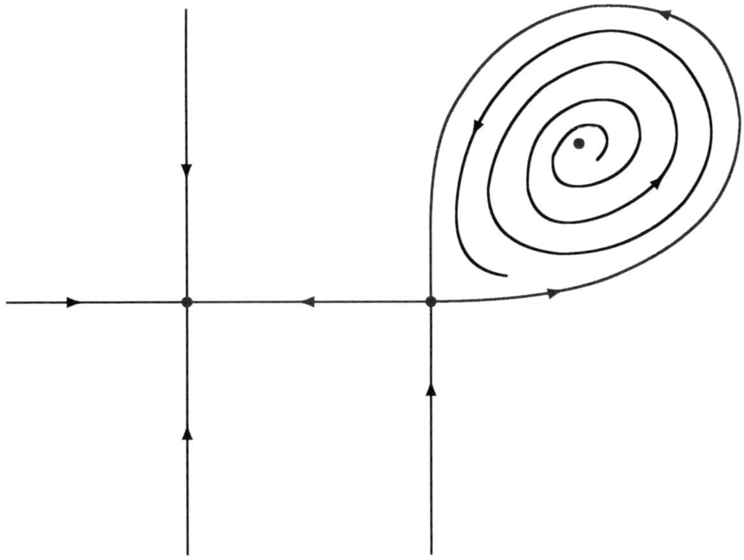

FIGURE 0.3. An attractor as union of unstable sets of equilibria.

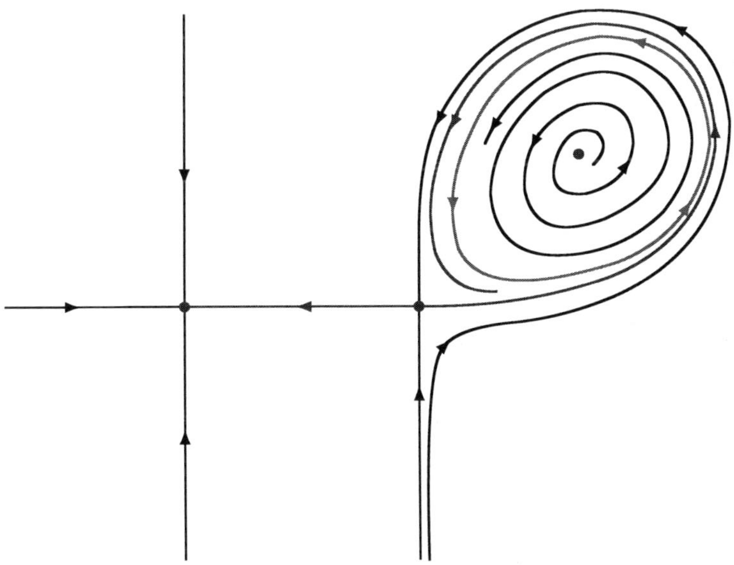

FIGURE 0.4. The perturbed attractor.

A possible perturbation of the attractor given in Figure 0.3 is the attractor given in Figure 0.4. Note that the characterization (0.4) does not hold with the isolated invariant sets given by the perturbations of the hyperbolic equilibria.

A natural question arises: is it possible, imposing some reasonable additional conditions, to ensure that $\mathcal{A}_\eta = \bigcup_{j=1}^{n} W^u(\Xi_{j,\eta})$? The answer to this question requires that homoclinic structures do not exist (that is, that the limit problem has the dynamical properties of a gradient semigroup). This is the subject of Chapter 3 (see also [**5**, **74**]).

To state this result we need to introduce a few notions.

DEFINITION 0.15 (Homoclinic structure). Let $\mathcal{T}$ be a semigroup in a metric space X with a global attractor $\mathcal{A}$ that contains a disjoint collection of isolated invariant sets $\mathbf{\Xi} = \{\Xi_1, \cdots, \Xi_n\}$.

A **homoclinic structure** relative to $\mathbf{\Xi}$ is a collection $\xi_1, \cdots, \xi_p \colon \mathbb{R} \to \mathcal{A}$ of global solutions and a subcollection $\{\Xi_{n_1}, \cdots, \Xi_{n_p}\}$ of $\mathbf{\Xi}$ such that

$$\Xi_{n_i} \overset{t\to-\infty}{\longleftarrow} \xi_i(t) \overset{t\to\infty}{\longrightarrow} \Xi_{n_{i+1}}, \quad 1 \leqslant i \leqslant p,$$

where we set $\Xi_{n_{p+1}} = \Xi_{n_1}$.

We are ready to define the *dynamically gradient semigroups* (which are *gradient*; see [**5**])

DEFINITION 0.16 (Dynamically gradient semigroups). Let X be a metric space and let $\mathcal{T}$ be a semigroup in X with a global attractor $\mathcal{A}$. Assume that

(G1) there is a disjoint collection of isolated invariant sets $\mathbf{\Xi} = \{\Xi_1, \cdots, \Xi_n\}$ in $\mathcal{A}$ with the property that any global solution $\xi \colon \mathbb{R} \to \mathcal{A}$ satisfies

$$\lim_{t\to-\infty} \operatorname{dist}_H(\xi(t), \Xi_i) = 0 \text{ and } \lim_{t\to\infty} \operatorname{dist}_H(\xi(t), \Xi_j) = 0,$$

for some $1 \leqslant i, j \leqslant n$, and

(G2) there is no homoclinic structure relative to $\mathbf{\Xi}$.

Then we say that $\mathcal{T}$ is a **dynamically gradient semigroup** relative to $\mathbf{\Xi}$.

THEOREM 0.17 ([**74**]). *Let $\{\mathcal{T}_\eta\}_{\eta\in[0,1]}$ be a continuous family of semigroups at* 0. *Suppose that $\mathcal{T}_\eta$ has a global attractor $\mathcal{A}_\eta$ for each $\eta \in [0,1]$ and also that*

(a) *there exists $\eta_0 \in (0,1]$ such that $\bigcup_{\eta\in[0,\eta_0]} \mathcal{A}_\eta$ is compact;*

(b) *there exists $n \in \mathbb{N}^*$ such that, for each $\eta \in [0,1]$, $\mathcal{A}_\eta$ has a disjoint collection of isolated invariant (uniformly in η) sets $\mathbf{\Xi}_\eta = \{\Xi_{1,\eta}, \cdots, \Xi_{n,\eta}\}$ and $\sup_{1\leqslant i\leqslant n} [\operatorname{dist}_H(\Xi_{i,\eta}, \Xi_{i,0}) + \operatorname{dist}_H(\Xi_{i,0}, \Xi_{i,\eta})] \overset{\eta\to0}{\longrightarrow} 0$;*

(c) *$\mathcal{T}_0$ is dynamically gradient relative to $\mathbf{\Xi}_0$.*

Then there exists $\eta_1 > 0 \in (0, \eta_0]$ such that, for all $\eta \in [0, \eta_1]$, $\mathcal{T}_\eta$ is dynamically gradient relative to $\mathbf{\Xi}_\eta$ and, consequently,

$$\mathcal{A}_\eta = \bigcup_{i=1}^{n} W^u(\Xi_{i,\eta}), \text{ for all } \eta \in [0, \eta_0].$$

Now consider the pictorial example given by Figure 0.5, with all equilibria e_1, e_2, e_3, e_4 and e_5 being hyperbolic. It is easy to deduce that the semigroup associated to this example is dynamically gradient.

If we make a regular perturbation (autonomous) of the semigroup associated to this example we obtain, from the previous theorem, that the perturbed semigroup is dynamically gradient. The resulting attractor can have the configuration represented in Figure 0.6.

Note that the characterization of the attractor as the union of the unstable manifolds of the hyperbolic equilibria is preserved as well as the properties that characterize a dynamically gradient semigroup. However, there is a loss of continuity in the connection between e_2 and e_3 that modifies the inner structure of the attractor.

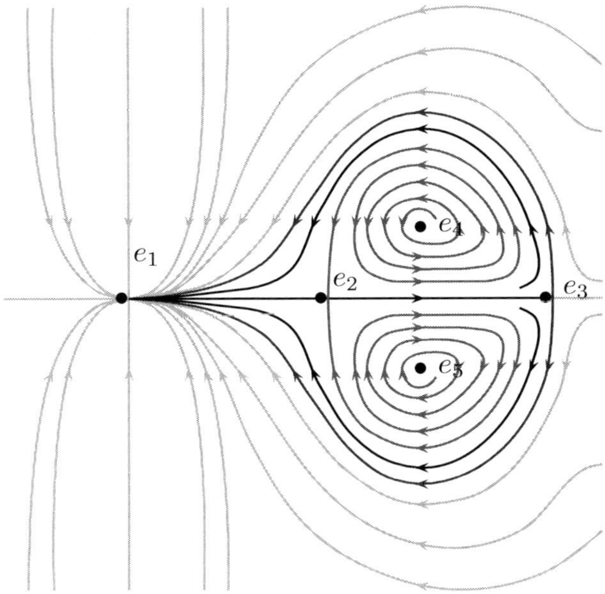

FIGURE 0.5. The *alien head*.

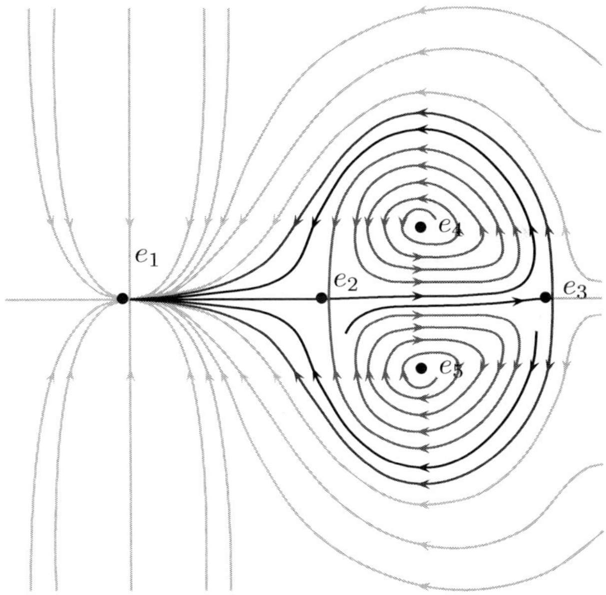

FIGURE 0.6. The perturbed *alien head*.

Having said that, it is natural to ask if it is possible to construct a class of semigroups for which the perturbations of the attractors would have attractors with the same inner structure. These will be the Morse-Smale semigroups, which are studied in Chapter 5. Next we briefly describe the results in that chapter.

First, for an isolated invariant set Ξ, we define its **stable set** by

$$W^s(\Xi) = \{x \in X \colon \lim_{t\to\infty} \mathrm{dist}_H(T(t)x, \Xi) = 0\}.$$

Also, for $\delta > 0$, we define the **(δ-)local stable set** by

$$W^s_\delta(\Xi) = \{x \in X \colon T(t)x \in \mathcal{O}_\delta(\Xi) \text{ for all } t \geqslant 0 \text{ and } \lim_{t\to\infty} \mathrm{dist}_H(T(t)x, \Xi) = 0\}.$$

Here we also can denote a local stable set of Ξ by $W^s_{loc}(\Xi)$.

DEFINITION 0.18. Let $\mathcal{T}$ be a semigroup in a Banach space X with a global attractor $\mathcal{A}$. We say that $\mathcal{T}$ is a **Morse-Smale semigroup** if:

(i) $\mathcal{T}$ is a dynamically gradient semigroup relative to the disjoint collection of isolated invariant sets $\boldsymbol{\Xi} = \{\Xi_i, \cdots, \Xi_n\}$, where each $\Xi \in \boldsymbol{\Xi}$ is either a hyperbolic equilibrium or a normally hyperbolic periodic orbit (notions that will be defined in Chapter 4);

(ii) $\mathcal{T}$ is **reversible**, that is, $T(t)|_{\mathcal{A}}$ is injective, $T(t)$ is differentiable in $\mathcal{A}$ and $D_xT(t)(z) \colon X \to X$ is an injective bounded linear operator, for each $t \geqslant 0$ and $z \in \mathcal{A}$;

(iii) $W^u_{loc}(\Xi)$ has finite dimension for each $\Xi \in \boldsymbol{\Xi}$;

(iv) given $\Xi, \hat{\Xi} \in \boldsymbol{\Xi}$ such that $W^u(\Xi) \cap W^s_{loc}(\hat{\Xi}) \neq \varnothing$, there is a point of *transversal intersection* (notion to be defined in Chapter 5) in $W^u(\Xi) \cap W^s_{loc}(\hat{\Xi})$.

Before presenting the main result of the autonomous part of this monograph, we would like to explain in a little more detail what this concept of *Morse-Smale semigroup* entails. More than just being a dynamically gradient semigroup, a Morse-Smale semigroup is injective and differentiable in its global attractor, and its derivative in any point of the attractor is an injective bounded linear operator. This property is required so that its *unstable set*, which is given as iterations of its *local unstable set*, is a differentiable manifold. The finite dimensionality of the local unstable sets allows us to define *levels of energy*, the higher the dimension, the more energy the invariant possesses. Finally, the transversality condition (together with the finite dimensionality of the local unstable sets) is there to ensure the *permanence* of intersections, since transversal intersection is, in this setting, an open property.

To state the main result on the continuity of the internal structure of the attractors for Morse-Smale semigroups, we present the notion of collective asymptotic compactness: a family $\{T_\eta\}_{\eta\in[0,1]}$ of semigroups will be called **collectively asymptotically compact at** $\eta = 0$ if given sequences $\{x_n\} \subset X$ bounded, $t_n \to \infty$ and $\eta_n \to 0$, the sequence $\{T(t_n)x_n\}$ has a convergent subsequence.

THEOREM 0.19. *Let $\{\mathcal{T}_\eta\}_{\eta\in[0,1]}$ be a family of reversible semigroups which is continuous and collectively asymptotically compact at $\eta = 0$, and such that*

(a) *$\mathcal{T}_\eta$ has a global attractor $\mathcal{A}_\eta$, and there is an $\eta_0 \in (0, 1]$ such that*

$$\bigcup_{\eta\in(0,\eta_0]} \mathcal{A}_\eta \text{ is bounded;}$$

(b) *for each bounded interval $J \subset \mathbb{R}^+$ and for some $\epsilon > 0$ we have*

$$\sup_{t \in J} \sup_{z \in \mathcal{O}_\epsilon(\mathcal{A}_0)} \|D_x T_\eta(t)(z) - D_x T(t)(z)\|_{\mathcal{L}(X)} \overset{\eta \to 0}{\to} 0.$$

(c) *$\mathcal{T}_0$ is a Morse-Smale semigroup with disjoint collection of isolated invariant sets (critical elements) $\Xi_0 = \{\Xi_{1,0}, \cdots, \Xi_{n,0}\}$.*

Then, there exists $\eta_0 > 0$ such that $\mathcal{T}_\eta$ is a Morse-Smale semigroup and there is a phase diagram isomorphism *between the $\mathcal{A}_\eta$ and $\mathcal{A}_0$, for each $\eta \in (0, \eta_0)$, that is, the perturbed and unperturbed attractors have the same structure of connections between the isolated invariant sets (equilibra and periodic orbits).*

To prove this result on *geometrical stability* of Morse-Smale semigroups (see Chapter 5) the λ-lemma of Palis (see [**180**]) plays a fundamental role. We present in Chapter 4 a detailed proof of an infinite-dimensional version of it. The concept of *phase diagram isomorpshim*, details introduced in Definition 5.13, roughly speaking means that, if there is a connection between two isolated invariant sets in the unperturbed (respectively, perturbed) attractor then it also exists between the corresponding isolated invariant sets of the perturbed (respectively, unperturbed) attractor, that is, the set of connections between isolated invariant sets is the same for the perturbed and unperturbed attractors.

We point out that the concept of *phase diagram isomorphism* is weaker than the C^0-conjugacy of flows, presented for instance in [**133**]. The C^0-conjugacy of flows (or A-stability) in this context means that there exists $\eta_0 > 0$ such that for each $0 < \eta < \eta_0$ there exists a homeomorphism $h_\eta \colon \mathcal{A}_\eta \to \mathcal{A}_0$ such that h_η takes orbits in $\mathcal{A}_\eta$ into orbits of $\mathcal{A}_0$, preserving the sense of time. We will not pursue the existence of such homeomorphisms in this monograph.

Part II is dedicated to non-autonomous evolution processes and it is organised in the same manner as Part I.

The attractors, when non-autonomous nonlinear terms are present in the equations, have been intensively studied in the last two decades. The first observation is that there exists some possible notions of an 'attractor' in the non-autonomous framework. Indeed, note that even the definition of *asymptotic dynamics* has to be clarified, as, due to the explicit dependence of the system of initial time s and final time t (differently from the autonomous case where it just depends on the elapsed time $t - s$), different dynamical scenarios emerge, in the sense that we can try to analyse asymptotic dynamics in t (and then uniformly or not in s) or in s, which, in general, will be unrelated. The pullback attractor, a family containing all bounded global solutions of the equations, emerges as one of the correct possible concepts to deal with the asymptotic regime of non-autonomous dynamical systems (see [**78**] and [**153**]). Much of the theory described above for global attractors of autonomous dynamical systems has been analogously developed up until the 'metric' of lower semicontinuity and topological stability for non-autonomous dynamical systems, the *evolution processes*, and their associated pullback attractors (see [**74, 75, 78**]).

In Chapter 6 we introduce the different notions of attractor for non-autonomous dynamical systems: pullback, attractor for the skew-product semiflow and the uniform attractor. We describe the relationship between these concepts. In Chapter 7 we obtain results on upper semicontinuity of pullback and uniform attractors and consider a non-autonomous perturbation of an autonomous differential equation. In Chapter 8 we discuss the topological stability of pullback and uniform attractors. In

Chapter 9 we study the global hyperbolic solutions, their permanence, their unstable manifolds and the continuity of these unstable manifolds under perturbations. Chapter 10 deals with the extension of the notion of Morse-Smale semigroups to the non-autonomous context with emphasis on a small non-autonomous perturbation of a semigroup, and also presents applications of the results to reaction-diffusion equations and to the damped wave equation. The approach followed here is that of [**31**].

We made an effort to ensure that Part II is completely analogous to Part I. Our approach to the non-autonomous dynamical system is centered in the skew product semiflow, which makes it possible that several results from the autonomous framework can be directly translated and re-interpreted to the non-autonomous context. Nonetheless, the central role played by the pullback attractors becomes clear, already in Chapter 8, in the description of the forward dynamics on the non-autonomous problem. For that reason, in several occasions, they will be treated with detail and separately. This allows the reader to view this book in the usual order:

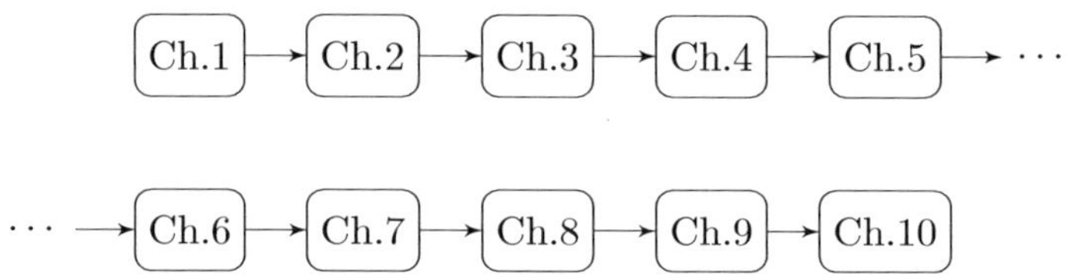

as well as switching between the autonomous and non-autonomous framework:

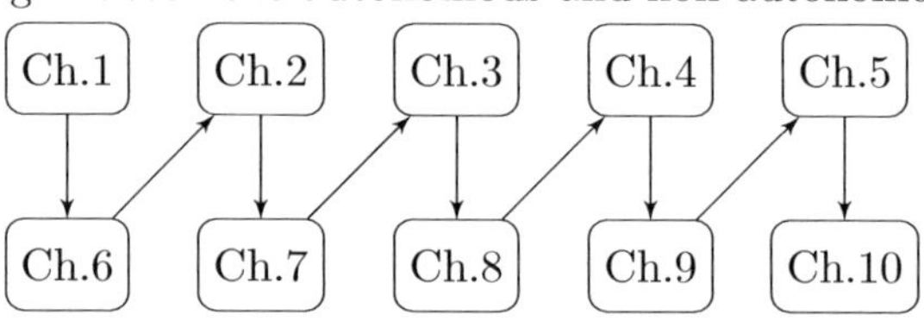

Lastly, many authors and in many references in the literature one can find mentions of the so-called λ-lemma (presented, for instance, in [**180**] for the finite-dimensional case, and mentioned in [**133**] for the infinite-dimensional case). We present a detailed proof of a more general version of this result found in [**180**]. It is worthwhile to point out that the version we present here is for an infinite-dimensional Banach space X and can also be used for the case of *center-stable manifolds*.

Our treatment of autonomous and non-autonomous dynamical systems, separate at first sight, but using the autonomous framework via the skew-product semiflow to deal with the non-autonomous setting, as well as the attention given to the matters of structural stability, are the main aspects that differentiate this monograph from that of [**78**].

Part 1

Autonomous Theory

CHAPTER 1

Semigroups and Global Attractors

To explain what a *semigroup* is (also known as *autonomous dynamical system*), imagine that we wish to predict the value of a variable x at future times. Denote the space of all possible values of this variable by X. The variable x can describe several distinct quantities from many areas of knowledge (Physics, Biology, Economics, etc.). As examples we mention the position of a body in the space (a vector in $\mathbb{R}^3$) or the temperature distribution in a body Ω (a function defined in Ω and taking values in $\mathbb{R}$). Consequently, the space X may be finite of infinite dimensional.

A semigroup will be a one-parameter family of operators $\mathcal{T} = \{T(t)\colon t \geqslant 0\}$, where $t \in \mathbb{R}$, in such a way that, given the initial value x of the variable, $T(t)x$ denotes the value of the variable after t instants of time. The knowledge of the semigroup allows us to predict the values of the variable that we know at the present time, after t instants of time, for each possible value of x in X and for each possible $t \geqslant 0$. Clearly, this family of operators must obey certain compatibility conditions which are: $T(0) = I_X$, where I_X is the identity operator in X, which means that, when we do not evolve our variable, we remain at the same state. Also, $T(t)T(s)x = T(t+s)x$ for all $t, s \geqslant 0$ and $x \in X$, that is, when evolving from the present time to a future time s, then evolving again from s to t, we arrive at the same state as we would when evolving directly from the present time to the future time $t+s$.

The space X is, in general, a metric space, and we denote by $d\colon X \times X \to \mathbb{R}^+$ its metric, where $\mathbb{R}^+ = [0, \infty)$. Usually, we also ask that the semigroup possesses some continuity property, that is, the map $\mathbb{R}^+ \times X \ni (t,x) \mapsto T(t)x \in X$ is continuous. This requirement means that the evolution of the variable depends continuously on time and on its initial value.

Collecting these initial thoughts we have the following definition for a semigroup in X.

Definition 1.1. Let X be a metric space, with metric d, and let $\mathcal{C}(X)$ be the set of continuous maps from X into itself. A **semigroup** is a family $\mathcal{T} = \{T(t)\colon t \geqslant 0\} \subset \mathcal{C}(X)$ with the properties:

(a) $T(0) = I_X$, with I_X being the identity in X,
(b) $T(t+s) = T(t)T(s)$, for all $t, s \geqslant 0$, and
(c) the map $\mathbb{R}^+ \times X \ni (t,x) \mapsto T(t)x \in X$ is continuous,

where $T(t)T(s)$ denotes the composition map $T(t) \circ T(s)$, for all $t, s \geqslant 0$.

Semigroups frequently appear associated to *autonomous differential equations*. These autonomous differential equations can be ordinary, partial, or functional evolving differential equations. Next we discuss examples that arise from ordinary differential equations.

EXAMPLE 1.2. Consider the initial value problem

$$\begin{cases} \dot{x} = f(x), \\ x(0) = x_0 \in \mathbb{R}^n, \end{cases} \tag{1.1}$$

where $f\colon \mathbb{R}^n \to \mathbb{R}^n$ is a *locally Lipschitz continuous function*, that is, given $x_0 \in \mathbb{R}^n$ there exists $\delta > 0$ and $L \geqslant 0$ such that for $\|x - x_0\| < \delta$ and $\|y - x_0\| < \delta$ we have

$$\|f(x) - f(y)\| \leqslant L\|x - y\|.$$

Under these assumptions, the following result holds.

THEOREM 1.3 ([**130**]). *For each $x_0 \in \mathbb{R}^n$, there exist $\tau > 0$ and a continuously differentiable function $[0, \tau) \ni t \mapsto \xi(t) \in \mathbb{R}^n$ such that $\xi(0) = x_0$ and $\dot{\xi}(t) = f(\xi(t))$ for all $t \in (0, \tau)$. This function is called a* solution *of the initial value problem* (1.1) *and has the following properties:*

1. *if there exist $\sigma > 0$ and a continuously differentiable function $\eta\colon [0, \sigma) \to \mathbb{R}^n$ such that $\eta(0) = x_0$ and $\dot{\eta}(t) = f(\eta(t))$ for all $t \in (0, \sigma)$, then $\xi(t) = \eta(t)$ for all $t \in [0, \min\{\sigma, \tau\})$;*
2. *there exist $\tau(x_0) > 0$ and a solution $[0, \tau_{x_0}) \ni t \mapsto x(t, x_0) \in \mathbb{R}^n$ of* (1.1) *such that either $\tau(x_0) = \infty$ or $\limsup_{t \to \tau(x_0)^-} \|x(t, x_0)\| = \infty$;*
3. *if $E = \{(t, x) \in \mathbb{R}^+ \times \mathbb{R}^n : 0 \leqslant t < \tau(x)\}$, the function $E \ni (t, x_0) \mapsto x(t, x_0) \in \mathbb{R}^n$ is continuous.*

We will denote by $x(t, x_0)$ the value of the solution of (1.1) at time t, $t \in [0, \tau(x_0))$. Assume also that there exists a constant $M > 0$ such that

$$f(x) \cdot x < 0, \text{ whenever } \|x\| \geqslant M, \tag{1.2}$$

where $\cdot$ denotes the usual scalar product in $\mathbb{R}^n$. Then, if $x = x(t, x_0)$ is the solution of (1.1), we have that

$$\frac{d}{dt}\|x\|^2 = 2f(x) \cdot x.$$

It follows from (1.2) and from Theorem 1.3 that $\tau(x_0) = \infty$ for all $x_0 \in \mathbb{R}^n$.

Under these conditions we define $T(t)\colon \mathbb{R}^n \to \mathbb{R}^n$ by

$$T(t)x_0 = x(t, x_0) \quad \text{for each } x_0 \in \mathbb{R}^n \text{ and } t \geqslant 0.$$

It is clear that, in view of Theorem 1.3, the only condition that we need to check to conclude that $\mathcal{T} = \{T(t) : t \geqslant 0\}$ is a semigroup (with $X = \mathbb{R}^n$) is condition (b) of Definition 1.1. This condition follows from the uniqueness of solutions of (1.1) given in Theorem 1.3 observing that $x(t, x(s, x_0))$ and $x(t+s, x_0)$ are both solutions of the following initial value problem:

$$\begin{cases} \dot{y} = f(y), \\ y(0) = x(s, x_0) \in \mathbb{R}^n. \end{cases}$$

After these preliminary considerations we already have a good idea of what a semigroup is, and we will seek to express what we will try to understand from it, and from here on, unless stated otherwise, $\mathcal{T} = \{T(t) : t \geqslant 0\}$ will be a semigroup in a metric space X.

Firstly, we insist that the semigroups studied here must include more general models than those coming from ordinary differential equations. Hence, we assume that X is a metric space or, in the cases where the structure of vector spaces are needed, a Banach space (that may be finite or infinite dimensional).

Secondly, given the nature of our state space (or phase space) X, we wish to study those semigroups that are *dissipative* (in the example above, all solutions eventually enter the ball or radius M). A subset B_0 of X with the property that, given any bounded subset B of X, *all solutions* with initial conditions in B enter B_0 after a given time τ_B (which may depend on B) is called *absorbing*. We require that the dynamical systems (or semigroups) that we consider have a bounded absorbing set. We also wish that the intersection of all absorbing sets be compact and we will call this intersection the *global attractor*. The global attractor will be the asymptotic set of states and the states outside the global attractor are called *transient states*.

Throughout this chapter we will seek conditions for the semigroups that ensure the existence of global attractors. It is already possible to see that, if $f\colon \mathbb{R}^n \to \mathbb{R}^n$ is a locally Lipschitz map such that (1.2) is satisfied, the semigroup associated to (1.1) has a global attractor.

In our study of semigroups, the global attractors will play an essential role. In this chapter we present the basic definitions and results that lead to the characterization of semigroups which possess a global attractor.

1. Global attractors for semigroups

Given a subset A of the metric space X and $r > 0$, the r-neighborhood of A in X is the set

$$\mathcal{O}_r(A) = \{x \in X\colon d(x,a) < r \text{ for some } a \in A\}.$$

The notion of *invariance* plays a fundamental role in the study of the asymptotic behavior of semigroups, and it is given as follows.

DEFINITION 1.4 (Invariance). Let A be a subset of X. We say that A is:

(i) **positively $\mathcal{T}$-invariant** if $T(t)A \subset A$ for all $t \geqslant 0$;
(ii) **negatively $\mathcal{T}$-invariant** if $T(t)A \supset A$ for all $t \geqslant 0$;
(iii) **$\mathcal{T}$-invariant** if $T(t)A = A$ for all $t \geqslant 0$, that is, A is positively and negatively $\mathcal{T}$-invariant.

If $A = \{a\}$ is an invariant unitary set we say that a is an **equilibrium point**, or simply an **equilibrium** for $\mathcal{T}$.

Now we define the notions of *attraction* and *absorption* under the action of the semigroup $\mathcal{T}$. To that end we recall the definition of the **Hausdorff semidistance** $\mathrm{dist}_H(A,B)$ between two nonempty subsets A and B of X, given by:

$$\mathrm{dist}_H(A,B) = \sup_{x\in A} \inf_{y\in B} d(x,y),$$

and we will denote by $\mathrm{dist}(A,B)$ the usual **distance** (which is not a distance in the sense of metric spaces) between A and B, that is,

$$\mathrm{dist}(A,B) = \inf_{x\in A} \inf_{y\in B} d(x,y).$$

REMARK 1.5. We point out that:

(a) $\mathrm{dist}_H(A,B) = 0$ if and only if $A \subset \overline{B}$;
(b) there exist two closed nonempty sets A and B such that $\mathrm{dist}(A,B) = 0$ and $A \cap B = \varnothing$.

DEFINITION 1.6 (Attraction and absorption). Let A and B be nonempty subsets of X. We say that A **$\mathcal{T}$-attracts** B if

$$\lim_{t\to\infty} \operatorname{dist}_H(T(t)B, A) = 0.$$

If there exists $t_0 \geqslant 0$ such that $T(t)B \subset A$ for all $t \geqslant t_0$, we say that A **$\mathcal{T}$-absorbs** B.

In particular, if A $\mathcal{T}$-absorbs B, then A $\mathcal{T}$-attracts B (the converse is not true in general). However, if A $\mathcal{T}$-attracts B, then given $r > 0$, the set $\mathcal{O}_r(A)$ $\mathcal{T}$-absorbs B.

We are now ready to define *global attractors* for semigroups. A global attractor for a semigroup, when it exists, is the set of all possible asymptotic states of the semigroup, as we shall see in the following definition.

DEFINITION 1.7 (Global attractor). A set $\mathcal{A}$ is called a **global attractor** for the semigroup $\mathcal{T}$ if it is compact, $\mathcal{T}$-invariant and $\mathcal{T}$-attracts all bounded subsets of X.

Note that the global attractor for a semigroup $\mathcal{T}$ is unique. In fact, if $\mathcal{A}_1$ and $\mathcal{A}_2$ are global attractors for this semigroup, we have

$$\operatorname{dist}_H(\mathcal{A}_1, \mathcal{A}_2) = \operatorname{dist}_H(T(t)\mathcal{A}_1, \mathcal{A}_2) \overset{t\to\infty}{\longrightarrow} 0,$$

and hence $\mathcal{A}_1 \subset \mathcal{A}_2$. Analogously $\mathcal{A}_2 \subset \mathcal{A}_1$, and the equality follows.

A very useful characterization of the global attractor of a semigroup $\mathcal{T}$ is given in terms of *global solutions*. To state this result, we introduce the following definition.

DEFINITION 1.8.

(a) A **global solution** of $\mathcal{T}$ is a continuous function $\phi\colon \mathbb{R} \to X$ such that $T(t)\phi(s) = \phi(t+s)$, for all $s \in \mathbb{R}$ and $t \geqslant 0$.
(b) If $x = \phi(0)$, we say that ϕ is a **global solution through** x of $\mathcal{T}$.
(c) If $\phi(\mathbb{R})$ is a bounded subset of X, we say that ϕ is a **bounded global solution** of $\mathcal{T}$.
(d) A constant global solution ϕ for $\mathcal{T}$ will be called a **stationary solution** for $\mathcal{T}$.

Note that since $T(t)$ is not necessarily injective, if a global solution through x exists, it does not need to be unique. Also, it is clear that if $\phi \equiv a$ is a stationary solution for $\mathcal{T}$, then a is an equilibrium point of $\mathcal{T}$.

We can now present a characterization result for global attractors using global solutions.

THEOREM 1.9. *Let $\mathcal{T}$ be a semigroup with a global attractor $\mathcal{A}$. Then, given $x \in \mathcal{A}$, there exists a bounded global solution $\phi\colon \mathbb{R} \to \mathcal{A}$ such that $\phi(0) = x$. Consequently,*

$$\mathcal{A} = \{x \in X\colon \text{ there exists a bounded global solution through } x \text{ of } \mathcal{T}\}.$$

PROOF. If $\mathcal{A}$ is the global attractor and $x \in \mathcal{A}$ for $\mathcal{T}$, then there is a global solution $\phi\colon \mathbb{R} \to \mathcal{A}$ through x. In fact, from the positive $\mathcal{T}$-invariance of $\mathcal{A}$, $\mathbb{R}^+ \ni t \mapsto \phi(t) = T(t)x \in \mathcal{A}$ is always uniquely defined. Now, since $x \in \mathcal{A} = T(1)\mathcal{A}$, there exists $x_{-1} \in \mathcal{A}$ such that $T(1)x_{-1} = x$ and we may construct a sequence $\{x_{-n}\}_{n\in\mathbb{N}}$

in $\mathcal{A}$ such that $x_0 = x$ and $T(1)x_{-n-1} = x_{-n}$ for all $n \in \mathbb{N}$ (recall that the sequence $\{x_{-n}\}_{n\in\mathbb{N}}$ does not need to be uniquely determined). Define

$$\phi(t) = \begin{cases} T(t)x, & t \geqslant 0, \\ T(j+t)x_{-j}, & t \in [-j,-j+1), j = 1,2,3,\cdots. \end{cases}$$

It follows that $\phi\colon \mathbb{R} \to \mathcal{A}$ is a global solution of $\mathcal{T}$ through x, and since $\phi(\mathbb{R}) \subset \mathcal{A}$, ϕ is bounded.

Conversely, given a bounded global solution $\phi\colon \mathbb{R} \to X$ of $\mathcal{T}$, using that $\phi(\mathbb{R})$ is a bounded $\mathcal{T}$-invariant set, and that $\mathcal{A}$ $\mathcal{T}$-attracts bounded sets, we obtain that $\phi(\mathbb{R}) \subset \mathcal{A}$, which concludes the proof. □

REMARK 1.10. Note that if A is any $\mathcal{T}$-invariant set, then there exists at least a global solution $\phi\colon \mathbb{R} \to A$ through x, for any given $x \in A$.

Our goal is to present results that give a characterization of the semigroups which possess a global attractor. With this in mind, we will note that the set of points where the orbit of a set B accumulates, called the *ω-limit set* of B, plays a fundamental role.

DEFINITION 1.11. Let $\mathcal{T}$ be a semigroup in X.

(a) If B is a subset of X, for each $t \geqslant 0$, we define the **image** of B by $T(t)$ by
$$T(t)B = \{T(t)x\colon x \in B\}.$$

(b) If B is a subset of X, we define for each $t \geqslant 0$ the **partial positive orbit** of B by
$$\gamma_t^+(B) = \bigcup_{s\geqslant t} T(s)B.$$
When $t = 0$, we call it simply the **positive orbit** of B, and denote it simply by $\gamma^+(B)$.

(c) When a global solution $\phi\colon \mathbb{R} \to X$ through x exists, we define the **entire orbit of x relative to the global solution ϕ** by
$$\gamma_\phi(x) = \{\phi(t)\colon t \in \mathbb{R}\}.$$
In this case, for each $t \in \mathbb{R}$, we define the **partial negative orbit** of x **relative to ϕ** by
$$(\gamma_\phi)_t^-(x) = \{\phi(s)\colon s \leqslant t\}.$$

With these definitions, we are able to define the ω and α-limit sets. As stated before, the ω-limit sets will play a key role in the characterization of semigroups with global attractors. As for the α-limit, it will be useful later to give a more detailed description of the internal structures of global attractors.

DEFINITION 1.12. The **ω-limit set** of $B \subset X$ is defined by
$$\omega(B) = \bigcap_{t\in\mathbb{R}^+} \overline{\gamma_t^+(B)}.$$

When a global solution $\phi\colon \mathbb{R} \to X$ through x exists, we also define the **α-limit set** of x relative to ϕ as
$$\alpha_\phi(x) = \bigcap_{t\in\mathbb{R}^-} \overline{(\gamma_\phi)_t^-(x)}.$$

The following characterization of the ω-limit sets and α-limit sets will be frequently used in the proofs of the results that follow.

PROPOSITION 1.13. *If $B \subset X$, then $\omega(B)$ is closed and*

$$\omega(B) = \{y \in X\colon \text{ there exist sequences } \{t_n\}_{n\in\mathbb{N}} \text{ in } \mathbb{R}^+ \text{ and } \{x_n\}_{n\in\mathbb{N}} \text{ in } B \text{ such that } t_n \overset{n\to\infty}{\longrightarrow} \infty \text{ and } y = \lim_{n\to\infty} T(t_n)x_n\}.$$

Also, if $\phi\colon \mathbb{R} \to X$ is a global solution of $\mathcal{T}$ through $x \in X$, then $\alpha_\phi(x)$ is closed and

$$\alpha_\phi(x) = \{v \in X\colon \text{ there exists a sequence } \{t_n\}_{n\in\mathbb{N}} \text{ in } \mathbb{R}^+ \text{ such that } t_n \overset{n\to\infty}{\longrightarrow} \infty \text{ and } v = \lim_{n\to\infty} \phi(-t_n)\}.$$

PROOF. The closedness of the sets is straightforward, since both are intersections of closed sets. Now, if $y \in \omega(B)$, then $y \in \bigcap_{t\in\mathbb{R}^+} \overline{\gamma_t^+(B)}$. Consequently $y \in \overline{\gamma_t^+(B)}$ for all $t \geqslant 0$, that is, for each $n \in \mathbb{N}$ there exists a sequence $\{y_k^n\}_{k\in\mathbb{N}} \subset \gamma_n^+(B)$ such that $y_k^n \overset{k\to\infty}{\longrightarrow} y$. Since $y_k^n \in \gamma_n^+(B)$ for all $n \in \mathbb{N}$ and $k \in \mathbb{N}$, there exists $\{x_k^n\}_{n,k\in\mathbb{N}} \subset B$ and $\{q_k^n\}_{n,k\in\mathbb{N}} \subset \mathbb{R}^+$ such that

$$y_k^n = T(n + q_k^n)x_k^n.$$

Also, given $n \in \mathbb{N}$ and $\epsilon > 0$, there is a $k(n,\epsilon) \in \mathbb{N}$ such that

$$d(y_k^n, y) < \epsilon, \text{ whenever } k \geqslant k(n,\epsilon),$$

that is, $d(T(n + q_k^n)x_k^n, y) < \epsilon$ if $k \geqslant k(n,\epsilon)$. Defining $t_n = n + q^n_{k(n,\frac{1}{n})}$ and $x_n = x^n_{k(n,\frac{1}{n})}$, we have

$$d(T(t_n)x_n, y) < \tfrac{1}{n} \overset{n\to\infty}{\longrightarrow} 0,$$

and therefore $y = \lim_{n\to\infty} T(t_n)x_n$, $t_n \overset{n\to\infty}{\longrightarrow} \infty$ and $x_n \in B$ for all $n \in \mathbb{N}$.

On the other hand, fix $y \in X$ such that there exist sequences $\{t_n\}_{n\in\mathbb{N}} \subset \mathbb{R}^+$, $\{x_n\}_{n\in\mathbb{N}} \subset B$ with $t_n \overset{n\to\infty}{\longrightarrow} \infty$ and $y = \lim_{n\to\infty} T(t_n)x_n$. Hence, if $\tau \geqslant 0$ we have $\{T(t_n)x_n\}_{t_n\geqslant\tau} \subset \gamma_\tau^+(B)$, and $y \in \overline{\gamma_\tau^+(B)}$. Consequently $y \in \omega(B)$ and the characterization of $\omega(B)$ is proved.

The proof of the characterization of $\alpha_\phi(x)$ is similar. □

The following proposition represents a topological property that will be useful for the theory that follows.

PROPOSITION 1.14. *Let K be a compact subset of X and $\{x_n\}_{n\in\mathbb{N}}$ a sequence in X such that*

$$\lim_{n\to\infty} \operatorname{dist}_H(x_n, K) = 0.$$

Then $\{x_n\}_{n\in\mathbb{N}}$ has a convergent subsequence to a point in K.

PROOF. Given $m \in \mathbb{N}$, there are $n_m \in \mathbb{N}$ and $y_{n_m} \in K$ such that $d(x_{n_m}, y_{n_m}) < \frac{1}{m}$. Since K is compact we may assume, passing to a subsequence if necessary, that $y_{n_m} \overset{m\to\infty}{\longrightarrow} y_0$ for some $y_0 \in K$. Hence, we obtain

$$d(x_{n_m}, y_0) \leqslant d(x_{n_m}, y_{n_m}) + d(y_{n_m}, y_0) \overset{n\to\infty}{\longrightarrow} 0,$$

that is, $\{x_n\}_{n\in\mathbb{N}}$ has a convergent subsequence to a point in K. □

LEMMA 1.15. *Let $\mathcal{T}$ be a semigroup in X. If $B \subset X$, then $\omega(B)$ is positively $\mathcal{T}$-invariant. If $B \subset X$ is such that $\omega(B)$ is compact and $\mathcal{T}$-attracts B, then $\omega(B)$ is $\mathcal{T}$-invariant.*

PROOF. If $\omega(B) = \varnothing$ there is nothing to prove. Assume then that $\omega(B) \neq \varnothing$ and fix $t \geqslant 0$. From Proposition 1.13, given $y \in \omega(B)$ there exist sequences $\{t_n\}_{n\in\mathbb{N}} \subset \mathbb{R}^+$ with $t_n \overset{n\to\infty}{\longrightarrow} \infty$ and $\{x_n\}_{n\in\mathbb{N}} \subset B$ such that $y = \lim_{n\to\infty} T(t_n)x_n$. It follows from the continuity of $T(t)$ that $T(t)y = \lim_{n\to\infty} T(t+t_n)x_n$ and hence $T(t)y \in \omega(B)$, using once more the characterization given in Proposition 1.13. Hence $T(t)\omega(B) \subset \omega(B)$, and since this holds for each $t \geqslant 0$, $\omega(B)$ is positively $\mathcal{T}$-invariant.

It remains to prove that if $\omega(B)$ is compact and $\mathcal{T}$-attracts B, then $\omega(B)$ is negatively $\mathcal{T}$-invariant. For $y \in \omega(B)$ there are sequences $t_n \to \infty$ and $x_n \in B$ such that $x = \lim_{n\to\infty} T(t_n)x_n$. For $t \geqslant 0$ fixed, since $t_n \to \infty$, there exists $n_0 \in \mathbb{N}$ such that $t_n > t$ for all $n \geqslant n_0$. Therefore $T(t)T(t_n - t)x_n = T(t_n)x_n \to x$ as $n \to \infty$. Since $\omega(B)$ is compact and attracts B we have that $\operatorname{dist}_H(T(t_n - t)x_n, \omega(B)) \overset{n\to\infty}{\longrightarrow} 0$. It follows from Proposition 1.14 that $\{T(t_n - t)x_n\}_{n\in\mathbb{N}}$ has a convergent subsequence (which we denote the same) to a point $x \in \omega(B)$. Hence, by continuity of $T(t)$, we have $T(t)x = y$. Therefore $\omega(B) \subset T(t)\omega(B)$ and $\omega(B)$ is negatively $\mathcal{T}$-invariant. This completes the proof. □

It follows immediately from Lemma 1.15 that, if $x \in X$, $\omega(x)$ attracts x and $\omega(x) = \{x^*\}$, then x^* is an equilibrium for $\mathcal{T}$. A similar result holds for $\alpha_\phi(x)$, as we will see below.

LEMMA 1.16. *Let $x \in X$ and assume that there is a global solution $\phi\colon \mathbb{R} \to X$ through x of $\mathcal{T}$ such that $\overline{\phi(\mathbb{R}^-)}$ is compact. Then $\alpha_\phi(x)$ is nonempty, compact, connected and $\mathcal{T}$-invariant.*

PROOF. From the definition of $\alpha_\phi(x)$, the compactness of $\overline{\phi(\mathbb{R}^-)}$ and from the finite intersection property, $\alpha_\phi(x)$ is nonempty and compact. Let us show that $\alpha_\phi(x)$ is $\mathcal{T}$-invariant.

Fix $t \geqslant 0$. From Proposition 1.13, given $y \in \alpha_\phi(x)$, there exists a sequence $t_n \overset{n\to\infty}{\longrightarrow} +\infty$ such that $\phi(-t_n) \overset{n\to\infty}{\longrightarrow} y$. From the continuity of $T(t)$ we obtain that $T(t)\phi(-t_n) = \phi(t - t_n) \overset{n\to\infty}{\longrightarrow} T(t)y$ and therefore $T(t)y \in \alpha_\phi(x)$. On the other hand, if $w \in \alpha_\phi(x)$, there exists a sequence $t_n \overset{n\to\infty}{\longrightarrow} \infty$ such that $\phi(-t_n) \overset{n\to\infty}{\longrightarrow} w$. Since $\{\phi(-t_n - t)\}_{n\in\mathbb{N}}$ is relatively compact, passing to a subsequence if necessary, there exists $z \in X$ such that $\phi(-t_n - t) \overset{n\to\infty}{\longrightarrow} z$ and $z \in \alpha_\phi(x)$. It follows from the continuity of $T(t)$ that $T(t)z = w$.

Finally, the connectedness follows noting that $\alpha_\phi(x) = \bigcap_{t\leqslant 0} \overline{\phi(\mathbb{R}_t^-)}$, which is an intersection of nested connected sets. □

As for connectedness of ω-limits, we have the following result.

LEMMA 1.17. *Let $\mathcal{T}$ be a semigroup in X and $B \subset X$ such that $\omega(B)$ is compact and $\mathcal{T}$-attracts B. If B is connected, then $\omega(B)$ is connected.*

PROOF. Note that $\overline{\gamma_t^+(B)}$ is connected for $t \geqslant 0$, since $[0,\infty) \times X \ni (s,x) \mapsto T(s)x \in X$ is continuous and takes the connected set $[t,\infty) \times B$ onto $\gamma_t^+(B)$. Since $\omega(B)$ attracts B, we can easily see that it is connected. □

The following definition is very important to study global attractors for semigroups.

DEFINITION 1.18. Let $\mathcal{T}$ be a semigroup. We say that $\mathcal{T}$ is **bounded dissipative** if there exists a bounded set $B_0 \subset X$ that $\mathcal{T}$-attracts bounded subsets of X and we say that $\mathcal{T}$ is **eventually bounded** if for each bounded subset B of X there exists $t_B > 0$ such that $\gamma^+_{t_B}(B)$ is bounded.

Note that in the above definition we can exchange the word *attracts* by the word *absorbs* without changing the meaning of the concept. We will now explore a crucial property for semigroups that possess a global attractor.

DEFINITION 1.19. A semigroup $\mathcal{T}$ is said to be **asymptotically compact** if, for each nonempty, closed, positively $\mathcal{T}$-invariant and bounded set $B \subset X$, there exists a compact set $J \subset B$ that $\mathcal{T}$-attracts B.

LEMMA 1.20. *Let $\mathcal{T}$ be an eventually bounded and asymptotically compact semigroup in X. Then for each nonempty bounded subset B of X we have $\omega(B)$ nonempty, compact, $\mathcal{T}$-invariant and $\mathcal{T}$-attracts B.*

PROOF. Since $\mathcal{T}$ is eventually bounded, if B is a bounded nonempty subset of X, there exists $t_0 \geqslant 0$ such that

$$T(t)B \subset B_0 \text{ for all } t \geqslant t_0.$$

Hence $\gamma^+_{t_0}(B) \subset B_0$ is bounded. Since $T(t)$ is continuous and $T(t)\gamma^+_{t_0}(B) \subset \gamma^+_{t_0}(B)$, it follows that $T(t)\overline{\gamma^+_{t_0}(B)} \subset \overline{\gamma^+_{t_0}(B)}$ for all $t \geqslant 0$. Thus $\overline{\gamma^+_{t_0}(B)}$ is a nonempty closed bounded and positively $\mathcal{T}$-invariant subset of X, and since $\mathcal{T}$ is asymptotically compact, there exists a compact set $J \subset \overline{\gamma^+_{t_0}(B)}$ that $\mathcal{T}$-attracts $\overline{\gamma^+_{t_0}(B)}$. Hence, there are sequences $\epsilon_n \to 0$ and $t_n \overset{n\to\infty}{\longrightarrow} \infty$ such that $T(t)(\overline{\gamma^+_{t_0}(B)}) \subset \mathcal{O}_{\epsilon_n}(J)$ for all $t \geqslant t_n$. So $\varnothing \neq \omega(B) \subset J$. Since $\omega(B)$ is closed and J is compact, we have $\omega(B)$ compact.

Now we show that $\omega(B)$ $\mathcal{T}$-attracts B. If this is not the case, there exist $\epsilon_0 > 0$ and sequences $\{x_n\}_{n\in\mathbb{N}} \subset B$ and $\{t_n\}_{n\in\mathbb{N}} \subset \mathbb{R}^+$ with $t_n \overset{n\to\infty}{\longrightarrow} \infty$ such that $\operatorname{dist}_H(T(t_n)x_n, \omega(B)) \geqslant \epsilon_0$. From the compactness of J and Proposition 1.14, we can extract subsequences $\{x_{n_j}\}_{j\in\mathbb{N}}$ of $\{x_n\}_{n\in\mathbb{N}}$, $\{t_{n_j}\}_{j\in\mathbb{N}}$ of $\{t_n\}_{n\in\mathbb{N}}$ and $z \in J$ such that $T(t_{n_j})x_{n_j} \overset{j\to\infty}{\longrightarrow} z$. It follows that $z \in \omega(B)$ and $\operatorname{dist}_H(z, \omega(B)) \geqslant \epsilon_0$, which is absurd, proving that $\omega(B)$ $\mathcal{T}$-attracts B.

Therefore, $\omega(B)$ is nonempty, compact and $\mathcal{T}$-attracts B. Finally, from Lemma 1.15, it follows that $\omega(B)$ is $\mathcal{T}$-invariant. □

LEMMA 1.21. *Let $\mathcal{T}$ be a bounded dissipative and asymptotically compact semigroup in X, and let B_0 be a bounded subset of X that $\mathcal{T}$-absorbs all bounded subsets of X. Then for each nonempty bounded subset B of X we have*

$$\omega(B) \subset \omega(B_0) \subset \overline{B_0}.$$

PROOF. Let B be a bounded subset of X. From Lemma 1.20, we know that $\omega(B)$ is nonempty, compact, $\mathcal{T}$-invariant and $\mathcal{T}$-attracts B. Moreover, from the fact that B_0 absorbs B, we know that $\gamma^+_{t_0}(B) \subset B_0$, for some $t_0 \geqslant 0$. Hence, if $y \in \omega(B)$, there exist sequences $\{x_n\}_{n\in\mathbb{N}}$ in B and $\{t_n\}_{n\in\mathbb{N}}$ in $\mathbb{R}^+$ with $t_n \overset{n\to\infty}{\longrightarrow} \infty$

and $y = \lim_{n\to\infty} T(t_n)x_n$. We can assume, without loss of generality, that $t_n \geqslant t_0$ for all $n \in \mathbb{N}$, and hence

$$T(t_n)x_n = T(t_n - t_0)T(t_0)x_n \subset T(t_n - t_0)T(t_0)B \subset T(t_n - t_0)B_0,$$

showing that $y \in \omega(B_0)$. Since B_0 is bounded, it $\mathcal{T}$-absorbs itself, hence we conclude that $\omega(B_0) \subset \overline{B_0}$. □

It is useful to have the following equivalent characterization of asymptotical compactness.

PROPOSITION 1.22. *Let $\mathcal{T}$ be a bounded dissipative semigroup in X. Then $\mathcal{T}$ is asymptotically compact iff it satisfies the following condition:*

$$\begin{cases} \text{for each bounded sequence } \{x_n\}_{n\in\mathbb{N}} \text{ in } X \text{ and } \{t_n\}_{n\in\mathbb{N}} \\ \text{with } t_n \overset{n\to\infty}{\longrightarrow} \infty \text{ we have } \{T(t_n)x_n\}_{n\in\mathbb{N}} \text{ relatively compact.} \end{cases} \tag{1.3}$$

PROOF. Let $B \subset X$ be nonempty, closed, bounded and positively $\mathcal{T}$-invariant. Using (1.3) we can easily prove that $\omega(B) \subset B$ is nonempty, compact, $\mathcal{T}$-invariant and $\mathcal{T}$-attracts B, simply rewriting the proof of Lemma 1.20 with (1.3) instead of the hypothesis of asymptotical compactness of $\mathcal{T}$. Therefore it follows that $\mathcal{T}$ is asymptotically compact.

Conversely, if $\mathcal{T}$ is a bounded dissipative semigroup and $\{x_n\}_{n\in\mathbb{N}}$ is a bounded sequence in X, there exists $t_0 \geqslant 0$ such that $B = \overline{\gamma^+_{t_0}(\{x_n\}_{n\in\mathbb{N}})}$ is a bounded set. Since B is positively $\mathcal{T}$-invariant and $\mathcal{T}$ is asymptotically compact, there exists a compact subset $J \subset B$ of X which $\mathcal{T}$-attracts B. In particular $\{T(t_n)x_n\}_{n\in\mathbb{N}}$ converges to J as $n \to \infty$ and therefore it is relatively compact. □

The following theorem characterizes the semigroups that have a global attractor.

THEOREM 1.23. *A semigroup $\mathcal{T}$ in X has a global attractor $\mathcal{A}$ iff $\mathcal{T}$ is bounded dissipative and asymptotically compact. Moreover, if B_0 is a bounded subset of X that $\mathcal{T}$-absorbs all bounded subsets of X, then*

$$\mathcal{A} = \omega(B_0).$$

PROOF. Assume first that $\mathcal{T}$ is bounded dissipative and asymptotically compact and let B_0 be a bounded subset of X that $\mathcal{T}$-absorbs all bounded subsets of X. Define $\mathcal{A} = \omega(B_0)$, which is nonempty, compact, $\mathcal{T}$-invariant and $\mathcal{T}$-attracts B_0, by Lemma 1.20.

Given B a nonempty bounded subset of X, Lemmas 1.20 and 1.21 show that $\omega(B)$ is nonempty, compact, $\mathcal{T}$-invariant, $\mathcal{T}$-attracts B and $\omega(B) \subset \mathcal{A}$. It follows that $\mathcal{A}$ $\mathcal{T}$-attracts B, and the proof that $\mathcal{A}$ is the global attractor of $\mathcal{T}$ is complete.

For the converse, if $\mathcal{T}$ has a global attractor $\mathcal{A}$, then it is clearly bounded dissipative, since $\mathcal{A}$ is a bounded subset of X that $\mathcal{T}$-attracts all bounded subsets of X. Using Proposition 1.22, the asymptotical compactness of $\mathcal{T}$ follows, and we conclude the proof. □

1.1. Asymptotic compactness in Banach spaces. In this short subsection we present a useful result to prove asymptotic compactness for semigroups in Banach spaces.

THEOREM 1.24. *Let X be a Banach space, with norm $\|\cdot\|$, and let $\mathcal{T}$ be a semigroup in X. Suppose that for each $t \geqslant 0$, we can write $T(t) = S(t) + K(t)$ and for each bounded subset B of X, there exists $t_B \geqslant 0$ such that:*

(i) *$K(t)B$ is relatively compact for all $t \geqslant t_B$.*

(ii) *$s_B(t) = \sup_{x \in B} \|S(t)x\|_X < \infty$ for all $t \geqslant t_B$ and $s_B(t) \stackrel{t\to\infty}{\longrightarrow} 0$.*

Then $\mathcal{T}$ is asymptotically compact.

PROOF. Let B be a nonempty, closed, bounded and positively $\mathcal{T}$-invariant subset of X. Given $\epsilon > 0$, we can choose $t_B \geqslant 0$ such that $s_B(t_B) < \epsilon/4$ and $K(t_B)B$ is relatively compact.

Hence, there exists $n_0 \in \mathbb{N}$ and points $y_1, \cdots, y_{n_0} \in X$ such that $K(t_B)B \subset \bigcup_{i=1}^{n_0} B_{\epsilon/4}(y_i)$. Since B is positively $\mathcal{T}$-invariant, we have

$$\begin{aligned}\omega(B) &\subset \bigcap_{t\in\mathbb{T}^+} \overline{T(t)B} \subset \overline{T(t_B)B} \subset \overline{S(t_B)B} + \overline{K(t_B)B} \\ &\subset B_{\epsilon/2}(0) + \bigcup_{i=1}^{n_0} B_{\epsilon/2}(y_i) \subset \bigcup_{i=1}^{n_0} B_{\epsilon}(y_i).\end{aligned}$$

Since $\epsilon > 0$ is arbitrary, we have $\omega(B)$ totally bounded. Thus $\omega(B)$ is closed and totally bounded in the Banach space X and therefore is compact.

Clearly $\omega(B)$ is nonempty, since for each sequence $\{x_n\}_{n\in\mathbb{N}}$ in B and $t_n \geqslant 0$ with $t_n \stackrel{n\to\infty}{\longrightarrow} \infty$, the sequence $\{T(t_n)x_n\}_{n\in\mathbb{N}} = \{K(t_n)x_n + S(t_n)x_n\}_{n\in\mathbb{N}}$ is totally bounded and therefore has a convergent subsequence.

It remains to prove that $\omega(B)$ $\mathcal{T}$-attracts B. Suppose that this is not the case. Then there exist $\epsilon_0 > 0$ and sequences $\{x_n\}_{n\in\mathbb{N}}$ in B, $\{t_n\}_{n\in\mathbb{N}}$ in $\mathbb{R}^+$ with $t_n \stackrel{n\to\infty}{\longrightarrow} \infty$, such that $\text{dist}_H(T(t_n)x_n, \omega(B)) \geqslant \epsilon_0$ for all $n \in \mathbb{N}$. As before, we know that the sequence $\{T(t_n)x_n\}_{n\in\mathbb{N}}$ has a convergent subsequence to a point $y \in X$. From this it follows that $y \in \omega(B)$ and $\text{dist}_H(y, \omega(B)) \geqslant \epsilon_0$. That leads us to a contradiction and proves that $\omega(B)$ attracts B.

This concludes the proof that $\mathcal{T}$ is asymptotically compact. □

COROLLARY 1.25. *If $\mathcal{T}$ is a bounded dissipative semigroup satisfying the conditions of Theorem 1.24, then $\mathcal{T}$ has a global attractor.*

2. Examples of semigroups with global attractors

In this section we present examples of semigroups with global attractors, using Theorem 1.23.

2.1. Diffusive coupling and tridiagonal matrices. Let $n \geqslant 3$ be an integer and consider positive real numbers ℓ_i, for $i = 0, \cdots, n-1$, and β_i, for $i = 0, \cdots, n-2$. Define $\alpha_0 = \beta_0$, $\alpha_i = \beta_i + \beta_{i-1}$, $1 \leqslant i \leqslant n-2$, $\alpha_{n-1} = \beta_{n-2}$, and

$$A = \begin{bmatrix} \begin{matrix} \alpha_0 & -\beta_0 & 0 & 0 \\ -\beta_0 & \alpha_1 & -\beta_1 & 0 \\ 0 & -\beta_1 & \alpha_2 & -\beta_2 \\ 0 & 0 & -\beta_2 & \alpha_3 \end{matrix} & & 0 \\ & \ddots & \\ 0 & & \begin{matrix} \alpha_{n-4} & -\beta_{n-4} & 0 & 0 \\ -\beta_{n-4} & \alpha_{n-3} & -\beta_{n-3} & 0 \\ 0 & -\beta_{n-3} & \alpha_{n-2} & -\beta_{n-2} \\ 0 & 0 & -\beta_{n-2} & \alpha_{n-1} \end{matrix} \end{bmatrix},$$

(1.4) $$\tilde{B} = A + \operatorname{diag}(\ell_0, \cdots, \ell_{n-1})I,$$

and $B = \operatorname{diag}(\ell_0^{-1}, \cdots, \ell_{n-1}^{-1})\tilde{B}$.

Tridiagonal matrices of the form (1.4) appear associated to the process of discretization of second-order differential operators (see [**38, 126**]) or as limiting operators for certain problems with varying diffusivity (see [**83, 123**]).

Consider $X = \mathbb{R}^n$ with the scalar product defined by

$$\langle \mathbf{x}, \mathbf{y} \rangle_X = \sum_{i=0}^{n-1} \ell_i x_i y_i \text{ for all } \mathbf{x} = (x_0, \cdots, x_{n-1}),\ \mathbf{y} = (y_0, \cdots, y_{n-1}) \in X,$$

and define the symmetric coercive bilinear form $\tilde{b}\colon X \times X \to \mathbb{R}$,

$$\tilde{b}(\mathbf{x}, \mathbf{y}) = \sum_{j=1}^{n-1} \beta_{j-1}(x_j - x_{j-1})(y_j - y_{j-1}) + \langle \mathbf{x}, \mathbf{y} \rangle_X,$$

for all $\mathbf{x} = (x_0, x_1, \cdots, x_{n-1})$ and $\mathbf{y} = (y_0, y_1, \cdots, y_{n-1})$ in X.

Thus for $\mathbf{w} = (w_0, w_1, \cdots, w_{n-1}) \in X$, the solution $\mathbf{u}$ of $B\mathbf{u} = \mathbf{w}$ is the solution of

$$\tilde{B}\mathbf{u} + \operatorname{diag}(\ell_0, \cdots, \ell_{n-1})\mathbf{u} = \operatorname{diag}(\ell_0, \cdots, \ell_{n-1})\mathbf{w},$$

and thus it is characterized as the minimizer of the functional $\mathbb{R}^n \ni \mathbf{x} \mapsto \frac{1}{2}\tilde{b}(\mathbf{x}, \mathbf{x}) - \langle \mathbf{w}, \mathbf{x} \rangle_X \in \mathbb{R}$, namely,

$$\begin{aligned} \mu &= \min_{\mathbf{x} \in \mathbb{R}^n} \left\{ \frac{1}{2}\tilde{b}(\mathbf{x}, \mathbf{x}) - \langle \mathbf{w}, \mathbf{x} \rangle_X \right\} \\ &= \min_{\mathbf{x} \in \mathbb{R}^n} \left\{ \sum_{j=1}^{n-1} \frac{\beta_{j-1}}{2}(x_j - x_{j-1})^2 + \sum_{j=0}^{n-1} \frac{\ell_j}{2} x_j^2 - \sum_{j=0}^{n-1} \ell_j w_j x_j \right\}. \end{aligned}$$

If $f\colon \mathbb{R} \to \mathbb{R}$ is a continuously differentiable function which satisfies

(1.5) $$\limsup_{|s| \to \infty} \frac{f(s)}{s} < 0,$$

the semigroup $\mathcal{T}$ defined in $\mathbb{R}^n$ by the system of ordinary differential equations of the form

(1.6) $$\frac{d\mathbf{u}}{dt} = -B\mathbf{u} + F(\mathbf{u}) \text{ for } t \in \mathbb{R},$$

where $F(\mathbf{u}) = (f(u_0), \cdots, f(u_{n-1}))^T$, has a global attractor which we will denote by $\mathbf{A}$. That can be achieved by multiplying the equation by $\operatorname{diag}(\ell_0, \cdots, \ell_{n-1})\mathbf{u}$ and using the dissipativity condition (1.5) to obtain that

$$\frac{d}{dt}\|\mathbf{u}\|_X^2 \leqslant -\theta\|\mathbf{u}\|_X^2 + C_\theta,$$

for some positive constants θ and C_θ, proving the existence of a bounded absorbing set. Now applying Theorem 1.23 we obtain the existence of the global attractor $\mathbf{A}$, recalling that the asymptotic compactness follows directly from the finite dimension of X and the compactness of balls. We remark that the flow of (1.6) is *structurally stable* in $\mathbf{A}$ (see [**124**]).

2.2. Ordinary differential equations modelling atmospheric problems. Consider the system of ordinary differential equations:

$$\left\{\begin{aligned} \dot u_0 &= f_0(u_0) + \alpha_0^{-1}\sum_{i=1}^{n}\alpha_i(u_i-u_0),\\ \dot u_1 &= f_1(u_1) + \alpha_1(u_0-u_1),\\ &\vdots\\ \dot u_n &= f_n(u_n) + \alpha_n(u_0-u_n), \end{aligned}\right. \tag{1.7}$$

where $f_i\colon \mathbb{R}\to\mathbb{R}$ is a continuously differentiable function and α_i is a positive real number, for each $i=0,\cdots,n$. This system of ordinary differential equations appears as a limiting problem for a system of parabolic problems with large diffusion modelling an atmospheric problem (see [**69, 197**]).

Consider $X=\mathbb{R}^{n+1}$ equipped with the inner product defined by

$$\langle \mathbf{u},\mathbf{v}\rangle_X = \alpha_0 u_0 v_0 + \sum_{i=1}^{n} u_i v_i, \text{ for } \mathbf{u}=(u_0,\cdots,u_n),\ \mathbf{v}=(v_0,\cdots,v_n)\in X,$$

which makes X a Hilbert space, and also consider $\|\cdot\|_X$ its associated norm.

For $\mathbf{u}\in\mathbb{R}^{n+1}$, let $\mathbf{f}(\mathbf{u})=(f_0(u_0),\cdots,f_n(u_n))$ and assume that there are positive constants M and θ such that

$$2\langle \mathbf{f}(\mathbf{u}),\mathbf{u}\rangle_X \leqslant -\theta\langle \mathbf{u},\mathbf{u}\rangle_X \quad \text{whenever } \|\mathbf{u}\|_X > M. \tag{1.8}$$

Under these assumptions, the semigroup associated to (1.7) has a global attractor. In fact, multiplying the first equation in (1.7) by $\alpha_0 u_0$, the others by u_i for $i=1,\ldots,n$ respectively and summing the results we get

$$\frac{d}{dt}\|\mathbf{u}\|_X^2 = 2\langle \mathbf{f}(\mathbf{u}),\mathbf{u}\rangle_X - 2\sum_{i=1}^{n}\alpha_i(u_i-u_0)^2.$$

Using the dissipativity condition from (1.8) we obtain

$$\frac{d}{dt}\|u\|_X^2 \leqslant -\theta\|u\|_X^2 + C_\theta.$$

This estimate proves that there exists a bounded absorbing set for the semigroup associated to (1.7) and Theorem 1.23 ensures the existence of a global attractor for it. Note also that (1.7) can be rewritten as

$$\dot{\mathbf{u}} = -A\mathbf{u} + \mathbf{f}(\mathbf{u}) + \mathbf{u},$$

where the operator $A\colon X\to X$ is given by

$$A = \begin{bmatrix} 1+\alpha_0^{-1}\sum_{i=1}^{n}\alpha_i & -\alpha_0^{-1}\alpha_1 & -\alpha_0^{-1}\alpha_2 & \cdots & -\alpha_0^{-1}\alpha_n \\ -\alpha_1 & 1+\alpha_1 & 0 & \cdots & 0 \\ -\alpha_2 & 0 & 1+\alpha_2 & \cdots & \vdots \\ \vdots & \vdots & \vdots & \ddots & 0 \\ -\alpha_n & 0 & 0 & \cdots & 1+\alpha_n \end{bmatrix}.$$

2.3. Scalar parabolic problems with global attractors. Let $\Omega \subset \mathbb{R}^N$ be an open bounded domain with smooth boundary Γ. Assume $\Gamma = \Gamma_0 \cup \Gamma_1$ is such that $\overline{\Gamma_0} \cap \overline{\Gamma_1} = \varnothing$. Let a, b, c be $C^1(\Omega)$ functions such that $a(x) \geqslant m > 0$ for all $x \in \Omega$. Consider the parabolic problem

$$(1.9) \qquad \begin{cases} u_t - \operatorname{div}(a(x)\nabla u) + c(x)u = f(x,u), & \text{in } \Omega, \\ u = 0, & \text{on } \Gamma_0, \\ a(x)\frac{\partial u}{\partial n} + b(x)u = g(x,u), & \text{on } \Gamma_1, \\ u(0) = u_0, \end{cases}$$

where $f(x,\cdot), g(x,\cdot)\colon \mathbb{R} \to \mathbb{R}$ are locally Lipschitz functions, uniformly for $x \in \overline{\Omega}$.

The well posedness of (1.9) has been considered in [**16**], and their results imply that, under suitable growth assumptions on the nonlinear terms, (1.9) can be solved, locally in time, with initial data in the phase spaces $X = L^q(\Omega)$ or $X = W^{1,q}(\Omega)$. The obtained solutions behave continuously with respect to time and initial conditions, and are classical solutions for positive times. The results also include the so-called *critical growth* conditions on the nonlinearities f and g. The growth assumptions, collectively called $(\mathrm{G})_X$, for the results in [**16**] are:

$(\mathrm{G})_X$ *Assume that:*

(1) *If $X = L^q(\Omega)$, there exist positive constants c_f, c_g and exponents ρ_f, ρ_g such that*

$$(1.10) \qquad |j(x,u) - j(x,v)| \leqslant c_j|u-v|(|u|^{\rho_j - 1} + |v|^{\rho_j - 1} + 1),$$

for $j = f, g$. Also for $N \geqslant 2$ (respectively, $N = 1$) we have

$$\rho_f \leqslant \rho_\Omega := 1 + \frac{2q}{N} \quad \text{and} \quad \rho_g \leqslant \rho_\Gamma := 1 + \frac{q}{N}$$

(respectively, $\rho_g < \rho_\Gamma := 1 + q$).

(2) *If $X = W^{1,q}_{\Gamma_0}(\Omega)$ is the space of functions in $W^{1,q}(\Omega)$ with zero trace in Γ_0, assume either*
(i) $q > N$,
(ii) $q = N$ and for every $\eta > 0$ there exists $c_\eta > 0$ such that

$$|j(x,u) - j(x,v)| \leqslant c_\eta\big(e^{\eta|u|^{\frac{N}{N-1}}} + e^{\eta|v|^{\frac{N}{N-1}}}\big)|u-v|,$$

for $j = f, g$, or
(iii) $1 < q < N$ and f, g satisfy (1.10), with exponents ρ_f, ρ_g, respectively, with

$$\rho_f \leqslant \rho_\Omega := 1 + \frac{2q}{N-q} \quad \text{and} \quad \rho_g \leqslant \rho_\Gamma := 1 + \frac{q}{N-q}.$$

Setting $\mathcal{E} = L^q(\Omega)$ or $W^{1,q}_{\Gamma_0}(\Omega)$ for some $1 < q < \infty$, the results from [**16**] can be summarized as follows.

THEOREM 1.26 ([**16**]). *If X is any space in the class $\mathcal{E}$ and f, g satisfy the growth assumption $(G)_X$, then for any $u_0 \in X$ there exists locally a unique (in a certain sense) mild solution $u(\cdot, u_0) \in C([0,\tau), X)$, of problem (1.9) satisfying $u(0, u_0) = u_0$ in X. This solution depends continuously jointly in time and on the initial data $u_0 \in X$ and it is a classical solution for $t > 0$. Also, the following regularizing effect takes place: if $u_0 \in X$, then $u(t, u_0) \in Y$ for any other space Y in the class $\mathcal{E}$, and $t \in (0,\tau)$.*

Under some additional conditions on the nonlinearities, given in [**17**], these solutions are shown to be defined for all $t \geqslant 0$. These conditions are the following:

(S) *Assume there exist* B_0, $C_0 \in \mathbb{R}$ *and* $B_1, C_1 \geqslant 0$ *such that*

$$u\, f(x,u) \leqslant -C_0 u^2 + C_1|u| \quad \text{and} \quad u\, g(x,u) \leqslant -B_0 u^2 + B_1|u|,$$

for $u \in \mathbb{R}$ *and* $x \in \overline{\Omega}$ *or* $x \in \Gamma_1$, *respectively.*

The following result on global existence holds.

THEOREM 1.27 ([**17**]). *Let* X *be any space in the class* $\mathcal{E}$. *Assume that the growth condition* $(G)_X$ *and the sign condition (S) hold. Then, for any* $u_0 \in X$ *the solution* $u(t, u_0)$ *of* (1.9) *starting at* u_0 *exists for all* $t \geqslant 0$. *Therefore we can define in* X *the semigroup* $\mathcal{T}$, *associated to* (1.9), *by* $T(t)u_0 = u(t, u_0)$ *for* $t \geqslant 0$. *Moreover, from the regularizing effect of Theorem* 1.26, $T(t)u_0 \in Y$ *for any other space* Y *in the class* $\mathcal{E}$ *and* $T(t)\colon X \to X$ *is compact for each* $t > 0$.

Of course, conditions $(G)_X$ and (S) may still allow solutions of (1.9) to grow exponentially. To obtain the dissipativity property, required for the existence of a global attractor, we still need an additional condition. This dissipativity condition, expressed in terms of a suitable eigenvalue problem, reflects a competition balance between the nonlinearities f and g, and are given by:

(D) *Assume that (S) holds and that, with* C_0 *and* B_0 *given in (S), the first eigenvalue* λ_1 *of the following problem is positive:*

$$\begin{cases} -\operatorname{div}(a(x)\nabla u) + (c(x) + C_0)u = \lambda u, & \text{in } \Omega, \\ u = 0, & \text{on } \Gamma_0, \\ a(x)\frac{\partial u}{\partial n} + (b(x) + B_0)u = 0, & \text{on } \Gamma_1. \end{cases}$$

The asymptotic compactness property follows trivially from the dissipativity of the semigroup associated to (1.9), which is compact, and from the fact that the time one evolution of the absorbing set is a relatively compact set that attracts bounded sets. With all these, the following result holds.

THEOREM 1.28 ([**17**]). *Let* X *be any space in the class* $\mathcal{E}$. *Assume that the growth condition* $(G)_X$ *and the dissipativity condition (D) hold. Then the semigroup* $\mathcal{T}$ *associated to* (1.9) *has a global attractor* $\mathcal{A}_X$ *in* X. *Moreover* $\mathcal{A}_X \subset Y$ *for any other space* Y *in the class* $\mathcal{E}$ *and it lies in a compact set of* Y.

With respect to the dependence of the attractor $\mathcal{A}_X$ on the phase space X, it is important to observe that if X and Y are spaces in the class $\mathcal{E}$, with $Y \hookrightarrow X$ (this is usually accomplished through a Sobolev type embedding), and $(G)_X$ holds, then $(G)_Y$ also holds. That is, the better the space the weaker the growth restriction on the nonlinearities. With this fact in mind and the regularizing effect of the semigroup $T(t)$ we can prove the following.

THEOREM 1.29 ([**17**]). *Let* X *be any space in the class* $\mathcal{E}$. *Assume that the growth condition* $(G)_X$ *and the dissipativity condition (D) hold. Then, for any space* Y *in the class* $\mathcal{E}$ *with* $Y \hookrightarrow X$ *we have the existence of the attractor* $\mathcal{A}_Y$. *Moreover* $\mathcal{A}_X = \mathcal{A}_Y$ *and it attracts bounded sets of* X *in the topology of* Y.

In [**17**] one can also find carefully obtained bounds on the attractors, which are uniform with respect to variations on the diffusion coefficient a and, in some cases, on the domain Ω. This can be very important to obtain upper semicontinuity of attractors in the next chapter.

PROPOSITION 1.30. *Let X be any space in the class $\mathcal{E}$. Assume that the growth restriction $(G)_X$ and the dissipative condition (D) hold. Denote by ϕ the solution of*

$$\begin{cases} -\operatorname{div}(a(x)\nabla\phi) + (c(x) + C_0)\phi = C_1, & \text{in } \Omega, \\ \phi = 0, & \text{on } \Gamma_0, \\ a(x)\frac{\partial\phi}{\partial n} + (b(x) + B_0)\phi = B_1, & \text{on } \Gamma_1. \end{cases}$$

Then $0 \leqslant \phi \in L^\infty(\Omega)$, $\limsup_{t\to\infty} |u(t,x,u_0)| \leqslant \phi(x)$, uniformly in $x \in \overline{\Omega}$ and for u_0 in bounded subsets of X. In particular, for any $v \in \mathcal{A}_X$, we have $|v(x)| \leqslant \phi(x)$.

In fact, in [**17**], the authors prove the following.

THEOREM 1.31. *Let X be any space in the class $\mathcal{E}$. Assume that the growth restriction $(G)_X$ and the dissipative condition (D) hold. Then*

(a) *there exists a constant $K_0 \geqslant 0$, depending on the domain Ω, λ_1, $m = \inf\{a(x)\colon x \in \Omega\} > 0$, $\sup\{c(x)\colon x \in \Omega\}$, $\sup\{b(x)\colon x \in \Gamma\}$ and on C_0, C_1, B_0 and B_1, such that*

$$\sup_{v\in\mathcal{A}_X} \{\|v\|_{L^\infty(\Omega)}\} \leqslant K_0.$$

Furthermore, if one of the following two conditions hold:

(i) there exists a $b_0 > 0$ with $b(x) + B_0 \geqslant b_0$;

(ii) $b(x) + B_0 \geqslant 0$, $B_1 = 0$ and there exists a $c_0 > 0$ such that $c(x) + C_0 \geqslant c_0 > 0$;

then K_0 will depend on $|\Omega|$ rather than on Ω;

(b) *for $x_0 \in \overline{\Omega}$ and $\rho_0 > 0$, there exists $\nu \in (0,1)$ such that*

$$\sup_{v\in\mathcal{A}_X} \{\|v\|_{C^\nu(\overline{\Omega}\cap B(x_0,\rho_0))}\} \leqslant K_\nu,$$

where K_ν depends on the same as K_0 and also on ν, on $M = \sup\{a(x)\colon x \in \overline{\Omega} \cap B(x_0, 2\rho_0)\}$, on $\sup\{\|f'(\psi)\|_{L^\infty(\Omega)}\colon \psi \in \mathcal{A}_X\}$, and on $\sup\{\|g'(\psi)\|_{L^\infty(\Gamma_1)}\colon \psi \in \mathcal{A}_X\}$.

2.4. Wave equations. For $\theta \in [0,1]$, $\eta > 0$, consider the family of problems of the form

$$\begin{cases} u_{tt} + \eta(-\Delta)^\theta u_t + (-\Delta)u = f(u), & t > 0,\ x \in \Omega, \\ u(0,x) = u_0(x),\ u_t(0,x) = v_0(x), & x \in \Omega, \\ u(t,x) = 0, & t \geqslant 0,\ x \in \partial\Omega, \end{cases} \tag{1.11}$$

where Ω is a bounded smooth domain in $\mathbb{R}^n$ for $n \geqslant 3$. Let $A = -\Delta$ be the negative Laplacian with Dirichlet boundary conditions. It is well known that A is a positive, self-adjoint operator with domain $D(A) = H^2(\Omega) \cap H_0^1(\Omega)$. Therefore $-A$ generates an analytic semigroup on $X = L^2(\Omega)$ (see [**78**, Section 6.5]). We denote by X^α the fractional power spaces associated to the operator A, that is, $X^\alpha = D(A^\alpha)$ endowed with the graph norm (see [**78**, Section 6.4]) and recall that $X^0 = X$, $X^{\frac{1}{2}} = H_0^1(\Omega)$ and $X^1 = H^2(\Omega) \cap H_0^1(\Omega)$.

Problem (1.11) can be rewritten as an evolution equation in the space $Y = X^{\frac{1}{2}} \times X$ as follows:

$$(1.12)\qquad \begin{cases} \dfrac{d}{dt}\begin{bmatrix} u \\ v \end{bmatrix} + \mathcal{A}_{(\theta)}\begin{bmatrix} u \\ v \end{bmatrix} = \mathcal{F}\left(\begin{bmatrix} u \\ v \end{bmatrix}\right), \text{ for } t > 0, \\ \begin{bmatrix} u(0) \\ v(0) \end{bmatrix} = \begin{bmatrix} u_0 \\ v_0 \end{bmatrix} \in Y, \end{cases}$$

with $\mathcal{A}_{(\theta)} \colon D(\mathcal{A}_{(\theta)}) \subset Y \to Y$ and $\mathcal{F}$ given by

$$D(\mathcal{A}_{(\theta)}) = Y^1_{(\theta)} = \left\{ \begin{bmatrix} \varphi \\ \psi \end{bmatrix} : \varphi \in X^{\frac{3}{2}-\theta},\ \psi \in X^{\frac{1}{2}},\ A^{1-\theta}\varphi + \eta\psi \in X^{\theta} \right\}$$

for $\theta \in [0, 1]$, and

$$\mathcal{A}_{(\theta)}\begin{bmatrix} \varphi \\ \psi \end{bmatrix} = \begin{bmatrix} 0 & -I \\ A & \eta A^{\theta} \end{bmatrix}\begin{bmatrix} \varphi \\ \psi \end{bmatrix} = \begin{bmatrix} -\psi \\ A^{\theta}(A^{1-\theta}\varphi + \eta\psi) \end{bmatrix} \quad \text{for} \quad \begin{bmatrix} \varphi \\ \psi \end{bmatrix} \in D(\mathcal{A}_{(\theta)}),$$

and

$$\mathcal{F}\left(\begin{bmatrix} u \\ v \end{bmatrix}\right) = \begin{bmatrix} 0 \\ F(u) \end{bmatrix},$$

where F is the Nemytskiĭ operator associated to f, that is, $F(u)(x) = f(u(x))$, $x \in \Omega$.

The linear problem associated to (1.12) in Y is

$$\begin{cases} \ddot{u} + \eta A^{\theta}\dot{u} + Au = 0,\ t > 0, \\ u(0) = u_0,\ \dot{u}(0) = v_0, \end{cases}$$

and corresponds to the damped wave equation when $\theta = 0$ or to the 'strongly damped wave equation' when $\theta \in (0, 1]$. For each $\theta \in [0, 1]$ the operator $-\mathcal{A}_{(\theta)}$ generates a strongly continuous semigroup $\{e^{-\mathcal{A}_{(\theta)}t} \colon t \geqslant 0\} \subset \mathcal{L}(Y)$. If $\theta = 0$, $\{e^{-\mathcal{A}_{(0)}t} \colon t \geqslant 0\}$ can be extended to a group. For $\theta > 0$, $\{e^{-\mathcal{A}_{(\theta)}t} \colon t \geqslant 0\}$ cannot be extended to a group and presents some regularization properties that can be described in the following way:

(i) If $\theta \in (0, \frac{1}{2})$, $-\mathcal{A}_{(\theta)}$ generates a semigroup which is of Gevrey class $\delta > \frac{1}{2\theta}$ and, consequently, is differentiable (see [**94**]).
(ii) If $\theta \in [\frac{1}{2}, 1]$, $-\mathcal{A}_{(\theta)}$ generates an analytic semigroup (see [**91, 92**]). In [**93**] the authors give a characterization of the domain of fractional powers of $\mathcal{A}_{(\theta)}$, that is, the fractional power spaces $Y^{\alpha}_{(\theta)}$, $\alpha \in [\frac{1}{2}, 1]$.

We will concentrate our attention on the Hilbert space case $Y = H^1_0(\Omega) \times L^2(\Omega)$ and describe the results in [**60, 64**], but for $\theta \in [\frac{1}{2}, 1]$ the problem can also be addressed in $W^{1,p}_0(\Omega) \times L^p(\Omega)$ (see [**63, 66**]).

Concerning the local well posedness of (1.11) the following results holds.

THEOREM 1.32 ([**61**]). *If f satisfies*

$$(1.13)\qquad |f(u) - f(u')| \leqslant c|u - u'|(1 + |u|^{\rho-1} + |u'|^{\rho-1})$$

with $\rho \leqslant \frac{n+2}{n-2}$ for $\theta \in [\frac{1}{2}, 1]$, and $\rho \leqslant \frac{n}{n-2}$ for $\theta \in [0, \frac{1}{2})$, then (1.11) is locally well posed in $H^1_0(\Omega) \times L^2(\Omega)$. If strict inequality holds in the conditions over ρ (subcritical growth), the solution of (1.11) either exists for all $t \geqslant 0$ or its $H^1_0(\Omega) \times L^2(\Omega)$ blows up in finite time.

From the results in [**60**], the semigroup associated to (1.11) has a global attractor under suitable assumptions. In particular, the following result holds.

THEOREM 1.33. *Assume that f satisfies* (1.13) *with $\rho < \frac{n+2}{n-2}$ for $\theta \in [\frac{1}{2}, 1]$, and $\rho < \frac{n}{n-2}$ for $\theta \in [0, \frac{1}{2})$ and the dissipativity condition*

$$\limsup_{|u|\to\infty} \frac{f(u)}{u} \leqslant 0. \tag{1.14}$$

Then problem (1.11) *has a global attractor.*

The critical growth case $\rho = \frac{n+2}{n-2}$ for $\theta = 1$, and $\rho = \frac{n}{n-2}$ for $\theta \in [0, \frac{1}{2})$ can be treated as in [**9**] (see also [**60**, Section 4.3]) under some additional assumptions that ensure the validity of a nonlinear variation of constants formula.

Under the assumptions of Theorem 1.33, the global solvability of (1.11) can be obtained noting that, for suitably regular initial data,

$$\frac{d}{dt}\left(\frac{1}{2}\|u_t\|^2_{L^2(\Omega)} + \frac{1}{2}\|\nabla u\|^2_{L^2(\Omega)} - \int_\Omega \int_0^u f(s)ds\right) = -\eta\|A^{\frac{\theta}{2}} u_t\|^2_{L^2(\Omega)},$$

which implies that, for all initial data,

$$\left\|\begin{bmatrix} u(t, u_0, v_0) \\ v(t, u_0, v_0) \end{bmatrix}\right\|_Y \leqslant c + c'\mathcal{L}\left(\begin{bmatrix} u_0 \\ v_0 \end{bmatrix}\right) \leqslant C\left(\left\|\begin{bmatrix} u_0 \\ v_0 \end{bmatrix}\right\|_Y\right), \tag{1.15}$$

where c, c' do not depend on η, β,

$$\mathcal{L}\left(\begin{bmatrix} w_1 \\ w_2 \end{bmatrix}\right) = \frac{1}{2}\|w_2\|^2_{L^2(\Omega)} + \frac{1}{2}\|\nabla w_1\|^2_{L^2(\Omega)} - \int_\Omega \int_0^{w_1} f(s)ds, \; w_1, w_2 \in Y, \tag{1.16}$$

and $C\colon \mathbb{R}^+ \to \mathbb{R}^+$ is a locally bounded function independent of η and β. In this case, we denote by $T(t)\colon Y \to Y$ the solution operator for (1.11).

From estimate (1.15) we also obtain that orbits of bounded subsets of Y are bounded. From the fact that $\rho < \frac{n+2}{n-2}$, the map $\mathcal{F}$ is compact. That together with the fact that the semigroup $\{e^{-\mathcal{A}_{(\theta)}t}\colon t \geqslant 0\}$ decays exponentially ensures the asymptotic compactness of $\mathcal{T} = \{T(t)\colon t \geqslant 0\}$ (see Theorem 1.24).

For the proof of the existence of the global attractor for $\mathcal{T}$ in Y, it suffices to prove that there is a bounded subset of Y that $\mathcal{T}$-attracts points. That follows from the fact that the functional $\mathcal{L}$ defined in (1.16) is decreasing along solutions and that the set of points where it is constant is bounded. This is done by showing that the solutions $\tilde{u}$ of

$$\begin{cases} -\Delta\tilde{u} = f(\tilde{u}) & \text{in } \Omega, \\ \tilde{u} = 0 & \text{on } \partial\Omega, \end{cases} \tag{1.17}$$

lie in a bounded set of $H^1_0(\Omega)$ and that follows by multiplying (1.17) by $\tilde{u}$, integrating and using (1.14).

Notes

In the last four decades the theory of semigroups and their global attractors has been intensively developed. Now we can count on several nice references from which one can learn the subject and its many applications to systems of differential or partial differential equations ([**22, 101, 103, 132, 142, 159, 189, 194, 200, 201**]). These references have settled what is now known as the 'classical theory' on the existence of global attractors. When the phase space is infinite dimensional, the finite fractal dimensionality of global attractors can be proved for most applications, as seen, for instance, in [**20, 78, 106, 120, 189, 200**], with the first results on this subject obtained by J. Mallet-Paret in the Hilbert space setting (see [**166**]) and by R. Mañe in the Banach space setting (see [**168**]). Some references for the characterization of global attractors and the robustness of its structure under perturbations are [**5, 78, 105, 133, 142, 180, 192**]. Other notions of attraction have been introduced, as the one of *exponential attractors* (see [**86, 87, 102, 115, 116**]), where the rate of attraction to a positively invariant set of finite fractal dimension is shown to be exponential, or the notion of inertial manifolds ([**119, 142, 167**]), for which the attractor is embedded into a finite-dimensional Lipschitz manifold, so that the asymptotic dynamics of the system in an infinite-dimensional phase space can be described by a system of ordinary differential equations. Applications of the theory of global attractors to models of differential and partial differential equations can be seen in a large number of papers, and the theory has also been enlarged to include, for instance, attractors for multivalued semigroups ([**23, 24, 171**]), delay differential equations ([**52**]) or even stochastic PDEs ([**108**]).

CHAPTER 2

Upper and Lower Semicontinuity

In this chapter we establish the *continuity* of global attractors relative to perturbations of a semigroup. As mentioned in the introduction, this study is essential to the mathematical modelling of phenomena using semigroups (or differential equations), due to the fact that mathematical models are always an approximation of the real evolution rule. Hence, to ensure that the result of the analysis reflects (in some way) the modelled phenomena, it is imperative that solutions of such models enjoy certain robustness properties relative to all sorts of perturbations.

In this section we address this aspect from the point of view of the asymptotic dynamics, giving conditions under which attractors behave *continuously* when we perform a perturbation on the semigroup.

DEFINITION 2.1. Let X and Λ be metric spaces and $\{\mathcal{A}_\lambda\}_{\lambda\in\Lambda}$ a family of subsets of X. We say that $\{\mathcal{A}_\lambda\}_{\lambda\in\Lambda}$ is

(1) **upper semicontinuous** at λ_0 if

$$\mathrm{dist}_H(\mathcal{A}_\lambda, \mathcal{A}_{\lambda_0}) = \sup_{x_\lambda\in\mathcal{A}_\lambda} \mathrm{dist}_H(x_\lambda, \mathcal{A}_{\lambda_0}) \to 0, \text{ as } \lambda\to\lambda_0,$$

(2) **lower semicontinuous** at λ_0 if

$$\mathrm{dist}_H(\mathcal{A}_{\lambda_0}, \mathcal{A}_\lambda) = \sup_{x\in\mathcal{A}_{\lambda_0}} \mathrm{dist}_H(x, \mathcal{A}_\lambda) \to 0, \text{ as } \lambda\to\lambda_0,$$

(3) **continuous** at λ_0 if it is upper and lower semicontinuous at λ_0, that is, if

$$\mathrm{dist}_H(\mathcal{A}_\lambda, \mathcal{A}_{\lambda_0}) + \mathrm{dist}_H(\mathcal{A}_{\lambda_0}, \mathcal{A}_\lambda) \to 0, \text{ as } \lambda\to\lambda_0.$$

The following characterization result is useful to prove upper and lower semicontinuity.

LEMMA 2.2. *Let X, Λ be metric spaces, $\lambda_0\in\Lambda$ and $\{\mathcal{A}_\lambda\}_{\lambda\in\Lambda}$ a family of compact subsets of X.*

(a) *If for $\lambda_n \overset{n\to\infty}{\longrightarrow} \lambda_0$, each sequence $\{x_n\}$ with $x_n\in\mathcal{A}_{\lambda_n}$ has a convergent subsequence with limit belonging to $\mathcal{A}_{\lambda_0}$, then $\{\mathcal{A}_\lambda\}_{\lambda\in\Lambda}$ is upper semicontinuous at λ_0. Conversely, if $\{\mathcal{A}_\lambda\}_{\lambda\in\Lambda}$ is upper semicontinuous at λ_0, $\lambda_n \overset{n\to\infty}{\longrightarrow} \lambda_0$, $\{x_n\}$ is such that $x_n\in\mathcal{A}_{\lambda_n}$, then $\{x_n\}$ has a convergent subsequence with limit belonging to $\mathcal{A}_{\lambda_0}$.*

(b) *If for each $x\in\mathcal{A}_{\lambda_0}$ and $\lambda_n \overset{n\to\infty}{\longrightarrow} \lambda_0$, there is a subsequence $\lambda_{n_k} \overset{k\to\infty}{\longrightarrow} \lambda_0$, and $\{x_k\}$ with $x_k\in\mathcal{A}_{\lambda_{n_k}}$ that converges to x, then $\{\mathcal{A}_\lambda\}_{\lambda\in\Lambda}$ is lower semicontinuous at λ_0. Conversely, if $\{\mathcal{A}_\lambda\}_{\lambda\in\Lambda}$ is lower semicontinuous at λ_0, $\lambda_n \overset{n\to\infty}{\longrightarrow} \lambda_0$ and $x\in\mathcal{A}_{\lambda_0}$, there is a subsequence $\lambda_{n_k} \overset{k\to\infty}{\longrightarrow} \lambda_0$, and $\{x_k\}$ with $x_k\in\mathcal{A}_{\lambda_{n_k}}$ that converges to x.*

PROOF. (a) If any sequence $\{x_n\}$ with $x_n \in \mathcal{A}_{\lambda_n}$, $\lambda_n \overset{n\to\infty}{\longrightarrow} \lambda_0$, has a convergent subsequence with limit belonging to $\mathcal{A}_{\lambda_0}$, and $\{\mathcal{A}_\lambda\}_{\lambda\in\Lambda}$ is not upper semicontinuous at λ_0, then there are $\epsilon > 0$ and a sequence $\{\lambda_n\}$ with $\lambda_n \overset{n\to\infty}{\longrightarrow} \lambda_0$ such that $\sup_{x\in\mathcal{A}_{\lambda_n}} \operatorname{dist}_H(x, \mathcal{A}_{\lambda_0}) \geqslant 2\epsilon$, $n \in \mathbb{N}$. Hence, for some $x_n \in \mathcal{A}_{\lambda_n}$, we have that $\operatorname{dist}_H(x_n, \mathcal{A}_{\lambda_0}) \geqslant \epsilon$, $n \in \mathbb{N}$. This contradicts the fact that $\{x_n\}$ has a convergent subsequence with limit belonging to $\mathcal{A}_{\lambda_0}$. The converse is left to the reader.

(b) If for any $x \in \mathcal{A}_{\lambda_0}$ and $\lambda_n \overset{n\to\infty}{\longrightarrow} \lambda_0$ there is a subsequence $\lambda_{n_k} \overset{k\to\infty}{\longrightarrow} \lambda_0$ and $\{x_k\}$ with $x_k \in \mathcal{A}_{\lambda_{n_k}}$ that converges to x and $\{\mathcal{A}_\lambda\}$ is not lower semicontinuous at λ_0, there are $\epsilon > 0$ and a sequence $\lambda_n \overset{n\to\infty}{\longrightarrow} \lambda_0$ such that $\sup_{x\in\mathcal{A}_{\lambda_0}} \operatorname{dist}_H(x, \mathcal{A}_{\lambda_n}) \geqslant 2\epsilon$, $n \in \mathbb{N}$. Hence, for some $x_n \in \mathcal{A}_{\lambda_0}$, we have $\operatorname{dist}_H(x_n, \mathcal{A}_{\lambda_n}) \geqslant 2\epsilon$, $n \in \mathbb{N}$. Since $\mathcal{A}_{\lambda_0}$ is compact we may assume that $\{x_n\}$ converges to some $x \in \mathcal{A}_{\lambda_0}$ and that $\operatorname{dist}_H(x, \mathcal{A}_{\lambda_n}) \geqslant \epsilon$, $n \in \mathbb{N}$. It follows, from the assumptions, that there exists a subsequence $\lambda_{n_k} \overset{k\to\infty}{\longrightarrow} \lambda_0$ and a sequence $y_k \in \mathcal{A}_{\lambda_{n_k}}$ such that $y_k \overset{k\to\infty}{\longrightarrow} x$ which leads to an absurdity. The converse is left to the reader. □

1. Upper semicontinuity

Before proving the upper semicontinuity for global attractors of autonomous perturbations of semigroups, we will need some definitions.

DEFINITION 2.3. Let X be a metric space and, for each $\eta \in [0, 1]$, we consider a semigroup $\mathcal{T}_\eta$ in X. We say that the family of semigroups $\{\mathcal{T}_\eta\}_{\eta\in[0,1]}$ is:

(i) **continuous** at $\eta = \eta_0$, for some $\eta_0 \in [0, 1]$, if
$$d(T_\eta(t)x, T_{\eta_0}(t)x) \to 0 \text{ as } \eta \to \eta_0,$$
uniformly for (t, x) in compact subsets $\mathbb{R}^+ \times X$;

(ii) **collectively asymptotically compact** if given sequences $\{t_k\}$ in $\mathbb{R}^+$ with $t_k \to \infty$, $\{\eta_k\}$ in $[0, 1]$ and $\{x_k\}$ bounded in X, then $\{T_{\eta_k}(t_k)x_k\}$ has a convergent subsequence in X. If, in the above condition, we additionally require that $\eta_k \to 0$, then we say that the family is **collectively asymptotically compact at** $\eta = 0$;

(iii) **collectively bounded dissipative** if there is a bounded subset B_0 of X which $\mathcal{T}_\eta$-attracts bounded subsets of X for each $\eta \in [0, 1]$.

At this point, this definition requires some important remarks. Firstly note that the definition of collectively asymptotically compactness, in particular, implies that $\mathcal{T}_\eta$ is an asymptotically compact semigroup, for each $\eta \in [0, 1]$. Since we will assume that $\mathcal{T}_\eta$ possesses a global attractor $\mathcal{A}_\eta$, for each $\eta \in [0, 1]$, this condition is not so restrictive. We point out, however, that this property will be important only for those sequences $\{\eta_k\}$ in $[0, 1]$ with $\eta_k \to 0$ and, of course, when the sequence $\{\eta_k\}$ is constant.

Secondly, note that the continuity of the family of semigroups is assumed *only* at $\eta = 0$, and finally, requiring that $\bigcup_{\eta\in[0,1]} \mathcal{A}_\eta$ is bounded implies that the family of semigroups $\{\mathcal{T}_\eta\}_{\eta\in[0,1]}$ is collectively bounded dissipative.

We will show that these three conditions are sufficient to ensure both the existence of a family of global attractors and their upper semicontinuity. We start with a lemma that plays a key role in much of what is done in the rest of this chapter. It is concerned with obtaining a global solution through a sequence of partial solutions which are defined in a sequence of finite intervals with length that goes to infinity.

LEMMA 2.4. *Let $\{\mathcal{T}_\eta\}_{\eta\in[0,1]}$ be a continuous and collectively asymptotically compact at $\eta = 0$ family of semigroups. Let $\eta_n \to 0$, $a_n < b_n < c_n$ such that $b_n - a_n$, $c_n - b_n \to \infty$, and set $\mathbb{J}_n = [a_n, c_n]$. Let $\xi_n\colon \mathbb{J}_n \to X$ be a solution of $\mathcal{T}_{\eta_n}$ and assume that*

$$\overline{\bigcup_{n\in\mathbb{N}} \xi_n(\mathbb{J}_n)} \text{ is bounded.}$$

Then there exists a subsequence $\{\xi_{n_k}\}$ of $\{\xi_n\}$ and a bounded global solution $\xi_0\colon \mathbb{R} \to X$ of $\mathcal{T}_0$ such that $\xi_{n_k}(t + b_{n_k}) \to \xi_0(t)$, uniformly for t in compact subintervals of $\mathbb{R}$.

PROOF. As a first step, we find a subsequence $\{n_{0,k}\}$ of $\mathbb{N}$ and $x_0 \in X$ such that

$$\xi_{n_{0,k}}(b_n) = T_{\eta_{n_{0,k}}}(b_n - a_n)\xi_{n_{0,k}}(a_n) \to x_0 \quad \text{as } k \to \infty,$$

using the collective asymptotically compactness at $\eta = 0$. Define $\xi_0\colon [0,\infty) \to X$ by $\xi_0(t) = T_0(t)x_0$ for $t \geqslant 0$. Then

$$\xi_0(t) = T_0(t)x_0 = \lim_{k\to\infty} T_{\eta_{n_{0,k}}}(t)\xi_{n_{0,k}}(b_n) = \lim_{k\to\infty} \xi_{n_{0,k}}(t + b_{n_k}),$$

with the convergence being uniform for t in compact subsets of $[0,\infty)$. In particular we obtain

$$\xi_0([0,\infty)) \subset \overline{\bigcup_{n\in\mathbb{N}} \xi_n(\mathbb{J}_n)}.$$

Suppose that we have already constructed a subsequence $\{n_{j,k}\}$ of $\{n_{j-1,k}\}$, for $j = 1,\cdots,m-1$, such that $\xi_{n_{j,k}}(\cdot + b_{n_{j,k}}) \to \xi_0(\cdot)$ as $k \to \infty$ uniformly on compact subintervals of $[-j,\infty)$, where $\xi_0\colon [-m+1,\infty) \to X$ is a solution of $\mathcal{T}_0$ that lies in $\overline{\bigcup_{n\in\mathbb{N}} \xi_n(\mathbb{J}_n)}$.

Now we find a subsequence $\{n_{m,k}\}$ of $\{n_{m-1,k}\}$ such that $\xi_{n_{m,k}}(-m+b_{n_{m,k}}) \to x_m$ for some $x_m \in X$, and extend the definition of $\xi_0(\cdot)$ to $[-m,\infty)$ by setting $\xi_0(s) = T_0(s+m)x_m$ for $s \in [-m,-m+1)$. Then

$$\begin{aligned} T_0(1)x_m &= \lim_{k\to\infty} T_{\eta_{n_{m,k}}}(1)\xi_{n_{m,k}}(-m + b_n) \\ &= \lim_{k\to\infty} \xi_{n_{m,k}}(-m + 1 + b_{n_{m,k}}) = \xi_0(-m+1), \end{aligned}$$

and so $\xi_0\colon [-m,\infty) \to X$ is a solution of $\mathcal{T}_0$. Clearly $\xi_{n_{m,k}}(\cdot + b_{n_{m,k}}) \to \xi_0(\cdot)$ uniformly on compact subintervals of $[-m,\infty)$, and $\xi_0([-m,\infty)) \subset \overline{\bigcup_{n\in\mathbb{N}} \xi_n(\mathbb{J}_n)}$.

Finally, if we let $n_k = n_{k,k}$, the sequence $\{\xi_{n_k}(\cdot + b_{n_k})\}$ and the corresponding global solution ξ_0 all have the properties stated in the theorem. □

Now we are able to prove the upper semicontinuity for a family of global attractors for semigroups.

THEOREM 2.5 (Upper semicontinuity). *Let $\{\mathcal{T}_\eta\}_{\eta\in[0,1]}$ be a collectively asymptotically compact at $\eta = 0$ and collectively bounded dissipative family of semigroups in X. Then $\mathcal{T}_\eta$ has a global attractor $\mathcal{A}_\eta$ for each $\eta \in [0,1]$, with $\bigcup_{\eta\in[0,1]} \mathcal{A}_\eta$ bounded in X.*

If, in addition, the family $\{\mathcal{T}_\eta\}_{\eta\in[0,1]}$ is continuous at $\eta = 0$, $\eta_k \to 0$, $\{s_k\}$ is a sequence in $\mathbb{R}$ and, for each $k \in \mathbb{N}$, ξ_k is a global solution of $\mathcal{T}_{\eta_k}$ in $\mathcal{A}_{\eta_k}$, then there exists a subsequence $\{\xi_{k_j}\}_{j\in\mathbb{N}}$ of $\{\xi_k\}_{k\in\mathbb{N}}$ such that $\{\xi_{k_j}(\cdot + s_{k_j})\}$ converges uniformly in bounded intervals of $\mathbb{R}$ to a global solution ξ_0 of $\mathcal{T}_0$ with $\xi_0(\mathbb{R}) \subset \mathcal{A}_0$. In particular, the family $\{\mathcal{A}_\eta\}_{\eta\in[0,1]}$ is upper semicontinuous at $\eta = 0$.

PROOF. From Theorem 1.23, since each semigroup $\mathcal{T}_\eta$ is asymptotically compact and bounded dissipative, it possesses a global attractor $\mathcal{A}_\eta$ for each $\eta \in [0,1]$. From Lemma 1.21, we have $\mathcal{A}_\eta \subset B_0$ for each $\eta \in [0,1]$, where B_0 is a bounded subset of X that comes from the definition of collective bounded dissipativity of $\{\mathcal{T}_\eta\}_{\eta\in[0,1]}$ (which does not depend on η). Hence $\bigcup_{\eta\in[0,1]} \mathcal{A}_\eta \subset B_0$ and therefore is bounded in X.

Now assume that the family $\{\mathcal{T}_\eta\}_{\eta\in[0,1]}$ is continuous at $\eta = 0$. Let $\eta_k \to 0$, and let $\{s_k\}$ be a sequence in $\mathbb{R}$ and ξ_k a global solution of $\mathcal{T}_{\eta_k}$ in $\mathcal{A}_{\eta_k}$ for each $k \in \mathbb{N}$. Using Lemma 2.4 with $a_n = -n + s_n$, $b_n = s_n$ and $c_n = n + s_n$, we obtain a bounded global solution ξ_0 of $\mathcal{T}_0$ such that $\xi_{n_k}(t + s_n) \to \xi_0(t)$, uniformly for t in compact subsets of $\mathbb{R}$, along a subsequence of $\mathbb{N}$. Since $\xi_0(\mathbb{R})$ is bounded and $\mathcal{T}_0$-invariant, we obtain $\xi_0(\mathbb{R}) \subset \mathcal{A}_0$.

Lastly assume $\eta_k \to 0$ and $x_k \in \mathcal{A}_{\eta_k}$ for each $k \in \mathbb{N}$. From Theorem 1.9, for each $k \in \mathbb{N}$ there exists a bounded global solution ξ_k of $\mathcal{T}_{\eta_k}$ through x_k. From what we have just proved, with $s_k = 0$ for each $k \in \mathbb{N}$, there exists a bounded global solution ξ_0 of $\mathcal{T}_0$ such that, along a subsequence, $\xi_k(t) \to \xi_0(t)$, uniformly for compact subintervals of $\mathbb{R}$. Thus, for $t = 0$ we obtain $x_k = \xi_k(0) \to \xi_0(0) \in \mathcal{A}_0$, and the upper semicontinuity of $\{\mathcal{A}_\eta\}_{\eta\in[0,1]}$ at $\eta = 0$ is proved. □

2. Examples with upper semicontinuity

2.1. Example of perturbation of an ODE. Let us start with a very simple example which is obtained as a perturbation of (1.1). Consider the initial value problem

$$\begin{cases} \dot{x} = f(x) + \eta g(x), \\ x(0) = x_0 \in \mathbb{R}^n, \end{cases} \tag{2.1}$$

where $f\colon \mathbb{R}^n \to \mathbb{R}^n$ is a locally Lipschitz continuous function, g is a globally bounded and locally Lipschitz function and $\eta \in [0,1]$. From Theorem 1.3, for each $x_0 \in \mathbb{R}^n$ and $\eta \in [0,1]$, there exists a unique and maximally defined solution $x_\eta(\cdot, x_0)\colon [0, \tau(x_0,\eta)) \to \mathbb{R}^n$ of (2.1).

Assume that

$$\limsup_{\|x\|\to\infty} \frac{f(x)\cdot x}{\|x\|} < 0. \tag{2.2}$$

Then, if $x(t,x_0)$ is the solution of (1.1), we have

$$\frac{d}{dt}\|x\|^2 = 2[f(x) + \eta g(x)]\cdot x.$$

Now, it is easy to see that there is an $\eta_0 \in (0,1]$ and $\delta > 0$ such that

$$\limsup_{\|x\|\to\infty} \frac{[f(x) + \eta g(x)]\cdot x}{\|x\|} < -\delta, \qquad \text{for all } \eta \in [0,\eta_0].$$

It follows from (2.2) and from Theorem 1.3 that $\tau(x_0) = \infty$ for all $x_0 \in \mathbb{R}^n$ and for all $\eta \in [0,\eta_0]$. Under these conditions, as before, we define $T_\eta(t)\colon \mathbb{R}^n \to \mathbb{R}^n$, $t \geqslant 0$, by

$$T_\eta(t)x_0 = x(t,x_0),\ x_0 \in \mathbb{R}^n \text{ and } \eta \in [0,\eta_0].$$

The continuity at $\eta = 0$ of the family $\{\mathcal{T}_\eta\}_{\eta\in[0,\eta_0]}$ follows from Theorem 1.3. It is easy to see that the family of semigroups $\{\mathcal{T}_\eta\}_{\eta\in[0,\eta_0]}$ is collectively asymptotically compact at $\eta = 0$ and collectively bounded dissipative. It follows from Theorem

2.5 that $\mathcal{T}_\eta$ has a global attractor $\mathcal{A}_\eta$, for each $\eta \in [0, \eta_0]$, and that the family $\{\mathcal{A}_\eta\}_{\eta\in[0,1]}$ is upper semicontinuous at $\eta = 0$.

Note that, if $f(u) = -u(|u| - 1)^2$ and $g\colon \mathbb{R}^n \to \mathbb{R}^n$ is globally bounded, locally Lipschitz and $g(u) \cdot u < 0$ for all $u \neq 0$, the conditions of this example are satisfied and the family of attractors $\{\mathcal{A}_\eta\}_{\eta\in[0,1]}$ is upper semicontinuous at $\eta = 0$. The attractor $\mathcal{A}_0$ is the unitary disk $\{x \in \mathbb{R}^n \colon \|x\| \leqslant 1\}$ and $\mathcal{A}_\eta = \{0\}$, for $\eta > 0$. This shows that the attractors may still implode under these types of perturbations, but they do not explode.

REMARK 2.6. The problems in Section 2.2 can be considered as a parameter dependent problem with parameter $\eta = (\alpha_0, \alpha_1, \cdots, \alpha_n)$. It is easy to see that the family of attractors $\{\mathcal{A}_\eta\}_\eta$ associated to the family of semigroups given by equation (1.7) is upper semicontinuous at $\eta = (1, 0, \cdots, 0)$ or at any other admissible value of η.

The same applies to the problems considered in Section 2.1 with α_i, ℓ_i, $0 \leqslant i \leqslant n-1$, and β_i, $0 \leqslant i \leqslant n-2$, as parameters.

2.2. Abstract semilinear parabolic problems. Let Z be a Banach space and consider a family of closed densely defined operators $B_\eta\colon D(B_\eta) \subset Z \to Z$ with $0 \in \rho(B_\eta)$ and a fixed domain $D = D(B_\eta)$ for all $\eta \in [0,1]$. Suppose that the graph norms in D are equivalent for all values of $\eta \in [0,1]$ and *with uniform equivalence constants*, that is, there exist constants $c_1, c_2 > 0$ such that

$$c_1(\|z\| + \|B_{\eta_1} z\|) \leqslant \|z\| + \|B_{\eta_2} z\| \leqslant c_2(\|z\| + \|B_{\eta_1} z\|),$$

for all $z \in D$ and $\eta_1, \eta_2 \in [0,1]$. Assume also that

$$\lim_{\eta\to 0} \|B_\eta^{-1} - B_0^{-1}\|_{\mathcal{L}(Z)} = 0, \tag{2.3}$$

and that the operators $-B_\eta$ are uniformly sectorial, that is, that there are constants $M, r > 0$ and $\phi \in (0, \frac{\pi}{2})$ such that $\Sigma_\phi = \{\lambda \in \mathbb{C}\colon \phi \leqslant \arg\lambda \leqslant \pi\} \cup B_r(0) \subset \rho(B_\eta)$ and

$$\|(\lambda - B_\eta)^{-1}\|_{\mathcal{L}(Z)} \leqslant \frac{M}{1 + |\lambda|} \quad \text{and} \quad \|B_\eta(\lambda - B_\eta)^{-1}\|_{\mathcal{L}(Z)} \leqslant M, \tag{2.4}$$

for all $\lambda \in \Sigma_\phi$ and for all $\eta \in [0,1]$. It is not difficult to see that, for all $\lambda \in \Sigma_\phi$, we have

$$(\lambda - B_\eta)^{-1} - (\lambda - B_0)^{-1} = B_\eta(\lambda - B_\eta)^{-1}(B_0^{-1} - B_\eta^{-1})B_0(\lambda - B_0)^{-1}$$

and, using (2.4), we obtain

$$\|(\lambda - B_\eta)^{-1} - (\lambda - B_0)^{-1}\|_{\mathcal{L}(Z)} \leqslant M^2 \|B_\eta^{-1} - B_0^{-1}\|_{\mathcal{L}(Z)}. \tag{2.5}$$

Let Γ be the boundary of Σ_ϕ, oriented in the direction of decreasing imaginary part. It is well known (see, for example, [**142**, Theorem 1.3.4]) that B_η generates an analytic linear semigroup $\{e^{B_\eta t}\colon t \geqslant 0\}$ with

$$e^{B_\eta t} = \frac{1}{2\pi i}\int_\Gamma (\lambda - B_\eta)^{-1} e^{\lambda t} d\lambda,$$

and for some $\omega > 0$ and $\bar{M} \geqslant 1$, independent of $\eta \in [0,1]$, for all $t \geqslant 0$ we have

$$\|e^{B_\eta t}\|_{\mathcal{L}(Z)} \leqslant \bar{M} e^{-\omega t} \quad \text{and} \quad \|B_\eta e^{B_\eta t}\|_{\mathcal{L}(Z)} \leqslant \bar{M} t^{-1} e^{-\omega t}. \tag{2.6}$$

THEOREM 2.7. *Assume that* (2.3) *and* (2.4) *are satisfied. Then there exists a constant* $C > 0$ *such that*

$$t^\gamma \|e^{B_\eta t} - e^{B_0 t}\|_{\mathcal{L}(Z)} \leqslant C\|B_\eta^{-1} - B_0^{-1}\|^\gamma_{\mathcal{L}(Z)} \quad \text{for any } \gamma \in [0,1], \tag{2.7}$$

and

$$t^{\gamma(1-\nu)+\nu}\|e^{B_\eta t} - e^{B_0 t}\|_{\mathcal{L}(Z,Z^\nu)} \leqslant C\|B_\eta^{-1} - B_0^{-1}\|^{\gamma(1-\nu)}_{\mathcal{L}(Z)} \quad \text{for any } \gamma,\nu \in [0,1]. \tag{2.8}$$

PROOF. From (2.5), we obtain

$$\begin{aligned}\|e^{B_\eta t} - e^{B_0 t}\|_{\mathcal{L}(Z)} &\leqslant Ct^{-1}\int_\Gamma e^{-\frac{1}{\cos\phi}|\mu|}d|\mu| \|B_\eta^{-1} - B_0^{-1}\|_{\mathcal{L}(Z)}\\ &\leqslant Ct^{-1}\|B_\eta^{-1} - B_0^{-1}\|_{\mathcal{L}(Z)}\end{aligned}$$

and (2.7) follows easily. Also, using (2.4) and proceeding as above, we have

$$\|e^{B_\eta t} - e^{B_0 t}\|_{\mathcal{L}(Z)} \leqslant C \text{ and } \|e^{B_\eta t} - e^{B_0 t}\|_{\mathcal{L}(Z,Z^1)} \leqslant Ct^{-1}.$$

From this, (2.7), and [**142**, Theorem 1.4.4], (2.8) follows. □

With this and (2.3) we can prove the convergence of the linear semigroups $\{e^{B_\eta t} \colon t \geqslant 0\}$.

Now denote by Z^ν the domain of the fractional power operator $(-B_0)^\nu$ with the graph norm and $\nu \in [0,1]$. For $\eta \in [0,1]$, consider the following family of semilinear problems:

$$\begin{cases}\dot{u} = B_\eta u + f_\eta(u),\\ u(0) = u_0,\end{cases} \tag{2.9}$$

where, for some $\nu \in [0,1)$, $f_\eta \colon Z^\nu \to Z$ is Lipschitz continuous in bounded subsets of Z^ν.

DEFINITION 2.8. A **solution** of (2.9) is a continuous function $u\colon [0,\tau) \to Z^\nu$, for some $\tau > 0$, such that

$$u(t) = e^{B_\eta t}u_0 + \int_0^t e^{B_\eta (t-s)} f_\eta(u(s))ds \quad \text{for all } t \in [0,\tau). \tag{2.10}$$

THEOREM 2.9 ([**142**, Section 3.3]). *Given* $u_0 \in Z^\nu$ *there exists a maximal time* $\tau_{u_0} > 0$ *and a unique solution* $u_\eta(\cdot, u_0)\colon [0,\tau_{u_0}) \to Z^\nu$ *of* (2.9). *If* $E = \{(t,z_0)\colon z_0 \in Z^\nu \text{ and } t \in [0,\tau_{z_0})\}$ *then* $u_\eta\colon E \to Z^\nu$ *is continuous, and if* $\tau_{u_0} < \infty$ *then* $\lim_{t\to\tau_{u_0}^-} \|u_\eta(t,u_0)\|_{Z^\nu} = \infty$.

Assume that for any $u_0 \in Z^\nu$ the Z^ν-norm of the solution of (2.9) does not blow up in finite time. As a consequence of Theorem 2.9 we obtain $\tau_{u_0} = \infty$ for all $u_0 \in Z^\nu$. In that case we define $T_\eta(t)u_0 = u_\eta(t,u_0)$ for all $t \geqslant 0$ and $u_0 \in Z^\nu$.

Suppose that

$$\lim_{\eta\to 0} \sup_{\|y\|_{Z^\nu}<r} \|f_\eta(y) - f_0(y)\|_Z = 0.$$

Using Theorem 2.7 and (2.10) we conclude that the family $\mathcal{T}_\eta = \{T_\eta(t)\colon t \geqslant 0\}$ is continuous at $\eta = 0$. If the semigroup $\mathcal{T}_\eta$ is bounded dissipative (usually obtained with the aid of some energy estimate) and $f_\eta\colon Z^\nu \to Z$ is compact (or B_η has compact resolvent), it follows that $\mathcal{T}_\eta$ is asymptotically compact (or compact). With all these hypotheses and results, $\mathcal{T}_\eta$ has a global attractor $\mathcal{A}_\eta$ in Z^ν.

If $\{\mathcal{T}_\eta\}_{\eta\in[0,1]}$ is collective bounded dissipative and either $f_\eta\colon Z^\nu \to Z$ is compact or B_η has compact resolvent, then

$$\overline{\bigcup_{\eta\in[0,1]} \mathcal{A}_\eta} \text{ is compact.}$$

As a consequence, the family $\{\mathcal{T}_\eta\}_{\eta\in[0,1]}$ is collective asymptotically compact at $\eta = 0$ and the family $\{\mathcal{A}_\eta\}_{\eta\in[0,1]}$ is upper semicontinuous at $\eta = 0$ in Z^ν.

2.3. Concrete semilinear parabolic problem. Let us now consider a concrete application of the above abstract example. Let Ω be a bounded smooth domain in $\mathbb{R}^N$, $N \geqslant 1$, and $\eta \in [0,1]$. Consider the parabolic problem

$$\begin{cases} u_t = \operatorname{div}(a_\eta \nabla u) - u + f(x,u), & x \in \Omega, \\ \dfrac{\partial u}{\partial n} = 0, \quad x \in \partial\Omega, \end{cases} \tag{2.11}$$

where $f(x,\cdot)\colon \mathbb{R} \to \mathbb{R}$ is a globally Lipschitz function with bounded derivatives up to second order in the second variable uniformly for $x \in \Omega$, and $a_\eta\colon \overline{\Omega} \to [1,2]$ is a continuously differentiable function.

Assume that $a_\eta \to a_0$ as $\eta \to 0$ in $L^1(\Omega)$ and let $A_\eta\colon D(A_\eta) \subset L^2(\Omega) \to L^2(\Omega)$ be the linear operator defined by

$$D(A_\eta) = \left\{u \in H^2(\Omega)\colon \frac{\partial u}{\partial n} = 0\right\} = H^2_n(\Omega),$$

with

$$A_\eta u = -\operatorname{div}(a_\eta \nabla u) + u, \text{ for all } u \in D(A_\eta).$$

Note that (2.4) follows immediately from the fact that A_η is positive and self-adjoint. Also, $D(A_\eta) = H^2_n(\Omega)$ is a fixed vector space and the graph norms in $H^2_n(\Omega)$ give equivalent norms for all values of $\eta \in [0,1]$, with uniform equivalence constants.

Let us verify condition (2.3). In order to prove that A_η^{-1} converges in the uniform operator topology to A_0^{-1} we note that $u_\eta = A_\eta^{-1} h_\eta$ means that u_η satisfies

$$\int_\Omega a_\eta \nabla u_\eta \nabla \phi + \int_\Omega u_\eta \phi = \int_\Omega h_\eta \phi \quad \text{for all } \phi \in H^1(\Omega). \tag{2.12}$$

It is easy to see that if $\|h_\eta\|_{L^2(\Omega)} \leqslant 1$ then there is a constant $c > 0$ such that $\|u_\eta\|_{H^1(\Omega)} \leqslant c$. From this we have that there is a subsequence (which we denote the same) such that $u_\eta \to u_0$ strongly in $L^2(\Omega)$ and weakly in $H^1(\Omega)$.

Also, since $a_\eta \to a_0$ in $L^1(\Omega)$ there is a subsequence which converges almost everywhere in Ω. From the Lebesgue Dominated Convergence Theorem, $a_\eta \nabla \phi$ converges to $a_0 \nabla \phi$ in $L^2(\Omega)$. Since the limiting is independent of the subsequence taken, the convergence follows. If $h_\eta \to h_0$ weakly in $L^2(\Omega)$, making $\eta \to 0$ in (2.12) we have

$$\int_\Omega a_0 \nabla u_0 \nabla \phi + \int_\Omega u_0 \phi = \int_\Omega h_0 \phi, \text{ for all } \phi \in H^1(\Omega) \tag{2.13}$$

and consequently $u_0 = A_0^{-1} h_0$.

REMARK 2.10. Asking that $a_\eta \to a_0$ uniformly in $\bar{\Omega}$ and that $h_\eta \to h_0$ weakly in $L^2(\Omega)$, we can also prove the convergence of the solutions u_η of (2.12) to the solution u_0 of (2.13) strongly in $H^1(\Omega)$. That is accomplished using the variational formulation of the solution.

Now we are able to prove:

THEOREM 2.11. $\|A_\eta^{-1} - A_0^{-1}\|_{\mathcal{L}(L^2(\Omega))} \to 0$.

PROOF. Suppose that is not the case. Then, there exist a sequence $h_n \in L^2(\Omega)$, $\|h_n\|_{L^2(\Omega)} = 1$ and $\epsilon_0 > 0$ such that

$$\|A_\eta^{-1}h_n - A_0^{-1}h_n\|_{L^2(\Omega)} \geqslant \epsilon_0. \tag{2.14}$$

Hence there is a subsequence, which we again denote by h_n, and a function h_0 such that $h_n \to h_0$ weakly in $L^2(\Omega)$. From our previous computation $A_\eta^{-1}h_n \to A_0^{-1}h_0$ in $L^2(\Omega)$, and that contradicts (2.14). □

From this (2.3) is satisfied and Theorem 2.7 holds.

Now, taking $\gamma < \frac{1}{2}$ and $\nu = \frac{1}{2}$, the Nemytskiĭ operator $f\colon H^1(\Omega) \to L^2(\Omega)$ ($H^1(\Omega) = Z^{\frac{1}{2}}$ with $Z = L^2(\Omega)$) is well defined, continuous and bounded. Thus the family of semigroups $\{\mathcal{T}_\eta\}_{\eta\in[0,1]}$, associated to (2.11), is well defined and continuous at $\eta = 0$. Here we use the uniform equivalent constants of the graph norms of $B_\eta^{\frac{1}{2}}$ and replace B_0 by $B_0^{\frac{1}{2}}$ in the second inequality in (2.6).

Since A_η^{-1} is compact and f is a bounded map, the semigroup $\mathcal{T}_\eta$ associated to (2.11) is compact and bounded dissipative, with a bounded absorbing set which can be chosen the same for all $\eta \in [0,1]$. That ensures the existence of a global attractor $\mathcal{A}_\eta$ for each $\eta \in [0,1]$ and that $\overline{\bigcup_{\eta\in[0,1]} \mathcal{A}_\eta}$ is compact.

Finally, the family $\{\mathcal{A}_\eta\}_{\eta\in[0,1]}$ is upper semicontinuous at $\eta = 0$ in $H^1(\Omega)$.

REMARK 2.12. Using the results of Section 2.3 and some other results on continuity of resolvents for elliptic operators one can also prove the upper semicontinuity of attractors with respect to domain perturbations, or with respect to blow up of the diffusivity in some regions of the domain (see [**8, 13, 17, 54**]).

REMARK 2.13. A similar scheme may be constructed when the semigroups generated by the operators B_η are only strongly continuous. In that case, the compactness usually comes from the compactness of the nonlinearity and the asymptotic compactness is obtained with the application of Theorem 1.24 and the continuity of the semigroups with the aid of the Trotter-Kato Theorem (see [**183**]). There are also situations where the perturbation changes the type of semigroup generated by the operators B_0 and B_η, $\eta > 0$ (see [**26, 37, 38, 67**]).

3. Lower semicontinuity

Given a semigroup $\mathcal{T}$ in a metric space X and a $\mathcal{T}$-invariant set A, we have already seen that there exists a global solution of $\mathcal{T}$ through x in A, for each $x \in A$ (cf. Remark 1.10). The simplest global solutions are the constant solutions, that is, the equilibrium points. The global solutions that converge to an equilibria y^* as t tends to $-\infty$ constitute an invariant set that we call the **unstable set** $W^u(y^*)$ of y^*, that is,

$$W^u(y^*) = \Big\{x \in X\colon \text{ there is a global solution } \xi \text{ of } \mathcal{T} \text{ through } x \\ \text{such that } \lim_{t\to-\infty} d(\xi(t), y^*) = 0\Big\}.$$

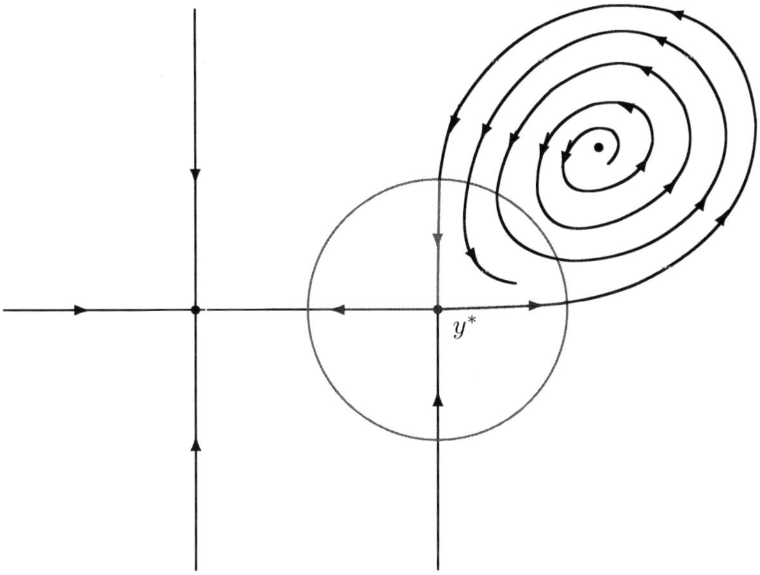

FIGURE 2.1. Local unstable set of y^*.

Given a neighborhood V of y^*, the set of points y of V such that there exists a global solution ξ of $\mathcal{T}$ through y with $\lim_{t\to-\infty} d(\xi(t), y^*) = 0$ and $\xi(t) \in V$ for all $t \in \mathbb{R}^-$ is called a *local unstable set* of y^* and is denoted by $W^u_{loc}(y^*)$. If $V = B_\delta(y^*)$, then we usually denote $W^u_{loc}(y^*)$ by $W^u_\delta(y^*)$.

The orbit $\gamma_\xi = \xi(\mathbb{R})$ of a global nonconstant solution ξ of $\mathcal{T}$ that tends to an equilibrium y^* as t tends both to $\pm\infty$ is called a *homoclinic orbit* in y^*. Observe that, when there exists a homoclinic orbit in y^*, $W^u_{loc}(y^*)$ may not coincide with the intersection $W^u(y^*)$ with a neighborhood of y^*, as Figure 2.1 shows.

In Figure 2.1, the region inside the circle represents the neighborhood V of y^*, the upper vertical segment inside the circle represents the part of $W^u(y^*)$ that *is not* in $W^u_{loc}(y^*)$ and the local unstable set is the horizontal segment inside the circle.

The idea used, in the seminal works of [**22, 135**], to prove lower semicontinuity (non-implosion) was to reproduce, in the perturbed attractor, the structures seen in the limiting attractor. Such an idea carries with it the assumption that we already understand the structures of the limiting attractor. Let us briefly describe the procedure used in these papers to obtain lower semicontinuity.

Consider a parameter dependent family $\{\mathcal{T}_\eta\}_{\eta\in[0,1]}$ of semigroups with global attractors $\{\mathcal{A}_\eta\}_{\eta\in[0,1]}$. We refer to $\mathcal{T}_\eta$ and $\{\mathcal{A}_\eta\}$ as the perturbed (limiting) semigroup and its global attractor if $\eta > 0$ (if $\eta = 0$). Suitable continuity properties with respect to the parameter η are also required (see Theorem 2.14).

The authors in [**22, 135**] require a context in which the limiting semigroup has a set of equilibria $\mathcal{E}_0$, and each equilibrium $y^* \in \mathcal{E}_0$ is *hyperbolic*. Hyperbolicity implies that each equilibria is isolated and that, together with the compactness of the attractor, there are finitely many of them (see Chapter 4). Moreover, they

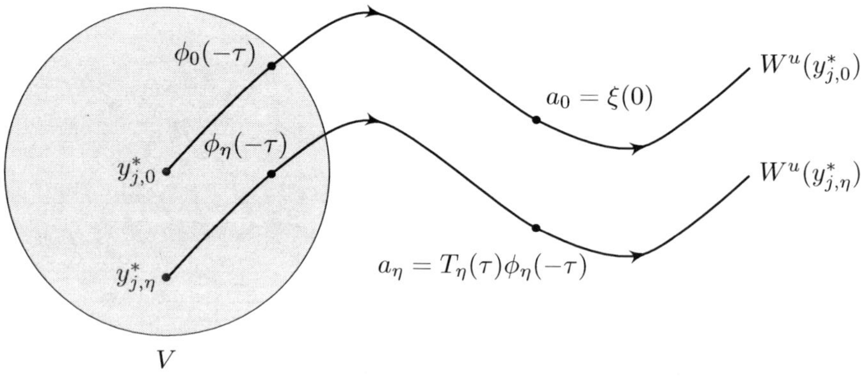

FIGURE 2.2. Lower semicontinuity explained.

require that the limiting attractor is the union of unstable sets of these equilibria $\mathcal{A}_0 = \bigcup_{j=1}^n W^u(y_j^*)$, where $\mathcal{E}_0 = \{y^*_{j,0}\}_{j=1}^n$. The hyperbolicity of equilibria $\{y^*_{j,0}\}_{j=1}^n$ ensures their permanence $\mathcal{E}_\eta = \{y^*_{j,\eta}\}_{j=1}^n$ under suitable small perturbations (see Chapter 6), and that their local unstable manifolds behave continuously under such perturbations (see Chapter 4).

The simple ideas that lead to lower semicontinuity are briefly explained next (see Figure 2.2): if $a_0 \in \mathcal{A}_0$, there is a global solution ϕ_0 of $\mathcal{T}_0$ through a_0, since $a_0 \in W^u(y^*_{j,0})$ for some $1 \leqslant j \leqslant n$, and a_0 can be driven by the flow (backwards in time) $-\tau$ units of time to the local unstable set, that is, $\phi_0(-\tau) \in W^u_{loc}(y^*_{j,0})$. Using the continuity of the local unstable sets, $\phi_0(-\tau)$ can be approximated by points $\phi_\eta(-\tau) \in W^u_{loc}(y^*_{j,\eta})$. Driving forward $\phi_\eta(-\tau)$ by the same amount of time τ, we produce an approximation of a_0 by points $a_\eta = T_\eta(\tau)\phi_\eta(-\tau)$ in the perturbed attractor $\mathcal{A}_\eta$. The lower semicontinuity at $\eta = 0$ then follows by Lemma 2.2.

Hyperbolicity of equilibria, in fact, is used to ensure their permanence and also the continuity of the local unstable sets under perturbation. These properties, however, can be assumed instead of assuming hyperbolicity of equilibria, and we are able to state the result for a general metric space. These ideas are made precise in the following theorem.

THEOREM 2.14 (Lower semicontinuity). *Let $\{\mathcal{T}_\eta\}_{\eta\in[0,1]}$ be a family of semigroups which is continuous at $\eta = 0$ and satisfies the following:*

(a) *$\mathcal{T}_\eta$ has a global attractor $\mathcal{A}_\eta$ for each $\eta \in [0, 1]$;*
(b) *there exist $n \in \mathbb{N}$, independent of $\eta \in [0, 1]$, and equilibria $\{y^*_{j,\eta}\}_{j=1}^n$ of $\mathcal{T}_\eta$ for each $\eta \in [0, 1]$;*
(c) *there exists $\delta > 0$ such that the family $\{W^u_\delta(y^*_{j,\eta})\}_{\eta\in[0,1]}$ is lower semicontinuous at $\eta = 0$;*
(d) *$\mathcal{A}_0 = \bigcup_{j=1}^n W^u(y^*_{j,0})$.*

Then the family of global attractors $\{\mathcal{A}_\eta\}_{\eta\in[0,1]}$ is lower semicontinuous at $\eta = 0$.

PROOF. From (d) it follows that if $u_0 \in \mathcal{A}_0$, then there exists a global solution ϕ_0 of $\mathcal{T}_0$ through u_0, $1 \leqslant j \leqslant n$ and $\tau \in \mathbb{R}^+$ such that $\phi_0(-\tau) \in W^u_\delta(y^*_{j,0})$. From (c), there is $u_\eta^{-\tau} \in W^u_\delta(y^*_{j,\eta})$ such that $u_\eta^{-\tau} \overset{\eta\to 0}{\longrightarrow} \phi_0(-\tau)$ and global solution ψ_η of

$\mathcal{T}_\eta$ through $u_\eta^{-\tau}$ such that $\psi_\eta(0) = u_\eta^{-\tau} \stackrel{\eta \to 0}{\longrightarrow} \phi_0(-\tau)$. It follows from (a) and from the continuity of the family $\{\mathcal{T}_\eta\}_{\eta \in [0,1]}$ at $\eta = 0$ that

$$\psi_\eta(\tau) = T_\eta(\tau)\psi_\eta(0) \to T_0(\tau)\psi_0(-\tau) = \psi_0(0) = u_0,$$

and the result now follows from Lemma 2.2. □

There are, of course, much more general situations for which the lower semicontinuity of attractors holds. Results in this direction can be achieved (see [**73, 74**]) by replacing the equilibria by more general *isolated invariant sets* and their unstable sets. As an example we mention the normally hyperbolic periodic orbits and their unstable sets (see Chapter 4).

We will present now some of these concepts, in order to present a more general version of Theorem 2.14.

DEFINITION 2.15. Let $\mathcal{T}$ be a semigroup in a metric space X and Ξ a $\mathcal{T}$-invariant set. Given an open neighborhood U of Ξ in X, we say that Ξ is the **maximal $\mathcal{T}$-invariant set in** U if given $B \subset U$ $\mathcal{T}$-invariant, then $B \subset \Xi$.

We also say that Ξ is an **isolated $\mathcal{T}$-invariant set** when Ξ is the maximal $\mathcal{T}$-invariant set in $\mathcal{O}_\delta(\Xi)$ for some $\delta > 0$.

DEFINITION 2.16. We say that $\mathbf{\Xi} = \{\Xi_1, \cdots, \Xi_n\}$ is a **disjoint collection of isolated $\mathcal{T}$-invariant sets** if each Ξ_i is an isolated $\mathcal{T}$-invariant set and there exists $\delta > 0$ such that $\mathcal{O}_\delta(\Xi_i) \cap \mathcal{O}_\delta(\Xi_j) = \varnothing$, $1 \leqslant i < j \leqslant n$.

When $\Xi = \{z^*\}$ is a unitary set we say that z^* is an **isolated stationary solution** (or **isolated equilibrium**) of $\mathcal{T}$.

REMARK 2.17. Note that a *center* is an example of isolated equilibrium which is not a unitary isolated invariant set. In fact, unitary isolated invariant sets (more generally, isolated invariant sets) have the topological hyperbolicity property stated in the next result. Typically an isolated invariant set will be a hyperbolic equilibrium or a normally hyperbolic periodic orbit.

LEMMA 2.18 (Topological hyperbolicity). *Let $\mathcal{T}$ be an asymptotically compact semigroup. Assume that U, V are open subsets of X with $\overline{U} \subset V$ and such that Ξ is the maximal $\mathcal{T}$-invariant subset of V. If there exists $t_0 > 0$ and $u_0 \in X$ such that $T(t)u_0 \in U$ for all $t \geqslant t_0$, then $\mathrm{dist}_H(T(t)u_0, \Xi) \stackrel{t \to \infty}{\longrightarrow} 0$. Furthermore, if ξ is a global solution and there exists $t_0 > 0$ such that $\xi((-\infty, -t_0)) \subset U$, then $\mathrm{dist}_H(\xi(t), \Xi) \stackrel{t \to -\infty}{\longrightarrow} 0$.*

PROOF. Let us prove the case $T(t)u_0 \in U$ for all $t \geqslant t_0$. The other case is completely similar. We argue by contradiction assuming that

$$\limsup_{t \to \infty} \mathrm{dist}_H(T(t)u_0, \Xi) > 0.$$

If this is the case, there exist $\epsilon_0 > 0$ and a sequence $t_n \to \infty$ such that $\mathrm{dist}_H(T(t_n)u_0, \Xi) \geqslant \epsilon_0$. From the asymptotic compactness of $\mathcal{T}$, this sequence has a convergent subsequence (which we denote the same). If $y = \lim_{n \to \infty} T(t_n)u_0$, then $\mathrm{dist}_H(y, \Xi) \geqslant \epsilon_0$ and y belongs to the ω-limit set of x_0. Hence $y \in \omega(x_0) \subset \bar{U}$, $\Xi \subsetneq \omega(x_0) \cup \Xi \subset V$ and $\omega(x_0) \cup \Xi$ is invariant. This contradicts the assumption that Ξ is the maximal $\mathcal{T}$-invariant subset of V and concludes the proof. □

Now we can define *unstable sets* for isolated $\mathcal{T}$-invariant sets, analogously to the ones defined for equilibria.

DEFINITION 2.19. Given an isolated invariant set Ξ we define the **unstable set** of Ξ by

$$W^u(\Xi) = \{x \in X\colon \text{ there is a global solution } \xi \text{ of } \mathcal{T} \text{ through } x \\ \text{such that } \operatorname{dist}(\xi(t), \Xi) \overset{t\to-\infty}{\longrightarrow} 0\}, \tag{2.15}$$

and a **local unstable set** of Ξ by

$$W^u_\delta(\Xi) = \{x \in W^u(\Xi)\colon \text{ there is a global solution } \xi \text{ of } \mathcal{T} \text{ through } x \\ \text{with } \operatorname{dist}_H(\xi(t), \Xi) \leqslant \delta \text{ for all } t \leqslant 0 \text{ and } \operatorname{dist}_H(\xi(t), \Xi) \overset{t\to-\infty}{\longrightarrow} 0\}. \tag{2.16}$$

With this we can state the main result of this section.

THEOREM 2.20. *Let $\{\mathcal{T}_\eta\}_{\eta\in[0,1]}$ be a collective asymptotically compact and continuous at $\eta = 0$, and collectively bounded dissipative family of semigroups, which satisfies the following:*

(a) *there exists a disjoint collection of isolated $\mathcal{T}_\eta$-invariant sets $\boldsymbol{\Xi}_\eta = \{\Xi_{1,\eta}, \cdots, \Xi_{n,\eta}\}$ for each $\eta \in [0,1]$;*
(b) $\operatorname{dist}_H(W^u_\delta(\Xi_{j,0}), W^u_\delta(\Xi_{j,\eta})) \overset{\eta\to 0}{\longrightarrow} 0$*;*
(c) $\mathcal{A}_0 = \bigcup_{j=1}^n W^u(\Xi_{j,0})$.

Then

$$\operatorname{dist}_H(\mathcal{A}_\eta, \mathcal{A}_0) + \operatorname{dist}_H(\mathcal{A}_0, \mathcal{A}_\eta) \overset{\eta\to 0}{\longrightarrow} 0.$$

The proof of this result follows, step by step, the proof of Theorem 2.14 and will be left to the reader.

In the applications of the above result the main steps to be checked are the characterization of the limiting attractor as the union of the unstable sets of the disjoint collection of isolated invariant sets (gradient structure obtained from the special form of the limiting problem), the permanence of the isolated invariant sets and their continuity under perturbations (usually a consequence of hyperbolicity) and, finally, the proof that the local unstable sets behave lower semicontinuously at $\eta = 0$ (usually a consequence of robustness of hyperbolicity).

4. Examples with lower semicontinuity

4.1. Example 2.1 - Continued. Let us revisit Example 2.1 and consider some additional structure that will allow us to talk about lower semicontinuity of global attractors. We start by assuming that all conditions imposed in Section 2.1 are satisfied, that $f, g\colon \mathbb{R}^n \to \mathbb{R}^n$ are continuously differentiable and $\mathbb{R}^n \ni x \mapsto g'(x) \in \mathcal{L}(\mathbb{R}^n)$ is bounded, and that each solution x^* of $f(x) = 0$ is hyperbolic, that is, the spectrum $\sigma(f'(x^*))$ of $f'(x^*)$ is disjoint from the imaginary axis. This condition along with the compactness of the attractor $\mathcal{A}_0$ implies that the number of equilibrium points is finite (see Chapter 4). Let $n \in \mathbb{N}^*$ be the number of solutions of $f(x) = 0$ and denote by $\mathcal{E} = \{x^*_{1,0}, \cdots, x^*_{n,0}\}$ the set of these solutions.

The results of Chapter 4 imply that there are exactly n hyperbolic solutions of $f(x) + \eta g(x) = 0$, namely $\mathcal{E}_\eta = \{x^*_{1,\eta}, \cdots, x^*_{n,\eta}\}$ for all suitably small η, and also we have

$$\lim_{\eta\to 0^+} \sup_{1\leqslant i\leqslant n} \|x^*_{i,\eta} - x^*_{i,0}\|_{\mathbb{R}^n} = 0.$$

It also follows from the results in Chapter 4 that the local unstable manifolds $W^u_{loc}(x^*_{i,\eta})$ behave continuously at $\eta = 0$. If we assume that

$$\mathcal{A}_0 = \bigcup_{i=1}^{n} W^u(x^*_{i,0}), \tag{2.17}$$

then the family of attractors $\{\mathcal{A}_\eta\}_{\eta\in[0,1]}$ is continuous at $\eta = 0$ as a result of Theorem 2.14.

We point out that if there exists a differentiable function $V\colon \mathbb{R}^n \to \mathbb{R}$ such that $\nabla V(x) = -f(x)$ for all $x \in \mathbb{R}^n$, then (2.17) holds as a consequence of Theorem 3.8.

4.2. Example 2.2 - Continued. Let us revisit Example 2.2 and consider some additional structure that will allow us to talk about lower semicontinuity of global attractors. We start by assuming that all conditions imposed in Example 2.2 are satisfied, that $f_\eta\colon \mathcal{Z}^\nu \to \mathcal{Z}$ is continuously differentiable for each $\eta \in [0,1]$, and that each solution u^* of

$$0 = B_0 u + f_0(u) \tag{2.18}$$

is hyperbolic, that is, the spectrum of $B_0 + f_0'(u^*)$ is disjoint from the imaginary axis. This condition together with the compactness of the attractor $\mathcal{A}_0$ implies that the number of equilibrium points is finite (see Chapter 4). Let $n \in \mathbb{N}^*$ be the number of solutions of (2.18) and denote by $\mathcal{E}_0 = \{u^*_{1,0}, \cdots, u^*_{n,0}\}$ the set of solutions of (2.18). Assume that, for each $r > 0$,

$$\sup_{\|y\|_{\mathcal{Z}^\nu}<r} \{\|f_\eta(y) - f_0(y)\|_{\mathcal{Z}} + \|f_\eta'(y) - f_0'(y)\|_{\mathcal{L}(\mathcal{Z}^\nu,\mathcal{Z})}\} \overset{\eta\to 0}{\longrightarrow} 0.$$

These conditions, along with the results of Chapter 4, imply that the solutions of

$$0 = B_\eta u + f_\eta(u) \tag{2.19}$$

are $\mathcal{E}_\eta = \{u^*_{1,\eta}, \cdots, u^*_{n,\eta}\}$ and moreover

$$\lim_{\eta\to 0} \sup_{1\leqslant i\leqslant n} \|u^*_{i,\eta} - u^*_{i,0}\|_{\mathcal{Z}^\nu} = 0.$$

Also, each $u^*_{i,\eta}$ is hyperbolic for all suitably small η and for all $i = 1, \cdots, n$. It also follows from the results in Chapter 4 that the local unstable manifolds $W^u_{loc}(u^*_{i,\eta})$ behave continuously at $\eta = 0$.

If we assume that

$$\mathcal{A}_0 = \bigcup_{i=1}^{n} W^u(u^*_{i,0}),$$

then the family of attractors $\{\mathcal{A}_\eta\}_{\eta\in[0,1]}$ is continuous at $\eta = 0$ as a result of Theorem 2.14.

4.3. Example 2.3 - Continued. Let us now consider the concrete application of the above abstract results as in Section 2.3. Assume that all conditions imposed in Section 2.2 hold, and that, if u^* is a solution of

$$\begin{cases} 0 = \operatorname{div}(a_0\nabla u) - u + f(x,u), & x \in \Omega, \\ \dfrac{\partial u}{\partial n} = 0, & x \in \partial\Omega, \end{cases} \tag{2.20}$$

then u^* is hyperbolic, that is, the spectrum $\sigma(A_0)$ of the operator $A_0\colon D(A_0) \subset L^2(\Omega) \to L^2(\Omega)$ defined by $D(A_0) = H^2_n(\Omega)$ and $A_0 v = \operatorname{div}(a_0\nabla v) - v + f'(u^*)v$

for all $v \in D(A_0)$ is disjoint from the imaginary axis (this condition is satisfied generically; see [**35**]).

This condition and the compactness of the attractor $\mathcal{A}_0$ for the semigroup associated to (2.11) with $\eta = 0$ implies that the number of equilibrium points is finite (see Section 6). Let $n \in \mathbb{N}^*$ be the number of solutions of (2.18) and denote by $\mathcal{E}_0 = \{u_{1,0}^*, \cdots, u_{n,0}^*\}$ the set of solutions of (2.20).

That and the results of Section 6 imply that the solutions of

$$\begin{cases} 0 = \operatorname{div}(a_\eta \nabla u) - u + f(x,u), & x \in \Omega, \\ \dfrac{\partial u}{\partial n} = 0, \quad x \in \partial\Omega, \end{cases} \tag{2.21}$$

are $\mathcal{E}_\eta = \{u_{1,\eta}^*, \cdots, u_{n,\eta}^*\}$ with

$$\lim_{\eta \to 0} \sup_{1 \leqslant i \leqslant n} \|u_{i,\eta}^* - u_{i,0}^*\|_{H^1(\Omega)} = 0,$$

and each $u_{i,\eta}^*$ is hyperbolic for all suitably small η and for all $i = 1, \cdots, n$. It also follows from the results in Chapter 4 that the local unstable manifolds $W_{loc}^u(u_{i,\eta}^*)$ behave continuously at 0.

We also have

$$\mathcal{A}_\eta = \bigcup_{i=1}^{n} W^u(u_{i,\eta}^*), \tag{2.22}$$

and the family of attractors $\{\mathcal{A}_\eta\}_{\eta \in [0,1]}$ is continuous at $\eta = 0$ as an application of Theorem 2.14.

The characterization (2.22) of the global attractor $\mathcal{A}_\eta$ is obtained from Theorem 3.8 noting that $V_\eta \colon H^1(\Omega) \to \mathbb{R}$ given by

$$V_\eta(u) = \frac{1}{2}\int_\Omega a_\eta |\nabla u|^2 + \int_\Omega F(u),$$

where $F(u)(x) = -\displaystyle\int_0^{u(x)} f(x,s)ds$, is a Lyapunov function for the semigroup associated to (2.11).

4.4. A planar initial value problem. For $\alpha \geqslant 0$ consider the following planar initial value problem:

$$\begin{cases} \dot{u} = -(u+1)u(u-1) - \alpha(u-v), \ t > 0, \\ \dot{v} = -(v+1)v(v-1) + \alpha(u-v), \ t > 0, \\ (u(0), v(0)) = (u_0, v_0) \in \mathbb{R}^2 \end{cases} \tag{2.23}$$

(for more details on this system, see [**65**]). It is easy to see that the function

$$\mathcal{V}_\alpha(u,v) = \frac{1}{4}(u^4 + v^4) - \frac{1}{2}(1-\alpha)(u^2+v^2) - \alpha uv \tag{2.24}$$

is a Lyapunov function for (2.23) and, if $(u(t), v(t))$ is a solution of (2.23), then

$$\frac{d}{dt}(u^2+v^2) \leqslant -2(u^2+v^2) + 4.$$

This estimate ensures that all solutions of (2.23) exist for all $t \geqslant 0$, that the ball $B_{\sqrt{2}}(0)$ of radius $\sqrt{2}$ around zero is an attracting set for the semigroup $\mathcal{T}_\alpha = \{T_\alpha(t) \colon t \geqslant 0\}$, independently of $\alpha \geqslant 0$.

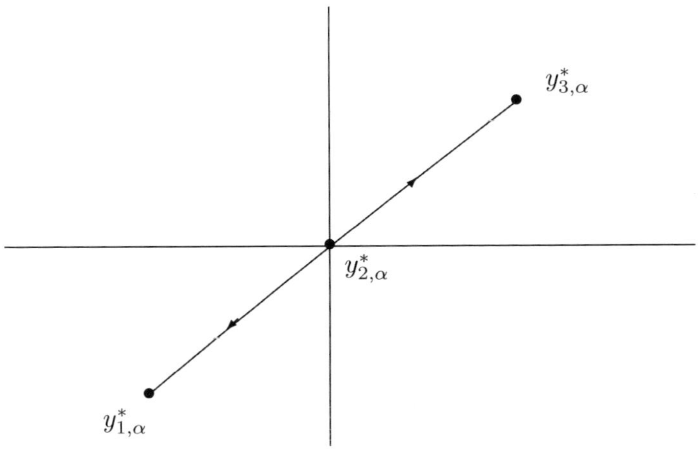

FIGURE 2.3. A representation of $\mathcal{A}_\alpha$ for $\alpha \geqslant \frac{1}{2}$.

It is immediate that the family of semigroups $\{\mathcal{T}_\alpha\}_{\alpha\geqslant 0}$ is collectively asymptotically compact and continuous at $\alpha = \alpha_0$ and collectively bounded dissipative. The associated family of attractors $\{\mathcal{A}_\alpha\}_{\alpha\geqslant 0}$ is such that $\bigcup_{\alpha\geqslant 0}\mathcal{A}_\alpha \subset B_{\sqrt{2}}(0)$ and $\{\mathcal{A}_\alpha\}_{\alpha\geqslant 0}$ is upper semicontinuous at $\alpha = \alpha_0 \geqslant 0$.

Also, the fact that we have a Lyapunov function $\mathcal{V}_\alpha$ for $\mathcal{T}_\alpha$ ensures that the global attractor is the unstable set $\mathcal{A}_\alpha = \bigcup_{(u^*_\alpha, v^*_\alpha)\in\mathcal{E}_\alpha} W^u((u^*_\alpha, v^*_\alpha))$, since the set of equilibria $\mathcal{E}_\alpha$ is always finite; see Theorem 3.8. There is only n_α ($n_\alpha = 3, 5, 7$ or 9) of them, $\mathcal{E}_\alpha = \{(u^*_{i,\alpha}, v^*_{i,\alpha})\colon i = 1, \cdots, n_\alpha\}$ and

$$\mathcal{A}_\alpha = \bigcup_{i=1}^{n_\alpha} W^u((u^*_{i,\alpha}, v^*_{i,\alpha})).$$

If all the equilibria in $\mathcal{E}_{\alpha_0}$ are hyperbolic, the local unstable sets behave continuously at $\alpha_0 \geqslant 0$ and, as a consequence of Theorem 2.14, the family of attractors $\{\mathcal{A}_\alpha\}_{\alpha\geqslant 0}$ is lower semicontinuous at $\alpha_0 \geqslant 0$. In fact, lower semicontinuity holds for all $\alpha_0 \geqslant 0$ (even if some of the equilibria are not hyperbolic; see **[65]**).

Let us briefly discuss the set $\mathcal{E}_\alpha$, $\alpha \geqslant 0$, to find the points where the hyperbolicity is not observed and where bifurcations occur. Clearly, the three points

$$y^*_{1,\alpha} = (-1,-1),\ y^*_{2,\alpha} = (0,0),\ y^*_{3,\alpha} = (1,1)$$

are equilibria, independently of $\alpha \geqslant 0$. Outside the diagonal, an equilibria (u, v) must satisfy

$$u^2 + uv + v^2 = 1 - 2\alpha, \tag{2.25}$$

and consequently, there will be no equilibria outside the diagonal for $\alpha \geqslant \frac{1}{2}$.

If $\alpha > \frac{1}{2}$ the equilibria $y^*_{1,\alpha}$, $y^*_{2,\alpha}$ and $y^*_{3,\alpha}$ are hyperbolic whereas 0 is in the spectrum of the linearization of (2.23) around $y^*_{1,\frac{1}{2}}$, that is, $y^*_{1,\frac{1}{2}}$ is not hyperbolic.

Nonetheless, the attractor $\mathcal{A}_{\frac{1}{2}}$ is the same as $\mathcal{A}_\alpha$ for $\alpha > \frac{1}{2}$ consisting of the three equilibria in the diagonal and the heteroclinic that connect them (see Figure 2.3).

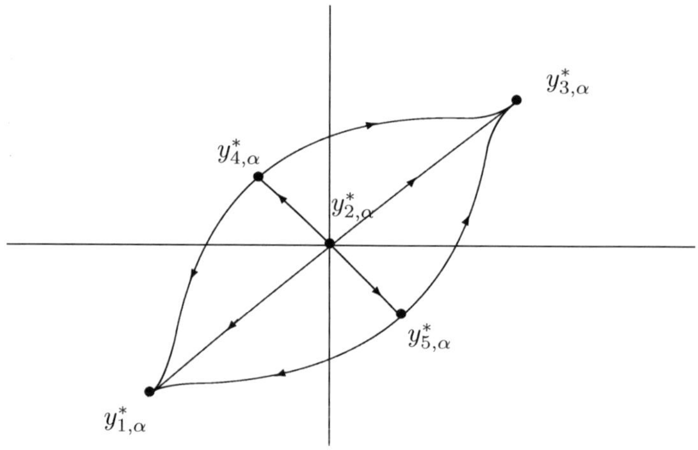

FIGURE 2.4. A representation of $\mathcal{A}_\alpha$ for $\alpha \in [\frac{1}{3}, \frac{1}{2})$.

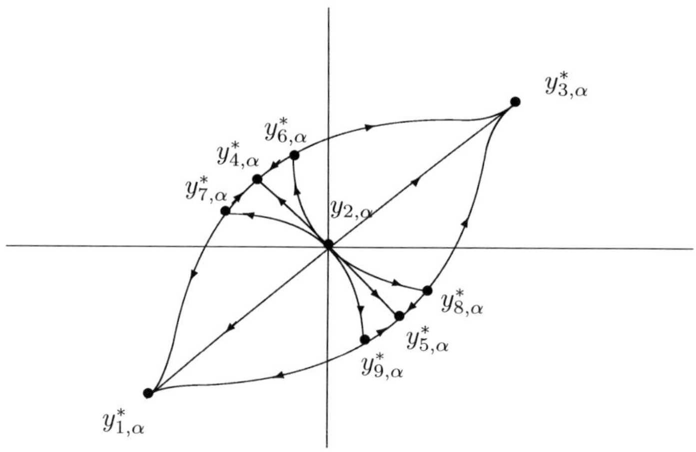

FIGURE 2.5. A representation of $\mathcal{A}_\alpha$ for $\alpha \in (0, \frac{1}{3})$ close to $\frac{1}{3}$.

A pitchfork bifurcation occurs when $\alpha = \frac{1}{2}$ and for $\alpha \in (\frac{1}{3}, \frac{1}{2})$ we observe that, making $u = -v$ in (2.25), two additional equilibria, namely

$$y^*_{4,\alpha} = (-\sqrt{1-2\alpha}, \sqrt{1-2\alpha}) \text{ and } y^*_{5,\alpha} = (\sqrt{1-2\alpha}, -\sqrt{1-2\alpha}), \tag{2.26}$$

appear and both of them are hyperbolic (see Figure 2.4).

To close the discussion, at $\alpha = \frac{1}{3}$ a secondary bifurcation occurs and, for $\alpha < \frac{1}{3}$, four new equilibria $\{y^*_{6,\alpha}, y^*_{7,\alpha}, y^*_{8,\alpha}, y^*_{9,\alpha}\}$ appear (see [**65**] for details) and they are given by (see Figure 2.5):

$$(u(v), v) \text{ where } v = \pm\sqrt{\frac{1-\alpha \pm \sqrt{1-2\alpha-3\alpha^2}}{2}},$$

and $u(v) = v - \frac{1}{\alpha}v + \frac{1}{\alpha}v^3$.

REMARK 2.21. We note that the lower semicontinuity also holds at $\alpha = \frac{1}{2}$ and $\alpha = \frac{1}{3}$ (see [**65**]). Also, the attractor $\mathcal{A}_0 = [-1,1]^2$ and the family of attractors $\{\mathcal{A}_\alpha\}_{\alpha \geqslant 0}$ behave continuously at 0.

Notes. Upper semicontinuity has been studied in a variety of problems and under very singular types of perturbations (see [**8, 14, 15, 18, 37, 57–59, 70–72, 78, 82–85, 114**]). The lower semicontinuity of global attractors include several singular perturbation problems, a more general class of limiting problems, with respect to the seminal works of [**22, 135**], which include also problems with normally hyperbolic periodic orbits (see [**8, 10, 11, 14, 29, 38, 55, 78**]). A theoretical equivalent condition for lower semicontinuity is the equi-attraction ([**145, 152, 164**]), where attraction to the global attractors is uniform with respect to the parameter, a property closely related to the rate of convergence of attractors under perturbation ([**22, 68**]). In applications, equi-attraction and the rate of convergence to the limit attractor is again associated to the inner structure of the attractors, in particular the rate of attraction to unstable manifolds ([**65**]).

CHAPTER 3

Topological Structural Stability of Attractors

Our goal in this chapter is, loosely speaking, to prove that a semigroup obtained as a perturbation (in some sense) of a *gradient semigroup* is still gradient (a property which will be called *topological structural stability of attractors*). Let us be more specific. We deal with the class of *gradient semigroups relative to an invariant set* (or a family of invariant sets) and having a global attractor. We will see that, under natural constraints, this type of semigroups can be characterized through its dynamical properties, introducing the auxiliary concept of *dynamically gradient semigroups* (those with the dynamical properties of a gradient semigroup but without requiring the existence of a Lyapunov function). Again, under natural assumptions, we obtain that a perturbation of a dynamically gradient semigroup is still a dynamically gradient semigroup. Finally we obtain that dynamically gradient semigroups have an associated Lyapunov function and therefore are gradient.

1. Gradient semigroups

We begin our chapter by considering the *gradient semigroups* relative to an invariant set. These semigroups appear naturally in several applications, and its inherent features allow us to describe with precision the inner structure of its attractors.

DEFINITION 3.1. A semigroup $\mathcal{T}$ in X with a closed $\mathcal{T}$-invariant set Ξ is said to be **gradient** relative to Ξ if there exists a continuous function $V\colon X \to \mathbb{R}$ such that:

(i) $\mathbb{R}^+ \ni t \mapsto V(T(t)x) \in \mathbb{R}$ is nonincreasing for each $x \in X$.
(ii) If x is such that $V(T(t)x) = V(x)$ for all $t \in \mathbb{R}^+$, then $x \in \Xi$.
(iii) V is constant in each connected component of Ξ.

A function $V\colon X \to \mathbb{R}$ with these properties is called a **Lyapunov function** for $\mathcal{T}$ relative to Ξ.

EXAMPLE 3.2 (Gradient vector fields). The notion of gradient semigroups in differential equations is associated to those vector fields which are the gradient of a potential, that is, if $\phi\colon \mathbb{R}^n \to \mathbb{R}$ is a C^1 potential with $\mathbb{R}^n \ni u \mapsto \nabla\phi(u) \in \mathbb{R}^n$ locally Lipschitz continuous, the system of differential equations

$$(3.1)\qquad \begin{cases} \dot{u} = -\nabla\phi(u), \ t > 0, \\ u(0) = u_0 \in \mathbb{R}^n \end{cases}$$

is called a *gradient system*.

The potential ϕ has the property that, if $[0, \tau_{u_0}) \ni t \mapsto u(t, u_0)$ is a solution of (3.1), then $[0, \tau_{u_0}) \ni t \mapsto \phi(u(t)) \in \mathbb{R}$ is nonincreasing, and consequently does not blow up as $t \to \tau_{u_0}^-$. If $\phi(x) \to \infty$ as $\|x\|_{\mathbb{R}^n} \to \infty$ we must have $\tau_{u_0} = +\infty$

for all $u_0 \in \mathbb{R}^n$, and we can define the semigroup $\mathcal{T} = \{T(t) : t \geqslant 0\}$ by setting $T(t)u_0 = u(t, u_0)$ for $t \geqslant 0$ and $u_0 \in \mathbb{R}^n$.

It is easy to see that $\phi(u(\cdot))$ is constant if and only if $u(\cdot)$ is constant and $\nabla\phi(u) = 0$, showing that $\phi : \mathbb{R}^n \to \mathbb{R}$ is a Lyapunov function for $\mathcal{T}$ relative to the set Ξ of equilibria for (3.1), that is, $\Xi = \{u \in \mathbb{R}^n : \nabla\phi(u) = 0\}$.

EXAMPLE 3.3 (A heat equation). Let Ω be a bounded smooth domain in $\mathbb{R}^N$, $N \geqslant 1$. Consider the following parabolic initial value problem with Dirichlet boundary condition:

$$\begin{cases} u_t = \Delta u + f(u), & x \in \Omega,\ t > 0, \\ u(x) = 0, & x \in \partial\Omega,\ t > 0, \\ u(0) = u_0 \in H_0^1(\Omega). \end{cases} \tag{3.2}$$

Assuming that $f : \mathbb{R} \to \mathbb{R}$ is a globally Lipschitz function with bounded derivatives up to second order, we can prove that there is a semigroup $\mathcal{T}$ in $H_0^1(\Omega)$ associated to (3.2).

Consider the functional $V : H_0^1(\Omega) \to \mathbb{R}$ given by

$$V(u) = \frac{1}{2}\int_\Omega |\nabla u|^2 - \int_\Omega F(u),$$

where $F(u)(x) = \int_0^{u(x)} f(s)ds$. It is easy to see that V is continuous and if $u : \mathbb{R}^+ \to H_0^1(\Omega)$ denotes the solution of (3.2), then

$$\frac{d}{dt}V(u(t)) = -\int_\Omega |u_t(t)(x)|^2 dx$$

and $V(u(\cdot))$ is constant if and only if $u(\cdot)$ is constant and $u \in H^2(\Omega) \cap H_0^1(\Omega)$ is a solution of

$$\begin{cases} 0 = \Delta u + f(u), & x \in \Omega, \\ u(x) = 0, & x \in \partial\Omega, \end{cases} \tag{3.3}$$

showing that $V : H_0^1(\Omega) \to \mathbb{R}$ is a Lyapunov function for $\mathcal{T}$ relative to the set Ξ of equilibria for (3.2), that is, $\Xi = \{u \in H^2(\Omega) \cap H_0^1(\Omega) : u \text{ satisfies } (3.3)\}$.

EXAMPLE 3.4 (A damped wave equation). Let Ω be a bounded smooth domain in $\mathbb{R}^N$, $N \geqslant 1$. For $\beta \geqslant 0$, consider the damped hyperbolic problem

$$\begin{cases} u_{tt} + 2\beta u_t = \Delta u + f(u), & x \in \Omega,\ t > 0, \\ u(x) = 0, & x \in \partial\Omega,\ t > 0, \\ u(0) = u_0 \in H_0^1(\Omega), \\ u_t(0) = v_0 \in L^2(\Omega). \end{cases} \tag{3.4}$$

This boundary-initial value problem can be written as an abstract ordinary differential equation in $H_0^1(\Omega) \times L^2(\Omega)$ in the following way: let $A : D(A) \subset L^2(\Omega) \to L^2(\Omega)$ be defined by $Au = \Delta u$ for all $u \in D(A) = H^2(\Omega) \cap H_0^1(\Omega)$ and $\Lambda : D(\Lambda) \subset H_0^1(\Omega) \times L^2(\Omega) \to H_0^1(\Omega) \times L^2(\Omega)$ is defined by

$$\Lambda \begin{bmatrix} u \\ v \end{bmatrix} = \begin{bmatrix} 0 & I \\ A & -2\beta \end{bmatrix} \begin{bmatrix} u \\ v \end{bmatrix} = \begin{bmatrix} v \\ Au - 2\beta v \end{bmatrix},$$

with $D(\Lambda) = D(A) \times L^2(\Omega)$. We may rewrite (3.4) as

$$\begin{cases} \dfrac{d}{dt}\begin{bmatrix} u \\ v \end{bmatrix} = \Lambda \begin{bmatrix} u \\ v \end{bmatrix} + \begin{bmatrix} 0 \\ f(u) \end{bmatrix}, \ t > 0, \\ \begin{bmatrix} u \\ v \end{bmatrix}(0) = \begin{bmatrix} u_0 \\ v_0 \end{bmatrix} \in H_0^1(\Omega) \times L^2(\Omega). \end{cases} \tag{3.5}$$

Assume that $f\colon \mathbb{R} \to \mathbb{R}$ is a globally Lipschitz function with bounded derivatives up to second order. Then, there is a semigroup $\mathcal{T}$ in $H_0^1(\Omega) \times L^2(\Omega)$ associated to (3.5). Consider now the functional $V\colon H_0^1(\Omega) \times L^2(\Omega) \to \mathbb{R}$ given by

$$V\left(\begin{bmatrix} u \\ v \end{bmatrix}\right) = \frac{1}{2}\int_\Omega |\nabla u|^2 + \beta \int_\Omega v^2 - \int_\Omega F(u),$$

where $F(u)(x) = \int_0^{u(x)} f(s)ds$.

It is easy to see that $V\left(\begin{bmatrix} u \\ v \end{bmatrix}(\cdot)\right)$ is continuous,

$$\frac{d}{dt}V\left(\begin{bmatrix} u(t) \\ u_t(t) \end{bmatrix}(\cdot)\right) = -\int_\Omega |u_t(t)(x)|^2 dx$$

(at least if $\begin{bmatrix} u_0 \\ v_0 \end{bmatrix} \in D(A) \times H_0^1(\Omega)$), showing that V is nonincreasing along solutions and is constant if and only if $\begin{bmatrix} u \\ v \end{bmatrix}(\cdot)$ is constant, $v = 0$ and $u \in H^2(\Omega) \cap H_0^1(\Omega)$ satisfies (3.3). Thus $V\colon H_0^1(\Omega) \times L^2(\Omega) \to \mathbb{R}$ is a Lyapunov function for $\mathcal{T}$ relative to the set Ξ of equilibria for (3.5), that is,

$$\Xi = \left\{ \begin{bmatrix} u \\ 0 \end{bmatrix} : u \in D(A) \text{ satisfies } (3.3) \right\}.$$

To begin our study of the inner structure of attractors for gradient semigroups relative to an invariant set, we present a result that allows us to characterize the omega and alpha limits of points.

LEMMA 3.5. *Let $\mathcal{T}$ be a gradient semigroup relative to an invariant set Ξ. Then $\omega(x) \subset \Xi$ for each $x \in X$, and if there is a global solution $\phi\colon \mathbb{R} \to X$ of $\mathcal{T}$ through x, then $\alpha_\phi(x) \subset \Xi$.*

Furthermore, if Ξ is a disjoint union of compact (or closed and positively separated) invariant sets $\Xi_1, \cdots, \Xi_n$, then $\omega(x) \subset \Xi_i$ for some $1 \leqslant i \leqslant n$, and if there is a global solution $\phi\colon \mathbb{R} \to X$ through x, then $\alpha_\phi(x) \subset \Xi_j$ for some $1 \leqslant j \leqslant n$.

PROOF. If $y \in \omega(x)$, there is a sequence $\{t_n\}$, with $t_n \to \infty$, such that $y = \lim_{n\to\infty} T(t_n)x$. From the continuity of V we have $V(y) = \lim_{n\to\infty} V(T(t_n)x)$, and since $\mathbb{R}^+ \ni t \mapsto V(T(t)x) \in \mathbb{R}$ is nonincreasing we have $V(y) = \lim_{t\to\infty} V(T(t)x)$. From this and the continuities of $\mathcal{T}$ and V we have

$$V(y) = \lim_{n\to\infty} V(T(t+t_n)x) = \lim_{n\to\infty} V(T(t)T(t_n)x) = V(T(t)y)$$

for each $t \geqslant 0$. From property (ii) in Definition 3.1, we have $y \in \Xi$, and since $y \in \omega(x)$ was arbitrarily chosen, we conclude that $\omega(x) \subset \Xi$.

Now suppose that there exists a global solution $\phi\colon \mathbb{R} \to X$ of $\mathcal{T}$ through x. With the same reasoning as above, if $y \in \alpha_\phi(x)$ we have $V(y) = \lim_{t\to\infty} V(\phi(-t))$.

From this and the continuities of $\mathcal{T}$ and V we have $V(T(t)y) = V(y)$ for each $t \geqslant 0$, therefore $y \in \Xi$ and we conclude that $\alpha_\phi(x) \subset \Xi$.

The last statement of the lemma follows from the fact that $\omega(x)$ and $\alpha_\phi(x)$ are connected (see Lemmas 1.16 and 1.17). □

Before stating the main result of this section, we will present some results that ensure the existence of global attractors for gradient semigroups.

LEMMA 3.6. *If $\mathcal{T}$ is an eventually bounded and asymptotically compact gradient semigroup relative to a bounded $\mathcal{T}$-invariant set Ξ, then there exists a positively $\mathcal{T}$-invariant and bounded set U that $\mathcal{T}$-absorbs all compact subsets of X.*

PROOF. For each $x \in X$ we have, from Lemma 1.16, that $\omega(x)$ is nonempty, compact, $\mathcal{T}$-invariant and $\mathcal{T}$-attracts x. From Lemma 3.5 we have $\omega(x) \subset \Xi$. Since Ξ is $\mathcal{T}$-invariant and $\mathcal{T}$ is eventually bounded and asymptotically compact we must have $\Xi = \omega(\Xi)$ compact.

It follows that, for any $\epsilon > 0$, $\mathcal{O}_\epsilon(\Xi)$ $\mathcal{T}$-absorbs points of X. Let $B = \{x \in \mathcal{O}_\epsilon(\Xi) \colon \gamma^+(x) \subset \mathcal{O}_\epsilon(\Xi)\}$. Since $\mathcal{O}_\epsilon(\Xi)$ $\mathcal{T}$-absorbs points, we have B nonempty. Clearly $\gamma^+(B) = B$, B is bounded and $\mathcal{T}$-absorbs points. Also, since $\mathcal{T}$ is eventually bounded and asymptotically compact, Lemma 1.20 implies that $\omega(B)$ is nonempty, compact, $\mathcal{T}$-invariant and $\mathcal{T}$-attracts B.

Let V be a neighborhood of $\omega(B)$ and $t_0 \in \mathbb{R}^+$ such that $\gamma^+_{t_0}(V)$ is bounded. Since $\omega(B)$ $\mathcal{T}$-attracts points of X and $T(t)$ is continuous, for each $x \in X$ there is a neighborhood $\mathcal{O}_x$ of x and $t_x > 0$ such that $T(t)(\mathcal{O}_x) \subset \gamma^+_{t_0}(V)$ for $t \geqslant t_x$, that is, $\gamma^+_{t_0}(V)$ $\mathcal{T}$-absorbs a neighborhood of x for each $x \in X$. From this it follows easily that $\gamma^+_{t_0}(V)$ $\mathcal{T}$-absorbs compact subsets of X. Hence $U = \gamma^+_{t_0}(V)$ has the required properties. □

LEMMA 3.7. *If $\mathcal{T}$ is an eventually bounded and asymptotically compact gradient semigroup relative to a bounded $\mathcal{T}$-invariant set Ξ, then $\mathcal{T}$ is bounded dissipative.*

PROOF. By Lemma 3.6, there exists a positively $\mathcal{T}$-invariant bounded set U that $\mathcal{T}$-absorbs all compact subsets of X. Since $\mathcal{T}$ is eventually bounded and asymptotically compact, Lemma 1.20 implies that $\omega(U)$ $\mathcal{T}$-attracts U and therefore it $\mathcal{T}$-attracts compact subsets of X.

Let B be a bounded subset of X. By Lemma 3.6, $\omega(B)$ is nonempty, compact, $\mathcal{T}$-invariant and $\mathcal{T}$-attracts B. Since $\omega(B)$ is compact and $\mathcal{T}$-invariant, and $\omega(U)$ $\mathcal{T}$-attracts all compact subsets of X, we must have $\omega(B) \subset \omega(U)$. Therefore we conclude that $\omega(U)$ $\mathcal{T}$-attracts B, which concludes the proof. □

With these lemmas we are able to state the main result of this section, which gives a characterization for the global attractor of a gradient semigroup.

THEOREM 3.8. *Let $\mathcal{T}$ be an eventually bounded, asymptotically compact, gradient semigroup relative to a bounded $\mathcal{T}$-invariant set Ξ. Then $\mathcal{T}$ has a global attractor $\mathcal{A}$. Moreover Ξ is compact, $\Xi \subset \mathcal{A}$ and $\mathcal{A} = W^u(\Xi)$, where*

$$W^u(\Xi) = \{y \in X \colon \text{ there is a global solution } \phi\colon \mathbb{R} \to X \text{ with } \phi(0) = y \text{ such that } \lim_{t \to -\infty} \mathrm{dist}_H(\phi(t), \Xi) = 0\}$$

is called the **unstable set** *of* Ξ. *If* $\Xi = \bigcup_{i=1}^{n} \Xi_i$, *where* $\boldsymbol{\Xi} = \{\Xi_1, \cdots, \Xi_n\}$ *is a disjoint collection of compact invariant sets, then* $\mathcal{A} = \bigcup_{i=1}^{n} W^u(\Xi_i)$. *Finally, if there is a bounded connected set* B *that contains* $\mathcal{A}$, *then* $\mathcal{A}$ *is connected.*

PROOF. From Lemma 3.7 we know that $\mathcal{T}$ is bounded dissipative and since $\mathcal{T}$ is asymptotically compact, Theorem 1.23 implies that $\mathcal{T}$ has a global attractor $\mathcal{A}$.

Let $x \in \mathcal{A}$. Theorem 1.9 implies that there exists a global solution $\phi\colon \mathbb{R} \to X$ of $\mathcal{T}$ through x. Since $\phi(\mathbb{R}) \subset \mathcal{A}$ is relatively compact we have $\alpha_\phi(x) \neq \varnothing$ and, from Lemma 3.5, $\alpha_\phi(x) \subset \Xi$. This shows that $\mathcal{A} \subset W^u(\Xi)$. If $x \in W^u(\Xi)$, there is a global solution $\phi\colon \mathbb{R} \to X$ through x and $\lim_{t\to\pm\infty} \mathrm{dist}_H(\phi(t), \Xi) = 0$, hence $\phi(\mathbb{R})$ is bounded. Since $\phi(\mathbb{R})$ is also $\mathcal{T}$-invariant we conclude that $\phi(\mathbb{R}) \subset \mathcal{A}$ and consequently $x \in \mathcal{A}$. This shows $\mathcal{A} \supset W^u(\Xi)$ and completes the proof that $\mathcal{A} = W^u(\Xi)$.

If $\Xi = \bigcup_{i=1}^{n} \Xi_i$ with $\{\Xi_1, \cdots, \Xi_n\}$ a disjoint collection of closed invariant sets, the equality $\mathcal{A} = \bigcup_{i=1}^{n} W^u(\Xi_i)$ follows immediately from Lemma 3.5.

Lastly, if $\mathcal{A}$ is contained in a bounded connected subset B of X, it follows from the $\mathcal{T}$-invariance of $\omega(B)$, from the fact that $B \supset \mathcal{A}$ and from Lemma 1.17 that $\mathcal{A} = \omega(B)$ is connected. □

If $\mathcal{T}$ is as in the previous theorem and $\Xi = \bigcup_{i=1}^{n} \Xi_i$, where $\boldsymbol{\Xi} = \{\Xi_1, \cdots, \Xi_n\}$ is a disjoint collection of compact invariant sets, we can use the Lyapunov function of $\mathcal{T}$ relative to Ξ to give a more detailed description of the attractor, in terms of *energy levels.*

To do that, we first need the following lemma, which is an immediate consequence of the continuity of semigroups. This lemma ensures that, given a $\mathcal{T}$-invariant set Ξ for the semigroup $\mathcal{T}$ and y close to Ξ, $\{T(s)y\colon 0 \leqslant s \leqslant t\}$ remains close to Ξ for large values of t.

LEMMA 3.9. *Let* $\mathcal{T}$ *be a semigroup and let* Ξ *be a compact* $\mathcal{T}$*-invariant set. Given* $t \geqslant 0$ *and* $\epsilon > 0$, *there exists* $\delta > 0$ *such that*

$$T(s)\mathcal{O}_\delta(\Xi) \subset \mathcal{O}_\epsilon(\Xi) \qquad \textit{for all } s \in [0, t].$$

PROOF. Suppose that there exist $t_0 \in (0, \infty)$ and $\epsilon_0 > 0$ such that, for each $k \in \mathbb{N}^*$ there are $x_k \in \mathcal{O}_{\frac{1}{k}}(\Xi)$ and $s_k \in [0, t_0]$ with $\mathrm{dist}_H(T(s_k)x_k, \Xi) \geqslant \epsilon_0$. We may assume that $s_k \to s_0$ for some $s_0 \in [0, t_0]$ and that $x_k \to y \in \Xi$. Since $\mathbb{R}^+ \times X \ni (t, x) \mapsto T(t)x \in X$ is continuous, we have $0 = \mathrm{dist}_H(T(s_0)y, \Xi) \geqslant \epsilon_0$ which is absurd, and concludes the proof. □

Now we can prove a more general description for attractors of gradient semigroups.

THEOREM 3.10. *Let* $\mathcal{T}$ *be a gradient semigroup relative to a* $\mathcal{T}$*-invariant set* Ξ, *with a global attractor* $\mathcal{A}$. *Suppose that* $\Xi = \bigcup_{i=1}^{n} \Xi_i$ *with* $\boldsymbol{\Xi} = \{\Xi_1, \cdots, \Xi_n\}$ *being a disjoint collection of compact invariant sets,* $n \in \mathbb{N}^*$, *and that the associated Lyapunov function is constant on each* Ξ_i, $1 \leqslant i \leqslant n$. *Let* $V(\Xi) = \{\mathfrak{n}_1, \cdots, \mathfrak{n}_p\}$ *with* $\mathfrak{n}_i < \mathfrak{n}_{i+1}$ *for* $1 \leqslant i \leqslant p-1$.

If $1 \leqslant j \leqslant p-1$ *and* $\mathfrak{n}_j \leqslant r < \mathfrak{n}_{j+1}$, *then* $X_r = \{z \in X\colon V(z) \leqslant r\}$ *is positively* $\mathcal{T}$*-invariant and* $\mathcal{T}_r$, *the restriction of* $\mathcal{T}$ *to* X_r, *has a global attractor* $\mathcal{A}^{(j)}$ *given by*

$$\mathcal{A}^{(j)} = \bigcup \{W^u(\Xi_\ell)\colon V(\Xi_\ell) \leqslant \mathfrak{n}_j\}.$$

In particular, $V(z) \leqslant \mathfrak{n}_j$ *for* $z \in \mathcal{A}^{(j)}$, $\mathfrak{n}_1 = \min\{V(x) \colon x \in X\}$ *and* $\mathcal{A}^{(1)} = \bigcup\{\tilde{\Xi} \in \boldsymbol{\Xi} \colon V(\tilde{\Xi}) = \mathfrak{n}_1\}$ *consists of all asymptotically stable* $\mathcal{T}$*-invariant sets, that is, for each* $\tilde{\Xi} \in \boldsymbol{\Xi}$ *with* $\tilde{\Xi} \subset \mathcal{A}^{(1)}$ *there is an* $\epsilon > 0$ *such that* $\lim_{t\to\infty} \mathrm{dist}_H(T(t)x, \tilde{\Xi}) = 0$ *whenever* $x \in \mathcal{O}_\epsilon(\tilde{\Xi})$.

PROOF. It is clear from the definition of a Lyapunov function that X_r is positively $\mathcal{T}$-invariant. To show the existence of a global attractor for $\mathcal{T}_r$ we note that the required properties for the existence of a global attractor are inherited from those of $\mathcal{T}$, namely, $\mathcal{T}_r$ is bounded dissipative and asymptotically compact. Hence $\mathcal{T}_r$ has a global attractor $\mathcal{A}^{(j)}$. The restriction V_r of V to X_r is a Lyapunov function for $\mathcal{T}_r$ associated with $\{\Xi_\ell \colon V(\Xi_\ell) \leqslant \mathfrak{n}_j\}$ and the characterization of $\mathcal{A}^{(j)}$ follows.

Let us now prove the last claim. Let $\delta_0 = \frac{1}{2}\min\{d(\tilde{\Xi}, \hat{\Xi})\} > 0$ where the minimum is taken over the sets $\tilde{\Xi}, \hat{\Xi} \in \boldsymbol{\Xi}$ such that $\tilde{\Xi}, \hat{\Xi} \subset \mathcal{A}^{(1)}$ and $\tilde{\Xi} \neq \hat{\Xi}$. Assume by contradiction that there exist $\boldsymbol{\Xi} \ni \tilde{\Xi} \subset \mathcal{A}^{(1)}$, $0 < \delta < \delta_0$ and sequences $\{x_k\}$ in X, $\{t_k\}$ in $\mathbb{R}^+$ with $x_k \to \Xi$, $\mathrm{dist}_H(T(t)x_k, \tilde{\Xi}) < \delta$ for $0 \leqslant t < t_k$ and $\mathrm{dist}_H(T(t_k)x_k, \tilde{\Xi}) = \delta$. From Lemma 3.9 we have $t_k \to \infty$ and the asymptotic compactness of $\mathcal{T}$ ensures that $\{T(t_k)x_k\}$ has a convergent subsequence, which we denote the same, and let y be its limit. It is simple to see that $\mathfrak{n}_1$ is the minimum value of V in X, and from this it is immediate that

$$\mathfrak{n}_1 = V(\tilde{\Xi}) = \lim_{k\to\infty} V(x_k) \geqslant \lim_{k\to\infty} V(T(t_k)x_k) = V(y) \geqslant \mathfrak{n}_1,$$

and

$$\mathfrak{n}_1 = V(\tilde{\Xi}) = \lim_{k\to\infty} V(x_k) \geqslant \lim_{k\to\infty} V(T(t+t_k)x_k) = V(T(t)y) \geqslant \mathfrak{n}_1 \text{ for each } t \geqslant 0.$$

Hence $V(y) = V(T(t)y) = \mathfrak{n}_1$ for all $t \in \mathbb{R}^+$ and therefore $y \in \Xi$. Since $\mathrm{dist}_H(y, \tilde{\Xi}) < \delta_0$ we have $y \in \tilde{\Xi}$. But $\mathrm{dist}_H(y, \tilde{\Xi}) = \delta > 0$, which gives us a contradiction. This proves that for each $\tilde{\Xi} \in \mathcal{A}^{(1)}$ and $0 < \delta < \delta_0$ there is a $0 < \delta' < \delta$ such that, for all $x \in \mathcal{O}_{\delta'}(\tilde{\Xi})$, we have $\gamma^*(x) \subset \mathcal{O}_\delta(\tilde{\Xi})$ and proves that $\mathcal{A}^{(1)}$ consists only of stable invariant sets. To conclude, we only need to note that, for each $x \in X$, it follows from Lemma 3.5 that $\lim_{t\to\infty} \mathrm{dist}_H(T(t)x, \omega(x)) = 0$, and that $\omega(x) \subset \tilde{\Xi}$ for some $\tilde{\Xi} \in \boldsymbol{\Xi}$. □

2. Dynamically gradient semigroups

In this section we present the concept of *dynamically gradient semigroups.* These semigroups have the dynamical properties of a gradient semigroup, but their definition does not require the existence of a Lyapunov function.

Under natural assumptions we show that dynamically gradient semigroups are gradient and that dynamically gradient semigroups are stable under perturbations. This stability of the gradient structure under perturbations (which we call *topological structural stability*) is a global property of attractors that is entirely obtained from the continuity properties of the semigroups. Our presentation is based on [**74**].

Let $\mathcal{T}$ be a semigroup with a global attractor $\mathcal{A}$, which contains a disjoint collection of isolated invariant sets $\boldsymbol{\Xi} = \{\Xi_1, \cdots, \Xi_n\}$. Let

$$\epsilon_0 = \epsilon_0(\boldsymbol{\Xi}) = \frac{1}{2} \min_{1\leqslant i<j\leqslant n} \mathrm{dist}(\Xi_i, \Xi_j) > 0. \tag{3.6}$$

We make the following definition.

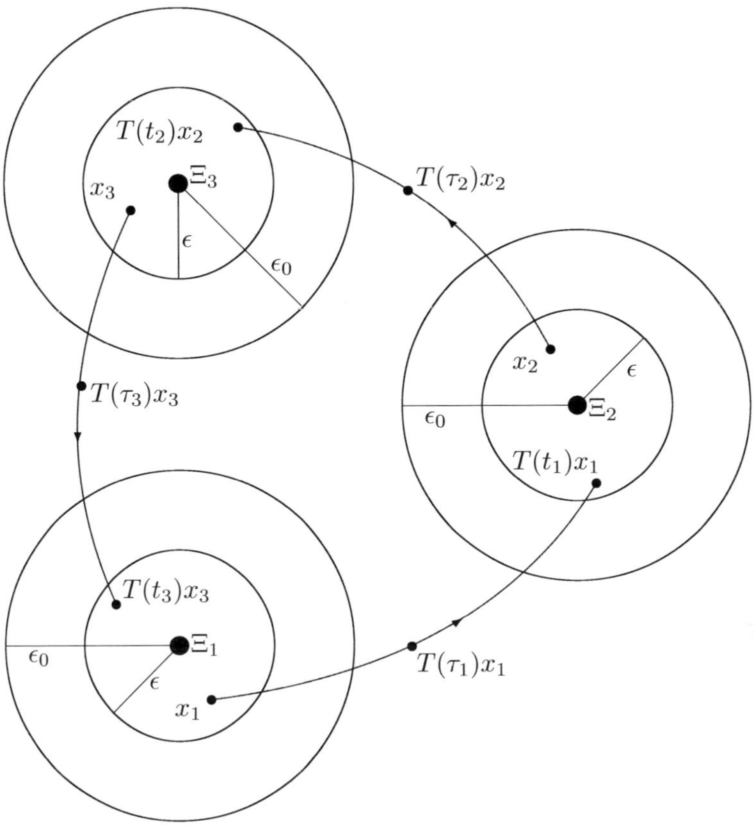

FIGURE 3.1. Example of an ϵ-chain.

DEFINITION 3.11. Let $\Xi^* \in \boldsymbol{\Xi}$ and $0 < \epsilon < \epsilon_0$. An **$\epsilon$-chain**[1] at Ξ^* consists of

(a) a subcollection $\{\Xi_{\ell_1}, \Xi_{\ell_2}, \cdots, \Xi_{\ell_k}\}$ of $\boldsymbol{\Xi}$ with $\Xi^* = \Xi_{\ell_1} =: \Xi_{\ell_{k+1}}$ and

$$\Xi_{\ell_i} \cap \Xi_{\ell_j} = \varnothing \quad \text{for} \quad 1 \leqslant i < j \leqslant k;$$

(b) a finite collection of points $\{x_1, \cdots, x_k\}$ in X with

$$\mathrm{dist}_H(x_i, \Xi_{\ell_i}) < \epsilon \quad \text{for} \quad 1 \leqslant i \leqslant k;$$

(c) finite sequences of nonnegative real numbers $\{t_1, \cdots, t_k\}$ and $\{\tau_1, \cdots, \tau_k\}$ with $\tau_i < t_i$, for all $i = 1, \cdots, k$, such that

$$\mathrm{dist}_H(T(t_i)x_i, \Xi_{\ell_{i+1}}) < \epsilon \quad \text{for} \quad 1 \leqslant i \leqslant k,$$

and also

$$\mathrm{dist}_H\Big(T(\tau_i)x_i, \bigcup_{j=1}^{k} \mathcal{O}_{\epsilon_0}(\Xi_{\ell_j})\Big) > 0 \quad \text{for} \quad 1 \leqslant i \leqslant k.$$

An isolated invariant set $\Xi^* \in \boldsymbol{\Xi}$ is said to be **chain recurrent** if there exists $\delta \in (0, \epsilon_0)$ and ϵ-chains at Ξ^*, for each $\epsilon \in (0, \delta)$.

[1] See Figures 3.1 and 3.2.

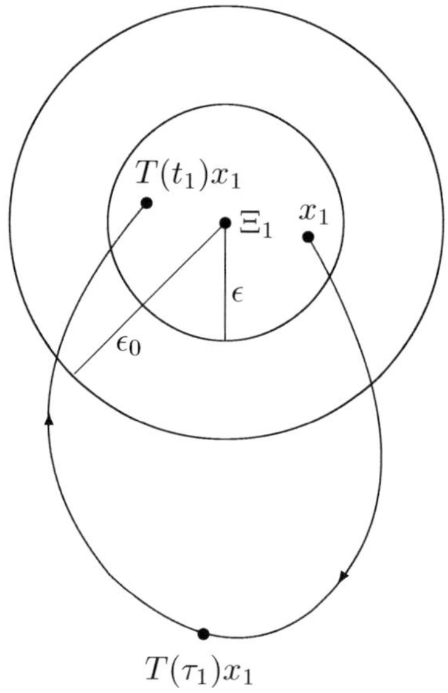

FIGURE 3.2. Example of an ϵ-chain when $k = 1$.

We are now ready to define the main concept of this section, the *dynamically gradient semigroups.*

DEFINITION 3.12. Let $\mathcal{T}$ be a semigroup in X with a global attractor $\mathcal{A}$ and a disjoint collection of isolated invariant sets $\Xi = \{\Xi_1, \cdots, \Xi_n\}$ in $\mathcal{A}$. We say that $\mathcal{T}$ is a **dynamically gradient semigroup** relative to Ξ if the following two conditions are satisfied:

(G1) Any global solution $\xi \colon \mathbb{R} \to X$ of $\mathcal{T}$ in $\mathcal{A}$ satisfies

$$\lim_{t\to-\infty} \mathrm{dist}_H(\xi(t), \Xi_i) = 0 \quad \text{and} \quad \lim_{t\to\infty} \mathrm{dist}_H(\xi(t), \Xi_j) = 0,$$

for some $1 \leqslant i, j \leqslant n$.

(G2) No set of Ξ is chain recurrent.

We note that condition (G1) is imposed only for solutions in the global attractor. However, as for gradient semigroups, we are able to prove that for a dynamically gradient semigroup the ω-limit set of each point is contained in one of the isolated invariant sets of Ξ.

The following lemmas will play an important role in the remaining results of the chapter. Our first lemma establishes that semigroups possessing a global attractor with a disjoint collection of isolated invariant sets and satisfying (G1) have the property that: given any suitably small neighborhood V of the union of the isolated invariant sets and any bounded set $B \subset X$, there is a time $t(V, B) > 0$ before which any solution starting at a point of B must visit the neighborhood V. The result also holds for families of semigroups.

LEMMA 3.13. *Let $\{\mathcal{T}_\eta\}_{\eta\in[0,1]}$ be a family of semigroups, continuous and asymptotically compact at $\eta = 0$, with a global attractor $\mathcal{A}_\eta$ and a disjoint collection of isolated invariant sets $\Xi_\eta = \{\Xi_{1,\eta}, \cdots, \Xi_{n,\eta}\}$, for each $\eta \in [0, 1]$, such that $\mathcal{T}_0$ satisfies* (G1)*. Given $0 < \delta < \epsilon_0(\Xi_0)$, with $\epsilon_0(\Xi_0)$ as in* (3.6)*, and a bounded* *set $B \subset X$, there exist positive numbers $t_0 = t_0(\delta, B)$ and $\eta_0 > 0$ such that $\{T_\eta(t)u_0 \colon 0 \leqslant t \leqslant t_0\} \cap \bigcup_{i=1}^n \mathcal{O}_\delta(\Xi_{i,0}) \neq \varnothing$ for all $u_0 \in B$ and $\eta \in [0, \eta_0]$.*

Moreover, if $\max_{j=1,\cdots,n} \operatorname{dist}_H(\Xi_{j,\eta}, \Xi_{j,0}) \to 0$, as $\eta \to 0$, we can replace $\Xi_{i,0}$ in the previous conclusion by $\Xi_{i,\eta}$.

PROOF. We argue by contradiction. Assume that there exist sequences $\{u_k\}$ in B, $\{\eta_k\}$ in $[0,1]$, with $\eta_k \to 0$, and a sequence of positive numbers $\{t_k\}$ with $t_k \to \infty$ such that $\{T_{\eta_k}(t)u_k \colon 0 \leqslant t \leqslant t_k\} \cap \bigcup_{i=1}^n \mathcal{O}_\delta(\Xi_{i,0}) = \varnothing$. Extracting subsequences, from Lemma 2.4 with $a_n = 0$, $b_n = \frac{t_n}{2}$ and $c_n = t_n$, we have a bounded global solution $\xi \colon \mathbb{R} \to X$ of $\mathcal{T}_0$ such that $T_{\eta_k}(t + \frac{t_k}{2})u_k \to \xi(t)$ uniformly in compact subsets of $\mathbb{R}$. Hence, $\xi(t) \notin \bigcup_{i=1}^n \mathcal{O}_\delta(\Xi_{i,0})$ for all $t \in \mathbb{R}$ and this contradicts (G1). □

Our next lemma establishes that dynamically gradient semigroups have the property that, given a neighborhood U of an isolated $\mathcal{T}$-invariant set Ξ, there is a smaller neighborhood V of Ξ such that, if a solution starts in V and, after some time, leaves U, then it never returns to V. The result also holds for families of semigroups.

LEMMA 3.14. *Let $\{\mathcal{T}_\eta\}_{\eta\in[0,1]}$ be a family of semigroups, continuous and collectively asymptotically compact at $\eta = 0$. Assume that for each $\eta \in [0,1]$ there exist a disjoint collection of isolated invariant sets $\boldsymbol{\Xi}_\eta = \{\Xi_{1,\eta}, \cdots, \Xi_{n,\eta}\}$ and a global attractor $\mathcal{A}_\eta$, and assume that $\max_{1\leqslant i\leqslant n} \operatorname{dist}_H(\Xi_{i,\eta}, \Xi_{i,0}) \to 0$ as $\eta \to 0$ and let $0 < \delta_0 < \epsilon_0(\boldsymbol{\Xi}_0)$. If $\mathcal{T}_0$ is dynamically gradient, then given $0 < \delta < \delta_0$, there exist $\eta_0 > 0$ and $\delta' > 0$ (independent from $\eta \in [0, \eta_0]$) such that, if $\eta \in [0, \eta_0]$, $\operatorname{dist}_H(z_0, \Xi_{i,\eta}) < \delta'$, $1 \leqslant i \leqslant n$, and for some $t_1 > 0$ we have $\operatorname{dist}_H(T_\eta(t_1)z_0, \Xi_{i,\eta}) \geqslant \delta$, then $\operatorname{dist}_H(T_\eta(t)z_0, \Xi_{i,\eta}) > \delta'$ for all $t \geqslant t_1$.*

PROOF. Assume that, for some $1 \leqslant i \leqslant n$, there exist a sequence $\{z_k\}$ in X, $\{\eta_k\} \subset [0,1]$ with $\eta_k \to 0$, $\operatorname{dist}_H(z_k, \Xi_{i,\eta_k}) < \frac{1}{k}$ and sequences $\sigma_k < \tau_k$ in $\mathbb{R}^+$ such that $\operatorname{dist}_H(T_{\eta_k}(\sigma_k)z_k, \Xi_{i,\eta_k}) \geqslant \delta$ and $\operatorname{dist}_H(T_{\eta_k}(\tau_k)z_k, \Xi_{i,\eta_k}) < \frac{1}{k}$. We claim that this contradicts property (G2) of $\mathcal{T}_0$.

First note that $\{\sigma_k\}$ can be chosen such that $T_{\eta_k}(s)u_k \in \mathcal{O}_\delta(\Xi_{i,\eta_k})$ for all $s < \sigma_k$ and, from Lemma 2.4, $\sigma_k \to \infty$. Taking subsequences we can construct a global solution $\xi_0 \colon \mathbb{R} \to \mathcal{A}$ of $\mathcal{T}_0$ such that $\lim_{t\to-\infty} \operatorname{dist}_H(\xi_0(t), \Xi_{i,0}) = 0$. From the fact that $\mathcal{T}_0$ is dynamically gradient there exists $j \neq i$ such that $\lim_{t\to\infty} \operatorname{dist}_H(\xi_0(t), \Xi_{j,0}) = 0$.

From the fact that $\lim_{t\to\infty} \operatorname{dist}_H(\xi_0(t), \Xi_{j,0}) = 0$ and from the construction of ξ_0, there exist a subsequence $\{\eta_{k_m}\}$ of $\{\eta_k\}$ and subsequences $\{\tau_{k_m}\}$, $\{t_{k_m}\}$ in $\mathbb{R}^+$, $\tau_{k_m} < \sigma_{k_m}$, such that

$$\operatorname{dist}_H(T_{\eta_{k_m}}(\tau_{k_m})z_{k_m}, \Xi_{j,\eta_{k_m}}) < \frac{1}{m} \quad \text{and} \quad \operatorname{dist}_H(T_{\eta_{k_m}}(\sigma_{k_m})z_k, \Xi_{j,\eta_{k_m}}) \geqslant \delta.$$

Proceeding exactly as in the previous step, we obtain a solution $\xi_1 \colon \mathbb{R} \to \mathcal{A}$ such that $\lim_{s\to-\infty} \operatorname{dist}_H(\xi_1(s), \Xi_{j,0}) = 0$. From the fact that $\mathcal{T}_0$ is dynamically gradient there exists $\Xi_{\ell,0} \in \boldsymbol{\Xi}_0$, $\Xi_{\ell,0} \notin \{\Xi_{i,0}, \Xi_{j,0}\}$, such that $\lim_{s\to\infty} \operatorname{dist}_H(\xi_1(s), \Xi_{\ell,0}) = 0$. In a finite number of steps, we construct a chain recurrent set in $\boldsymbol{\Xi}_0$, which leads to a contradiction with (G2). □

A first consequence of Lemmas 3.14 and 3.13 is given in the next result, which establishes that any solution of a dynamically gradient semigroup must converge to an isolated invariant set.

LEMMA 3.15. *Suppose that $\mathcal{T}$ is a dynamically gradient semigroup with global attractor $\mathcal{A}$ and a disjoint collection of isolated invariant sets $\Xi = \{\Xi_1, \cdots, \Xi_n\}$. Given $x \in X$ there exists $1 \leqslant j \leqslant n$ such that*

$$\lim_{t\to\infty} \mathrm{dist}_H(T(t)x, \Xi_j) = 0,$$

that is, $\omega(x) \subset \Xi_j$.

PROOF. It follows from Lemma 3.14 that given $\delta \in (0, \delta_0)$ there is a $\delta' \in (0, \delta)$ such that if $\mathrm{dist}_H(y, \Xi_i) < \delta'$ and for some $t_{y,\delta} > 0$, $\mathrm{dist}_H(T(t_{y,\delta})y, \Xi_i) \geqslant \delta$, then $\mathrm{dist}_H(T(t)v, \Xi_i) > \delta'$ for all $t \geqslant t_{y,\delta}$. On the other hand, using that $\gamma^+(x)$ is bounded, it follows from Lemma 3.13 that, given δ' there exists $t_{\delta'} = t_{\delta'}(\gamma^+(x)) \geqslant 0$ such that, for each $y \in \gamma^+(x)$, we have

$$\{T(t)y \colon 0 \leqslant t \leqslant t_{\delta'}\} \cap \bigcup_{i=1}^{n} \mathcal{O}_{\delta'}(\Xi_i) \neq \varnothing.$$

From the fact that Ξ is finite, there exist $\Xi_j \in \Xi$ and, for each $\delta \in (0, \delta_0)$, a $s_\delta \geqslant 0$ such that $T(s)x \in \mathcal{O}_\delta(\Xi_j)$ for all $s \geqslant s_\delta$. This, together with another application of Lemma 3.14, completes the proof. □

Later in this chapter, in Section 4, we will prove the topological structural stability of dynamically gradient semigroups, that is, suitably small autonomous perturbations of dynamically gradient semigroups are still dynamically gradient semigroups. To that end, the following notion is, sometimes, a convenient replacement for condition (G2) of Definition 3.12.

DEFINITION 3.16. Let $\mathcal{T}$ be a semigroup with global attractor $\mathcal{A}$ and a disjoint collection of isolated invariant sets $\Xi = \{\Xi_1, \cdots, \Xi_n\}$. Let $0 < \delta < \epsilon_0$, where ϵ_0 is given in (3.6). A **homoclinic structure** in $\mathcal{A}$ is a disjoint subcollection $\{\Xi_{\ell_1}, \cdots, \Xi_{\ell_k}\}$ of Ξ and a finite number of bounded global solutions $\xi_i \colon \mathbb{R} \to X$, for $1 \leqslant i \leqslant k$, such that, with $\Xi_{\ell_{k+1}} = \Xi_{\ell_1}$,

$$\lim_{t\to-\infty} \mathrm{dist}_H(\xi_i(t), \Xi_{\ell_i}) = 0 \quad \text{and} \quad \lim_{t\to+\infty} \mathrm{dist}_H(\xi_i(t), \Xi_{\ell_{i+1}}) = 0$$

for $1 \leqslant i \leqslant k$, and

$$\min_{1\leqslant i\leqslant k} \sup_{t\in\mathbb{R}} \mathrm{dist}_H\Big(\xi_i(t), \bigcup_{i=1}^{n} \mathcal{O}_\delta(\Xi_i)\Big) > 0. \tag{3.7}$$

REMARK 3.17. Condition (3.7) has a technical nature and it is used only when $k = 1$ to ensure that the global solution $\xi \colon \mathbb{R} \to \mathcal{A}$ is not entirely contained in the isolated invariant set, that is, $\xi(t) \in \Xi_i$ for all $t \in \mathbb{R}$ and some $i \in \{1, \cdots, n\}$ *does not* make a homoclinic structure. A pictorial example of a homoclinic structure can be seen in Figure 3.3.

In what follows we prove that, if (G1) is satisfied, we can replace (G2) in Definition 3.12 by the nonexistence of homoclinic structures.

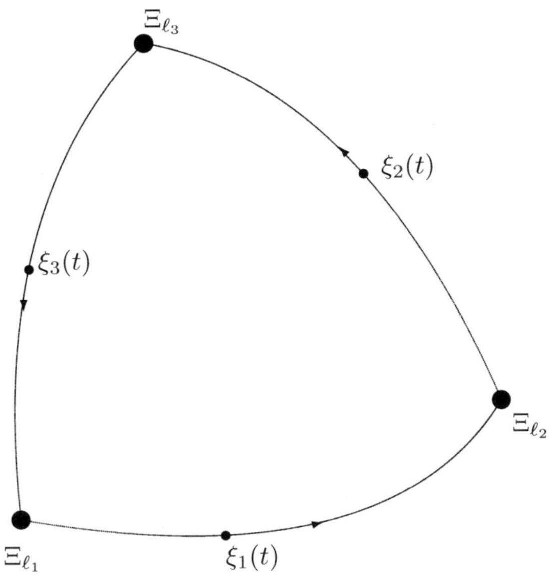

FIGURE 3.3. Example of a homoclinic structure.

PROPOSITION 3.18. *If $\mathcal{T}$ is a semigroup with a disjoint collection of isolated invariant sets $\boldsymbol{\Xi} = \{\Xi_1, \cdots, \Xi_n\}$ which has a global attractor $\mathcal{A}$ and satisfies* (G1), *then* (G2) *is satisfied if and only if $\mathcal{A}$ does not have any homoclinic structure.*

PROOF. Assuming that $\mathcal{A}$ has a homoclinic structure it is easy to see that any of the isolated invariant sets in it is chain recurrent. On the other hand, if one of the isolated invariant sets $\Xi \in \boldsymbol{\Xi}$ is chain recurrent, there exist a subset $\{\Xi_{k_1}, \cdots, \Xi_{k_\ell}\}$, $\Xi = \Xi_{k_1} =: \Xi_{k_{\ell+1}}$ which we will denote (after a reordering of $\boldsymbol{\Xi}$) by $\{\Xi_1, \cdots, \Xi_\ell\}$ and, for each positive integer k, points $x_1^k, \cdots, x_\ell^k$ and positive numbers $t_1^k, \cdots, t_\ell^k$ such that

$$\mathrm{dist}_H(x_i^k, \Xi_i) < \frac{1}{k} \quad \text{and} \quad \mathrm{dist}_H(T(t_i^k)x_i^k, \Xi_{i+1}) < \frac{1}{k}, \quad 1 \leqslant i \leqslant \ell,$$

and we may assume that $\overline{\bigcup_{k=1}^\infty \bigcup_{t\in[0,t_i^k]} T(t)x_i^k} \cap \Xi_j \neq \varnothing$ if j is different from i and $i+1$ (adding more isolated invariant sets to the $\frac{1}{k}$ chains if needed).

Consider ϵ_0 as in (3.6), and choose $0 < \delta < \epsilon_0$ and $\tau_i^k > 0$ such that $\mathrm{dist}_H(T(t)x_i^k, \Xi_i) < \delta$ for all $0 \leqslant t < \tau_i^k$ and $\mathrm{dist}_H(T(\tau_i^k)x_i^k, \Xi_i) = \delta$. It is clear from the continuity of $\mathcal{T}$ and from the invariance of Ξ_i that $\tau_i^k \to \infty$ as $k \to \infty$. For $t \in [-\tau_i^k, \infty)$ let $\xi_i^k(t) = T(\tau_i^k + t)x_i^k$. Taking subsequences we define the global solutions $\xi_i \colon \mathbb{R} \to X$ by $\xi_i(t) = \lim_{k\to\infty} \xi_i^k(t)$. Since each $\xi_i(t)$ must converge to an isolated invariant set as $t \to +\infty$ and as $t \to -\infty$ and since $\xi_i(t) \in \mathcal{O}_\delta(\Xi_i)$ for all $t \leqslant 0$ we have that $\lim_{t\to-\infty} \mathrm{dist}_H(\xi_i(t), \Xi_i) = 0$. Since $\overline{\bigcup_{k=1}^\infty \bigcup_{t\in[0,t_i^k]} T(t)x_i^k} \cap \Xi_j \neq \varnothing$, j different from i and $i+1$, we must have $\lim_{t\to\infty} \mathrm{dist}_H(\xi_i(t), \Xi_{i+1}) = 0$.

The collection of isolated invariant sets $\{\Xi_1, \cdots, \Xi_\ell\} \subset \mathbf{\Xi}$ and the set of global solutions $\{\xi_i \colon \mathbb{R} \to X,\ 1 \leqslant i \leqslant \ell\}$ are such that $\Xi = \Xi_1 = \Xi_{\ell+1}$ and

$$\lim_{t\to-\infty} \mathrm{dist}_H(\xi_i(t), \Xi_i) = 0 \quad \text{and} \lim_{t\to\infty} \mathrm{dist}_H(\xi_i(t), \Xi_{i+1}) = 0,$$

for $1 \leqslant i \leqslant \ell$. Hence $\mathcal{A}$ has a homoclinic structure. □

COROLLARY 3.19. *If $\mathcal{T}$ is a dynamically gradient semigroup relative to a disjoint collection of isolated invariant sets $\mathbf{\Xi} = \{\Xi_1, \cdots, \Xi_n\}$ and $\mathcal{A}$ is its global attractor, then there are isolated invariant sets Ξ_α and Ξ_ω such that Ξ_α has trivial stable set in $\mathcal{A}$, that is, $W^s(\Xi_\alpha) \cap \mathcal{A} = \Xi_\alpha$, and Ξ_ω has trivial unstable set, that is, $W^u(\Xi_\omega) = \Xi_\omega$.*

PROOF. Let us prove the existence of at least one isolated invariant set Ξ_ω with a trivial unstable set. If such a set does not exist, then for each Ξ_i, $1 \leqslant i \leqslant n$, there is a global solution $\xi_i \colon \mathbb{R} \to \mathcal{A}$ such that $\lim_{t\to-\infty} \mathrm{dist}_H(\xi_i(t), \Xi_i) = 0$ and $\xi_i(\mathbb{R}) \not\subset \Xi_i$. Choose $\xi_1 \colon \mathbb{R} \to \mathcal{A}$ such that $\lim_{t\to-\infty} \mathrm{dist}_H(\xi_1(t), \Xi_1) = 0$ and $\xi_1(\mathbb{R}) \not\subset \Xi_1$. Set $\Xi_{\ell_1} = \Xi_1$. From (G1), there exists $\Xi_{\ell_2} \in \mathbf{\Xi}$ with $\ell_2 \neq \ell_1$ such that $\lim_{t\to\infty} \mathrm{dist}_H(\xi_1(t), \Xi_{\ell_2}) = 0$. Choose $\xi_2 \colon \mathbb{R} \to X$ such that $\lim_{t\to-\infty} \mathrm{dist}_H(\xi_2(t), \Xi_{\ell_2}) = 0$ and $\xi_2(\mathbb{R}) \not\subset \Xi_{\ell_2}$. Let $\ell_3 \notin \{\ell_1, \ell_3\}$ be such that $\lim_{t\to\infty} \mathrm{dist}_H(\xi(t), \Xi_{\ell_3}) = 0$. In a finite number of steps we arrive at a contradiction. This proves the existence of an isolated invariant set with trivial unstable set.

The proof of the existence of an isolated invariant set Ξ_α with trivial stable set in $\mathcal{A}$ is similar and it is left to the reader. □

REMARK 3.20. We point out that in the next two chapters we will deal with *discrete semigroups*. But for a dynamically gradient semigroup $\mathcal{T}$ with a global attractor $\mathcal{A}$, if we consider the associated discrete semigroup $\{T^n \colon n \in \mathbb{N}\}$, where $T = T(1)$, then it is easy to see that $\mathcal{A}$ is also the global attractor of $\{T^n \colon n \in \mathbb{N}\}$ (see [**28**, Remark 3.3] for details).

2.1. Examples.

EXAMPLE 3.21. If $\mathcal{T}$ is a gradient semigroup relative to a finite set of equilibria $\mathbf{\Xi} = \{x_1^*, \cdots, x_n^*\}$ and has a global attractor, it is not difficult to prove that $\mathcal{T}$ is dynamically gradient relatively to $\mathbf{\Xi}$, using Lemma 3.5.

EXAMPLE 3.22. Consider the semigroup in $\mathbb{R}^2$ which is pictorially represented by Figure 3.4.

If x_1^* is the stable node, x_2^* is the saddle and x_3^* is the unstable focus, choose $\Xi_1 = \{x_1^*\}$, $\Xi_2 = W^u(x_2^*) \cup \{x_2^*\}$ and $\Xi_3 = \{x_3^*\}$. It follows that the semigroup $\mathcal{T}$ associated to Figure 3.4 is dynamically gradient relative to the disjoint family of isolated invariant sets $\mathbf{\Xi} = \{\Xi_1, \Xi_3, \Xi_3\}$.

EXAMPLE 3.23. Consider the following autonomous equation in polar coordinates:

$$\begin{cases} \dot{r} = -r(r-1)(r-2), \\ \dot{\theta} = 1, \end{cases} \tag{3.8}$$

which has, in the (x, y)-plane, $\sigma_0 = (0, 0)$ as a fixed point and two periodic orbits

$$P_{r_0} = \{(r_0 \cos\theta, r_0 \sin\theta) \colon \theta \in [0, 2\pi]\},$$

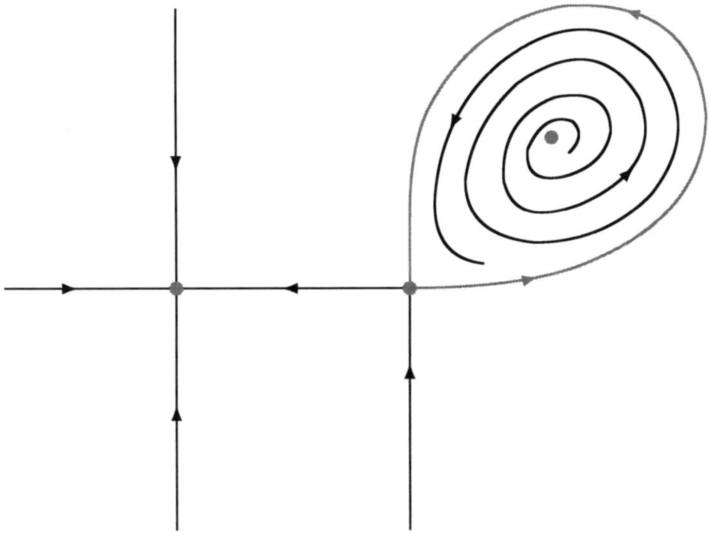

FIGURE 3.4. Isolated invariant sets: the connected sets formed by the three equilibria and the homoclinic.

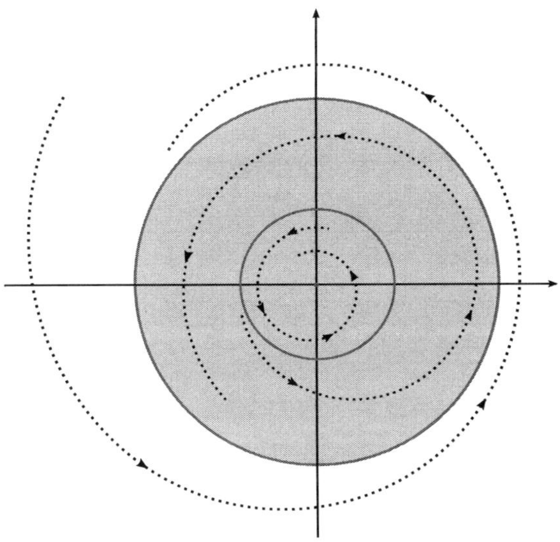

FIGURE 3.5. Global attractor.

with $r_0 = 1$ or 2. We can see that $(0, 0)$ is stable, the periodic orbit P_1 is unstable and P_2 is stable.

It is easy to see that (3.8) has a global attractor $\mathcal{A}_0 = \{(x, y) \in \mathbb{R}^2 \colon x^2 + y^2 \leqslant 4\}$ and that (3.8) generates a dynamically gradient semigroup with the family of isolated invariant sets $\mathbf{P} = \{\sigma_0, P_1, P_2\}$. Its internal dynamics can be seen in Figure 3.5.

EXAMPLE 3.24. Assume that $f \colon \mathbb{R} \to \mathbb{R}$ and $g \colon \mathbb{R}^2 \to \mathbb{R}$ are continuously differentiable functions, and for $h(\cdot) = f(\cdot)$ or $h(\cdot) = g(u, \cdot)$, for each $u \in \mathbb{R}$, we have

$$\limsup_{|s|\to\infty} \frac{h(s)}{s} \leqslant -\delta < 0, \tag{3.9}$$

for some $\delta > 0$.

Consider the initial boundary value problem

$$\begin{cases} \dot{u} = f(u), \\ \dot{v} = g(u,v), \\ u(0) = u_0 \in \mathbb{R}, \\ v(0) = v_0 \in \mathbb{R}, \end{cases} \tag{3.10}$$

which defines a semigroup $\mathcal{T}$ in $\mathbb{R}^2$ with a global attractor $\mathcal{A}$.

Assume that for all u^* such that $f(u^*) = 0$ we have $f'(u^*) \neq 0$ and $g'(u^*, v^*) \neq 0$ for those v^* such that $g(u^*, v^*) = 0$. It is clear that, if $(u(\cdot), v(\cdot)) \colon \mathbb{R} \to \mathbb{R}^2$ is a global solution of (3.10) its first coordinate converges (backwards and forwards; see Lemma 3.5) to $u^*_\pm$ such that $f(u^*_\pm) = 0$. It follows that $v(\cdot) \colon \mathbb{R} \to \mathbb{R}$ is a solution of the asymptotically autonomous (backwards and forwards) problem $\dot{v} = g(u(t), v)$ and (as a consequence of the results in Chapter 8) it converges (backwards and forwards) to $v^*_\pm$ satisfying $g(u^*_\pm, v^*_\pm) = 0$. This reasoning ensures that the global solutions of (3.10) satisfy (G1). The fact that (G2) is satisfied is immediate, since it has to be satisfied for the first coordinate.

REMARK 3.25. We are not able to exhibit a Lyapunov function for (3.10), but we will show in Theorem 3.41 that such a function exists.

The same reasoning above can be applied to general cascade examples in $\mathbb{R}^n$ or to some cascade-like examples that arise in partial differential equations such as the following: let Ω be a bounded smooth domain in $\mathbb{R}^n$ and $P_0, P_1 \in \overline{\Omega}$. Assume that $f, g \in C^2(\mathbb{R}, \mathbb{R})$ and satisfy (3.9). Consider the initial boundary value problem

$$\begin{cases} u_t = \Delta u + f(u) & \text{in } \Omega, \\ \frac{\partial u}{\partial n} = 0 & \text{in } \partial\Omega, \\ u(0) = u_0 \in C(\overline{\Omega}), & \\ v_t = v_{xx} + g(v) & \text{in } (0,1), \\ v(t,0) = u(t, P_0), \; v(t,1) = u(t, P_1), & \\ v(0,\cdot) = v_0 \in H^1(0,1), & \end{cases} \tag{3.11}$$

which defines a semigroup $\mathcal{T}$ in $W^{1,p}(\Omega) \times W^{1,p}(0,1)$ with a global attractor $\mathcal{A}$ (see, for example, [**12**]).

Assume that (3.11) has a finite number of equilibria and each of them is hyperbolic. It is clear that (G2) is satisfied and, as before, the solutions in $\mathcal{A}$ are backwards and forwards asymptotic to equilibria (satisfy (G1)). Consequently, $\mathcal{T}$ is dynamically gradient.

3. Dynamically gradient semigroups are gradient

In this section we exhibit a Morse decomposition for a dynamically gradient semigroup and, using that, prove the existence of a Lyapunov function for it. With this we will have proved that a dynamically gradient semigroup is, in fact, gradient. This result allows us to conclude that the gradient structure of semigroups is robust under autonomous perturbations, that is, a *small perturbation of a gradient semigroup is still a gradient semigroup*, by a simple application of Theorem 3.42. The material presented here is adapted from [**5**].

3.1. Morse decomposition for dynamically gradient semigroups. Now we will introduce the notion of a *Morse decomposition* for a global attractor $\mathcal{A}$ of a dynamically gradient semigroup $\mathcal{T}$. We will define firstly the notion of *attractor-repeller pairs*.

DEFINITION 3.26. Let $\mathcal{T}$ be a semigroup in a metric space X with a global attractor $\mathcal{A}$. We say that a nonempty subset Ξ of $\mathcal{A}$ is a **local attractor** if there exists an $\epsilon > 0$ such that $\omega(\mathcal{O}_\epsilon(\Xi)) = \Xi$. The **repeller** Ξ^* associated to a local attractor Ξ is the set defined by

$$\Xi^* = \{x \in \mathcal{A} \colon \omega(x) \cap \Xi = \varnothing\}.$$

The pair (Ξ, Ξ^*) is called an **attractor-repeller pair** for $\mathcal{T}$.

LEMMA 3.27. *Let $\mathcal{T}$ be a semigroup in a metric space X with a global attractor $\mathcal{A}$. If Ξ is a local attractor, then there exists $\epsilon > 0$ such that $\Xi^* \cap \mathcal{O}_\epsilon(\Xi) = \varnothing$. Moreover, Ξ^* is closed and $\mathcal{T}$-invariant.*

PROOF. Since Ξ is a local attractor, by definition there exists $\epsilon > 0$ such that $\omega(\mathcal{O}_\epsilon(\Xi)) = \Xi$. Hence if $x \in \mathcal{O}_\epsilon(\Xi)$, then $\omega(x) \cap \Xi \neq \varnothing$ and therefore $x \notin \Xi^*$. Now, since $\omega(T(t)x) = \omega(x)$ for each $t \geqslant 0$ we have $x \in \Xi^*$ if and only if $T(t)x \in \Xi^*$ and the $\mathcal{T}$-invariance follows. To see that Ξ^* is closed note that, since the closure of a $\mathcal{T}$-invariant is $\mathcal{T}$-invariant and since Ξ^* does not intersect $\mathcal{O}_\epsilon(\Xi)$, all points in the closure of Ξ^* are also in Ξ^*. □

DEFINITION 3.28. An ordered n-tuple $\boldsymbol{\Xi} = \{\Xi_1, \cdots, \Xi_n\}$ of isolated invariant sets is called a **Morse decomposition** of $\mathcal{A}$ if, whenever $\phi \colon \mathbb{R} \to \mathcal{A}$ is a global solution of $\mathcal{T}$, then either

$$\lim_{t\to-\infty} \operatorname{dist}_H(\phi(t), \Xi_j) = 0 \quad \text{and} \quad \lim_{t\to\infty} \operatorname{dist}_H(\phi(t), \Xi_i) = 0 \quad \text{with } i < j,$$

or $\phi(\mathbb{R}) \subset \Xi_i$ for some $1 \leqslant i \leqslant n$.

Observe that, from Lemma 1.20, Ξ is a local attractor if and only if it is compact, $\mathcal{T}$-invariant and $\mathcal{T}$-attracts $\mathcal{O}_\epsilon(\Xi)$ for some $\epsilon > 0$.

REMARK 3.29. We stress the fact that the above definition differs slightly from the usual definition since the local attractor is required to $\mathcal{T}$-attract a neighborhood of Ξ in X and not in $\mathcal{A}$ as in [**105, 192**]. We will prove next that both definitions coincide.

LEMMA 3.30. *Let $\mathcal{T}$ be a semigroup in a metric space X with a global attractor $\mathcal{A}$. If Ξ is compact, $\mathcal{T}$-invariant and there is an $\epsilon > 0$ such that Ξ $\mathcal{T}$-attracts $\mathcal{O}_\epsilon(\Xi) \cap \mathcal{A}$, then given $\delta > 0$ there is a $\delta' > 0$ such that $\gamma^+(\mathcal{O}_{\delta'}(\Xi)) \subset \mathcal{O}_\delta(\Xi)$, where $\gamma^+(\mathcal{O}_{\delta'}(\Xi)) = \bigcup_{x\in\mathcal{O}_{\delta'}(\Xi)} \bigcup_{t\geqslant 0} T(t)x$.*

PROOF. Given $0 < \delta < \epsilon$ if there is no $\delta' > 0$ such that $\gamma^+(\mathcal{O}_{\delta'}(\Xi)) \subset \mathcal{O}_\delta(\Xi)$, there are $x \in \Xi$, $X \ni x_n \to x$ and, from Lemma 3.9, $\mathbb{R} \ni t_n \to \infty$ such that $\operatorname{dist}_H(T(t_n)x_n, \Xi) = \delta$ and $T(t)x_n \in \mathcal{O}_\delta(\Xi)$, $t \in [0, t_n]$. Since $\mathcal{T}$ has a global attractor it is not difficult to see that there is a global solution $\xi \colon \mathbb{R} \to X$ of $\mathcal{T}$ such that $\xi_{n_k} \colon [-t_{n_k}, \infty) \to X$ given by $\xi_{n_k}(t) = T(t_{n_k} + t)x_{n_k}$ satisfies $\xi_{n_k}(t) \to \xi(t)$ for each $t \in \mathbb{R}$ and some subsequence $n_k \to \infty$. Clearly $\xi(t) \in \overline{\mathcal{O}_\delta(\Xi)} \cap \mathcal{A} \subset$

$\mathcal{O}_\epsilon(\Xi) \cap \mathcal{A}$ for all $t \leqslant 0$, and $\mathrm{dist}_H(\xi(0), \Xi) = \delta$, and consequently Ξ cannot $\mathcal{T}$-attract $\mathcal{O}_\epsilon(\Xi) \cap \mathcal{A}$. □

LEMMA 3.31. *If $\mathcal{T}$ is a semigroup in X with a global attractor $\mathcal{A}$ and $S(t) = T(t)|_{\mathcal{A}}$, clearly $\mathcal{S}$ is a semigroup in the metric space $\mathcal{A}$. If Ξ is a local attractor for $\mathcal{S}$ in $\mathcal{A}$ and K is a compact subset of $\mathcal{A}$ such that $K \cap \Xi^* = \varnothing$, then Ξ attracts K. Furthermore Ξ is a local attractor for $\mathcal{T}$ in X.*

PROOF. Let K be a compact subset of $\mathcal{A}$ such that $K \cap \Xi^* = \varnothing$. If Ξ does not $\mathcal{T}$-attract K, there exist a $\delta > 0$, $t_n \to \infty$, $x \in K$ and $K \ni x_n \to x$ such that $\mathrm{dist}_H(T(t_n)x_n, \Xi) \geqslant \delta$. From Lemma 3.30 there exists $0 < \delta' < \delta$ such that $\mathrm{dist}_H(T(t)x_n, \Xi) \geqslant \delta'$ for all $t \in [0, t_n]$. This implies that $\mathrm{dist}_H(T(t)x, \Xi) \geqslant \delta'$ for all $t \geqslant 0$ and $\omega(x) \cap \Xi = \varnothing$, which is a contradiction with $K \ni x \notin \Xi^*$, proving the first part of the result.

For the remaining part note that, from Lemma 3.30, there exists $\delta' > 0$ such that $\omega(\mathcal{O}_{\delta''}(\Xi)) \cap \Xi^* = \varnothing$ for all $\delta'' < \delta'$. From the $\mathcal{S}$-invariance of $\omega(\mathcal{O}_{\delta'}(\Xi))$ and the property that Ξ $\mathcal{S}$-attracts compact subsets of $\mathcal{A}$ that do not intersect Ξ^*, we must have that $\omega(\mathcal{O}_{\delta''}(\Xi)) \subset \Xi$. Since $\omega(\mathcal{O}_{\delta''}(\Xi))$ $\mathcal{T}$-attracts $\mathcal{O}_{\delta''}(\Xi)$, we have the result. □

LEMMA 3.32. *Let $\mathcal{T}$ be a semigroup in X with a global attractor $\mathcal{A}$ and an attractor-repeller pair (Ξ, Ξ^*). A global solution $\xi \colon \mathbb{R} \to X$ of $\mathcal{T}$ with the property that, for some $\delta > 0$ such that $\mathcal{O}_\delta(\Xi^*) \cap \Xi = \varnothing$, $\xi(t) \in \mathcal{O}_\delta(\Xi^*)$ for all $t \leqslant 0$ must satisfy $\lim\limits_{t \to -\infty} \mathrm{dist}_H(\xi(t), \Xi^*) = 0$.*

PROOF. If the conclusion is false there exist a $\delta' > 0$ and a sequence $t_n \to \infty$ such that $\mathrm{dist}_H(\xi(-t_n), \Xi^*) \geqslant \delta'$. This leads to a contradiction with the fact that Ξ must attract $K = \{z \in \mathcal{A} \colon \mathrm{dist}_H(z, \Xi^*) \geqslant \delta'\}$. □

LEMMA 3.33. *Let $\mathcal{T}$ be a semigroup in X with a global attractor $\mathcal{A}$ and an attractor-repeller pair (Ξ, Ξ^*). If $\xi \colon \mathbb{R} \to X$ is a global bounded solution for $\mathcal{T}$ through $x \notin \Xi \cup \Xi^*$, then $\lim\limits_{t \to \infty} \mathrm{dist}_H(\xi(t), \Xi) = 0$ and $\lim\limits_{t \to -\infty} \mathrm{dist}_H(\xi(t), \Xi^*) = 0$. Furthermore, if $x \in X \setminus \mathcal{A}$, then $\lim\limits_{t \to \infty} \mathrm{dist}_H(T(t)x, \Xi \cup \Xi^*) = 0$.*

PROOF. If $\xi \colon \mathbb{R} \to X$ is a global bounded solution through $x \notin \Xi \cup \Xi^*$ it follows from Lemma 3.31 that $\lim\limits_{t \to \infty} \mathrm{dist}_H(\xi(t), \Xi) = 0$.

Assume now that $\xi(t)$ does not converge to Ξ^* as $t \to -\infty$. We divide this part of the proof of the result into two cases:

Case 1: $\overline{\xi(\mathbb{R})} \cap \Xi^* = \varnothing$. In this case, from Lemma 3.31, $\overline{\xi(\mathbb{R})}$ is attracted by Ξ. The invariance of $\overline{\xi(\mathbb{R})}$ implies that $\overline{\xi(\mathbb{R})} \subset \Xi$ which is a contradiction with $\xi(0) = x \notin \Xi \cup \Xi^*$.

Case 2: $\overline{\xi(\mathbb{R})} \cap \Xi^* \neq \varnothing$. In this case, there exist $\delta > 0$ and sequences $t_n, \tau_n \to \infty$ such that $\mathrm{dist}_H(\xi(-t_n), \Xi^*) = \delta$ and $\xi(-t_n + \tau_n) \to z \in \Xi^*$ and $\mathrm{dist}_H(\xi(-t_n + t), \Xi) \leqslant \delta$ for all $0 \leqslant t \leqslant \tau_n$. From this we obtain a global solution $\zeta \colon \mathbb{R} \to \mathcal{A}$ of $\mathcal{T}$ with the property that $\mathrm{dist}_H(\zeta(t), \Xi^*) \leqslant \delta$ for all $t \geqslant 0$. Therefore we have $\omega(\zeta(0)) \notin \Xi$, which implies that $\zeta(0) \in \Xi^*$, leading to a contradiction.

For the last part of the result, let $x \in X \setminus \mathcal{A}$. If $\overline{\gamma^+(x)} \cap \Xi \neq \varnothing$ we have $\lim\limits_{t \to \infty} \mathrm{dist}_H(T(t)x, \Xi) = 0$. Otherwise, there exists $\delta > 0$ with $\gamma^+(z) \cap \mathcal{O}_\delta(\Xi) = \varnothing$, and in this case we claim that $\lim\limits_{t \to \infty} \mathrm{dist}_H(T(t)x, \Xi^*) = 0$. If the claim is false, there are $\epsilon > 0$ and a sequence $t_n \to \infty$ such that $\mathrm{dist}_H(T(t_n)x, \Xi^*) \geqslant \epsilon$. Considering the sequence of functions $\xi_n \colon [-t_n, \infty) \to X$ defined by $\xi_n(t) = T(t+t_n)x$ for $t \geqslant -t_n$, we construct a global solution $\xi \colon \mathbb{R} \to \mathcal{A}$ of $\mathcal{T}$ such that $\mathrm{dist}_H(\xi(0), \Xi^*) \geqslant \epsilon$ and $\mathrm{dist}_H(\xi(t), \Xi) \geqslant \delta$ for all $t \in \mathbb{R}$. Hence $\omega(\xi(0)) \cap \Xi = \varnothing$ and $\xi(0) \notin \Xi^*$ which is a contradiction. □

LEMMA 3.34. *Let $\mathcal{T}$ be a dynamically gradient semigroup in a metric space X with a global attractor $\mathcal{A}$ and a disjoint collection of isolated invariant sets $\boldsymbol{\Xi} = \{\Xi_1, \cdots, \Xi_n\}$. Then there exists $1 \leqslant i \leqslant n$ such that Ξ_i is a local attractor.*

PROOF. Assume that there are no local attractors in $\boldsymbol{\Xi}$. Choose $0 < \epsilon < \epsilon_0(\boldsymbol{\Xi})$. It follows from Lemma 3.14 that, for each suitably small $\delta \in (0, \epsilon)$ there exist an $x_{i,\delta} \in \mathcal{O}_\delta(\Xi_i)$ and a $\tau_{i,\delta} > 0$ such that $\mathrm{dist}_H(T(\tau_{i,\delta})x_{i,\delta}, \Xi_i) \geqslant \epsilon$ and $\mathrm{dist}_H(T(t+\tau_{i,\delta})x_{i,\delta}, \Xi_i) \geqslant \delta$ for all $t \geqslant 0$. It follows that $\lim_{t\to\infty} \mathrm{dist}_H(T(t)x_{i,\delta}, \Xi_j) = 0$, for some $j \neq i$. That produces a chain recurrent isolated invariant set and leads to a contradiction. □

Next we describe the construction of a Morse decomposition of the attractor of a dynamically gradient semigroup. Let $\mathcal{T}$ be a dynamically gradient semigroup with associated disjoint collection of isolated invariant sets $\boldsymbol{\Xi} = \{\Xi_1, \cdots, \Xi_n\}$. If (after possible reordering) Ξ_1 is a local attractor and

$$\Xi_1^* = \{x \in \mathcal{A} \colon \omega(x) \cap \Xi_1 = \varnothing\},$$

then each Ξ_i, $i > 1$, is contained in Ξ_1^* and that for any $x \notin \mathcal{A} \setminus \{\Xi_1 \cup \Xi_1^*\}$ and global solution $\phi \colon \mathbb{R} \to \mathcal{A}$ of $\mathcal{T}$ through x we have

$$\lim_{t\to-\infty} \mathrm{dist}_H(\phi(t), \Xi_1^*) = 0 \quad \text{and} \quad \lim_{t\to\infty} \mathrm{dist}_H(\phi(t), \Xi_1) = 0.$$

We already know that Ξ_1^* is $\mathcal{T}$-invariant and, hence, we can consider the restriction $\mathcal{T}_1$ of $T(t)$ to $X_{1,0} := \Xi_1^*$. Note that, making this restriction, we have removed all possible global solutions $\phi \colon \mathbb{R} \to \mathcal{A}$ of $\mathcal{T}$ such that $\phi(t) \to \Xi_1$ as $t \to \infty$, that is, if $\phi_1 \colon \mathbb{R} \to \mathcal{A}$ is a global solution of $\mathcal{T}_1$ and $\phi_1(t) \to \Xi_i$, then $i > 1$. Also, it is simple to see that $\mathcal{T}_1$ is a dynamically gradient semigroup in $X_{1,0}$ with disjoint collection of isolated invariant sets $\{\Xi_2, \cdots, \Xi_n\}$, and this implies that if a global solution $\phi \colon \mathbb{R} \to \mathcal{A}$ of $\mathcal{T}$ satisfies $\Xi_j \overset{t\to-\infty}{\longleftarrow} \phi(t) \overset{t\to\infty}{\longrightarrow} \Xi_1$, then we must have $j \geqslant 1$.

We may assume, without loss of generality, that Ξ_2 is a local attractor for the semigroup $\mathcal{T}_1$ in $X_{1,0}$. If $X_{2,1} =$ is the repeller associated to the isolated invariant set Ξ_2 for $\mathcal{T}_1$ in $X_{1,0}$ we may proceed and consider the restriction $\mathcal{T}_2$ of the semigroup $\mathcal{T}_1$ to $X_{2,1}$ and $\mathcal{T}_2$ is a dynamically gradient semigroup in $X_{2,1}$ with associated disjoint collection of isolated invariant sets $\{\Xi_3, \cdots, \Xi_n\}$. Note that in this case, we have removed all the global solutions $\phi \colon \mathbb{R} \to X$ of $\mathcal{T}$ such that $\phi(t) \to \Xi_2$ as $t \to \infty$ and therefore, if $\Xi_j \overset{t\to-\infty}{\longleftarrow} \phi(t) \overset{t\to\infty}{\longrightarrow} \Xi_2$, then $j \geqslant 2$.

Proceeding with this until all isolated invariant sets are exhausted we obtain a reordering of $\{\Xi_1, \cdots, \Xi_n\}$ in such a way that Ξ_j is a local attractor for the restriction of $\mathcal{T}$ to $X_{j-1,j-2}$, where $X_{0,-1} = \mathcal{A}$.

With the construction above, if a global solution $\phi \colon \mathbb{R} \to \mathcal{A}$ of $\mathcal{T}$ satisfies

$$\Xi_\ell \overset{t\to-\infty}{\longleftarrow} \xi(t) \overset{t\to\infty}{\longrightarrow} \Xi_k, \tag{3.12}$$

then $\ell \geqslant k$. This proves that this reordering of $\boldsymbol{\Xi} = \{\Xi_1, \cdots, \Xi_n\}$ (which we denote the same) is a Morse decomposition to $\mathcal{A}$.

This proves the following result.

THEOREM 3.35. *If $\mathcal{T}$ is a dynamically gradient semigroup with disjoint collection of isolated invariant sets $\boldsymbol{\Xi} = \{\Xi_1, \cdots, \Xi_n\}$, then there is a reordering of $\boldsymbol{\Xi}$ that it is a Morse decomposition to $\mathcal{A}$.*

Now we will prove that, for a suitably chosen sequence of local attractors $\varnothing = A_0 \subsetneq A_1 \subsetneq A_2 \subsetneq \cdots \subsetneq A_n = \mathcal{A}$, we have that $\Xi_j = A_j \cap A_{j-1}^*$, $1 \leqslant j \leqslant n$, and that we have a Morse decomposition in the sense of [**105, 192**]. To this end, using

$$W^u(\Xi_i) = \{x \in X\colon \text{ there exists a global solution } \xi\colon \mathbb{R} \to X \text{ of } \mathcal{T}$$
$$\text{through } x \text{ such that } \xi(t) \stackrel{t\to-\infty}{\to} \Xi_i\},$$

we can define $A_0 = \varnothing$, $A_1 = \Xi_1$ and for $j = 2, 3, \ldots, n$

$$A_j = \bigcup_{i=1}^{j} W^u(\Xi_i). \tag{3.13}$$

It is clear that $A_n = \mathcal{A}$.

With this construction, we can prove that each A_j is a local attractor, $j = 0, \ldots, n$, and if we consider the correspondent repellers A_j^*, then the sequence of attractor-repeller pairs $\{(A_j, A_j^*)\}_{j=0,\cdots,n}$ generates exactly the Morse decomposition given by $\boldsymbol{\Xi}$; that is, $\Xi_j = A_j \cap A_{j-1}^*$ for all $j = 1, \cdots, n$ and $\bigcap_{j=0}^n (A_j \cup A_j^*) = \bigcup_{i=1}^n \Xi_i$.

LEMMA 3.36. *The A_j defined in (3.13) is compact, for each $j = 0, \cdots, n$.*

PROOF. We prove this result using a finite induction argument. Clearly the result is true for $j = 0$ and $j = 1$. Now assume that A_j is compact for $1 \leqslant j < n$ and we prove that A_{j+1} is also compact. Since $A_{j+1} \subset \mathcal{A}$, it suffices to show that A_{j+1} is closed. To this end let $\{x_k\}_{k\in\mathbb{N}}$ be a sequence in A_{j+1} with $x_k \to x \in \mathcal{A}\backslash A_{j+1}$. If there exists a subsequence of $\{x_k\}_{k\in\mathbb{N}}$ in $A_j \cup \Xi_{j+1}$, the compactness of $A_j \cup \Xi_{j+1}$ completes the proof. If this is not the case, then there exists $k_0 \in \mathbb{N}$ such that $x_k \in W^u(\Xi_{j+1})\backslash\Xi_{j+1}$ for all $k \geqslant k_0$. Thus, for each such k, there exists a global solution $\xi_k\colon \mathbb{R} \to \mathcal{A}$ of $\mathcal{T}$ through x_k such that $\text{dist}_H(\xi_k(t), \Xi_{j+1}) \to 0$ as $t \to -\infty$.

From these, we can construct a sequence of global solutions $\{\zeta_1, \cdots, \zeta_{k-1}\}$ connecting isolated invariant sets $\{\Xi_{\ell_1}, \cdots, \Xi_{\ell_p}\}$ with $\Xi_{\ell_1} = \Xi_{j+1}$,

$$\Xi_{\ell_i} \stackrel{t\to-\infty}{\longleftarrow} \zeta_i(t) \stackrel{t\to\infty}{\longrightarrow} \Xi_{\ell_{i+1}}, \ 1 \leqslant i \leqslant p-1,$$

and $\zeta_{i_0}(0) = x$ for some $1 \leqslant i_0 \leqslant p-1$. Since x is not in the unstable manifolds of Ξ_ℓ, $1 \leqslant \ell \leqslant j+1$, we have $\text{dist}_H(\zeta_{i_0}(t), \Xi_m) \stackrel{t\to-\infty}{\longrightarrow} 0$ for some $j+1 < m \leqslant n$, which contradicts (3.12). □

THEOREM 3.37. *Let $\mathcal{T}$ be a dynamically gradient semigroup with disjoint collection of isolated invariant sets $\boldsymbol{\Xi} = \{\Xi_1, \cdots, \Xi_n\}$, reordered in such a way that it is a Morse decomposition. Then A_j defined in (3.13) is a local attractor for $\mathcal{T}$ in X and*

$$\Xi_j = A_j \cap A_{j-1}^* \quad \textit{for } j = 1, \cdots, n.$$

PROOF. From Lemma 3.31, it is sufficient to prove that $A_j = A_{j-1} \cup W^u(\Xi_j)$ is a local attractor for $\mathcal{T}$ restricted to $\mathcal{A}$.

Claim: there exists $\delta_0 > 0$ such that $\mathcal{O}_{\delta_0}(A_j) \cap (\bigcup_{i=j+1}^n \Xi_i) = \varnothing$.

Indeed, assume that there exists a sequence $\{x_k\}$ with $x_k \in \bigcup_{i=j+1}^n \Xi_i$ and $x_k \to A_j$, as $k \to \infty$. We can assume that $x_k \in \Xi_{i_0}$, for all $k \in \mathbb{N}$, for some fixed

$j+1 \leqslant i_0 \leqslant n$. By Lemma 3.36, we can assume that $x_k \to x \in A_j \cap \Xi_{i_0} = \varnothing$, which is a contradiction and proves our claim.

If there are $\delta < \delta_0$ and $\delta' < \delta$ such that $\gamma^+(\mathcal{O}_{\delta'}(A_j)) \subset \mathcal{O}_\delta(A_j)$, then $\omega(\mathcal{O}_{\delta'}(A_j))$ $\mathcal{T}$-attracts $\mathcal{O}_{\delta'}(A_j)$ and (as $\omega(\mathcal{O}_{\delta'}(A_j))$ is $\mathcal{T}$-invariant) is contained in A_j proving that A_j is a local attractor. If that is not the case, for $\delta < \delta_0$, there is a sequence $\{x_k\}$ in $\mathcal{A}\backslash A_j$ with $\operatorname{dist}_H(x_k, A_j) \to 0$, through each x_k a global solution $\xi_k\colon \mathbb{R} \to \mathcal{A}$ of $\mathcal{T}$ and a sequence $t_k \to \infty$, such that $\operatorname{dist}_H(\xi_k(t), A_j) \leqslant \delta$ for all $t \in [0, t_k]$. In this way, using Lemma 2.4, we construct a global solution $\xi\colon \mathbb{R} \to \mathcal{A}$ such that $\operatorname{dist}_H(\xi_k(t_k), A_j) = \delta$ and $\xi(0) \notin A_j$ and $\operatorname{dist}_H(\xi(t), A_j) \leqslant \delta$ for all $t \leqslant 0$. This and the fact that $\mathcal{T}$ is a dynamically gradient semigroup gives us a contradiction, since $\xi(t) \to \Xi_k$ as $t \to -\infty$ for some $1 \leqslant k \leqslant j$, which implies that $\xi(0) \in A_j$.

To prove that $\Xi_j = A_j \cap A^*_{j-1}$ note that

$$A_j = \bigcup_{i=1}^{j} W^u(\Xi_i)$$

and $A^*_{j-1} = \{z \in \mathcal{A}\colon \omega(z) \cap A_{j-1} = \varnothing\}$. Hence, given $z \in A_j \cap A^*_{j-1}$ we have that the global solution $\xi\colon \mathbb{R} \to \mathcal{A}$ through z must satisfy that

$$\{\Xi_i\colon 1 \leqslant i \leqslant j\} \overset{t\to-\infty}{\longleftarrow} \xi(t) \overset{t\to\infty}{\longrightarrow} \{\Xi_i\colon j \leqslant i \leqslant n\}.$$

As a consequence, together with the fact that $\mathcal{T}$ is a dynamically gradient semigroup with disjoint collection of isolated invariant sets $\{\Xi_1, \cdots, \Xi_n\}$, for which any global solution $\xi\colon \mathbb{R} \to \mathcal{A}$ of $\mathcal{T}$ satisfies $\Xi_\ell \overset{t\to-\infty}{\longleftarrow} \xi(t) \overset{t\to-\infty}{\longrightarrow} \Xi_k$ with $\ell \geqslant k$, we obtain that $z \in \Xi_j$. This shows that $A_j \cap A^*_{j-1} \subset \Xi_j$. The other inclusion is immediate from the definition of A_j and A^*_{j-1}. □

PROPOSITION 3.38. *Let $\mathcal{T}$ be a dynamically gradient semigroup with associated disjoint collection of isolated invariant sets $\boldsymbol{\Xi} = \{\Xi_1, \cdots, \Xi_n\}$ reordered in such a way that it is a Morse decomposition. Then,*

$$\bigcap_{j=0}^{n} (A_j \cup A^*_j) = \bigcup_{i=1}^{n} \Xi_i.$$

PROOF. If $z \in \bigcup_{i=1}^n \Xi_i$, let $k \in \{1, 2, \cdots, n\}$ be such that $z \in \Xi_k = A_k \cap A^*_{k-1}$. Hence $z \in A_k \subset A_{k+1} \subset \cdots \subset A_n$ and $z \in A^*_{k-1} \subset A^*_{k-2} \subset \cdots \subset A^*_0$. Thus

$$z \in \bigcap_{j=k}^{n} A_j \cap \bigcap_{j=1}^{k-1} A^*_j \subset \bigcap_{j=k}^{n} (A_j \cup A^*_j) \cap \bigcap_{j=0}^{k-1} (A_j \cup A^*_j) = \bigcap_{j=0}^{n} (A_j \cup A^*_j),$$

proving the inclusion $\bigcup_{i=1}^n \Xi_i \subset \bigcap_{j=0}^n A_j \cup A^*_j$.

Now, let $z \in \bigcap_{j=0}^n (A_j \cup A^*_j)$ and $I := \{i_1, i_2, \cdots, i_k\}$, $J := \{j_1, j_2, \cdots, j_l\}$ such that $I \cup J = \{0, 1, \cdots, n\}$ with $I \cap J = \varnothing$ and $z \in A_i$ for all $i \in I$ and $z \in A^*_j$ for all $j \in J$. Clearly, if $i := \min I$, necessarily $I = \{i, i+1, i+2, \cdots, n\}$ and $J = \{0, 1, \cdots, i-1\}$, consequently $z \in A_i$ and $z \in A^*_{i-1}$. So, $z \in A_i \cap A^*_{i-1} = \Xi_i$, from which we conclude that $\bigcap_{j=0}^n (A_j \cup A^*_j) \subset \bigcup_{i=1}^n \Xi_i$ and the proof is completed. □

3.2. Lyapunov functions for dynamically gradient semigroups. Inspired by the results in the work of Conley [**105**] (see also the work of Ribakowski [**192**]) we present in this section the equivalence between gradient semigroups and dynamically gradient semigroups. The approach used here is that of [**5, 78**], which does not assume backwards uniqueness of solutions when backwards solutions exist.

The key result in this section is Theorem 3.41, which states the equivalence between the concepts of dynamically gradient semigroups and gradient semigroups. The main task is to construct a Lyapunov function for a dynamically gradient semigroup, and to give an idea of such construction, we will state two auxiliary results, which will play a crucial role in the proof of such equivalence.

LEMMA 3.39. *If $\mathcal{T}$ is a semigroup with global attractor $\mathcal{A}$, the map $h\colon X \to \mathbb{R}$ given by*

$$h(x) = \sup_{t\geqslant 0} \operatorname{dist}_H(T(t)x, \mathcal{A}), \qquad \text{for each } x \in X$$

is well defined, continuous, nonincreasing along solutions of $\mathcal{T}$ and $h^{-1}(0) = \mathcal{A}$.

PROOF. Indeed, by Lemma 3.31, given $\varepsilon > 0$ let $0 < \varepsilon' < \varepsilon$ such that $\gamma^+(\mathcal{O}_{\varepsilon'}(\mathcal{A})) \subset \mathcal{O}_\varepsilon(\mathcal{A})$. This and the $\mathcal{T}$-invariance of $\mathcal{A}$ show the continuity of h on each $x \in \mathcal{A}$.

Let $x_0 \in X \backslash \mathcal{A}$ be given. We have $h(x_0) \geqslant \operatorname{dist}_H(x_0, \mathcal{A}) > 0$ and we consider $0 < \mu < \operatorname{dist}_H(x_0, \mathcal{A})$. From the continuity of the map $X \ni x \mapsto \operatorname{dist}_H(x, \mathcal{A}) \in [0, \infty)$, we can choose a bounded neighborhood V of x_0 in X such that $\operatorname{dist}_H(z, \mathcal{A}) > \mu$ if $z \in V$. Now, let $\tau > 0$ be such that $\gamma^+(T(\tau)V) \subset \mathcal{O}_\mu(\mathcal{A})$. Hence,

$$h(z) = \sup_{0\leqslant s\leqslant \tau} \operatorname{dist}_H(T(s)z, \mathcal{A}), \qquad \text{for each } z \in V,$$

and from the continuity properties of the $\mathcal{T}$, it follows that h is continuous in x_0.

To see that h is nonincreasing along solutions of $\mathcal{T}$ note that, if $z \in X$ and $t_1 > 0$, then

$$\begin{aligned} h(T(t_1)z) &= \sup_{t\geqslant 0} \operatorname{dist}_H(T(t)T(t_1)z, \mathcal{A}) = \sup_{t\geqslant 0} \operatorname{dist}_H(T(t+t_1)z, \mathcal{A}) \\ &= \sup_{t\geqslant t_1} \operatorname{dist}_H(T(t)z, \mathcal{A}) \leqslant \sup_{t\geqslant 0} \operatorname{dist}_H(T(t)z, \mathcal{A}) = h(z). \end{aligned}$$

□

PROPOSITION 3.40. *Let $\mathcal{T}$ be a semigroup in a metric space X with global attractor $\mathcal{A}$ and an attractor-repeller pair (Ξ, Ξ^*). Then there exists a function $f\colon X \to \mathbb{R}$ satisfying the following:*

(i) *f is continuous in X.*
(ii) *f is nonincreasing along solutions of $\mathcal{T}$.*
(iii) *$f^{-1}(0) = \Xi$ and $f^{-1}(1) \cap \mathcal{A} = \Xi^*$.*
(iv) *Given $z \in X$, if $f(T(t)z) = f(z)$ for all $t \geqslant 0$, then $z \in \Xi \cup \Xi^*$.*

PROOF. First note that Ξ and Ξ^* are disjoint compact subsets of X. With the convention that $\operatorname{dist}_H(z, \varnothing) = 1$ for each $z \in X$, define the function $l\colon X \to [0,1]$ associated to (Ξ, Ξ^*) by

$$l(z) = \frac{\operatorname{dist}_H(z, \Xi)}{\operatorname{dist}_H(z, \Xi) + \operatorname{dist}_H(z, \Xi^*)} \qquad \text{for each } z \in X,$$

which is the canonical Urysohn function if both Ξ and Ξ^* are nonempty. Clearly l is well defined. Setting $d_0 = \operatorname{dist}(\Xi, \Xi^*) > 0$, it holds that $|l(z) - l(w)| \leqslant \frac{2}{d_0} d(z, w)$ for $z, w \in X$, that is, l is uniformly continuous. Moreover, $l^{-1}(0) = \Xi$ and $l^{-1}(1) = \Xi^*$.

Define also the function $k\colon X \to \mathbb{R}$ by

$$k(z) = \sup_{t\geqslant 0} l(T(t)z) \qquad \text{for each } z \in X. \tag{3.14}$$

We now show that k is continuous and nonincreasing along solutions of $\mathcal{T}$, $k(X) \subset [0,1]$ (with equality when X is connected and Ξ and Ξ^* are nonempty), $k^{-1}(0) = \Xi$ and $k^{-1}(1) \cap \mathcal{A} = \Xi^*$.

Clearly $k(X) \subset [0,1]$ follows and $t \geqslant 0$. If X is connected and Ξ, Ξ^* are nonempty, due to the fact that $\mathcal{A}$ is connected and l is continuous, we have $k(X) = [0,1]$.

To prove that $[0,\infty) \ni t \mapsto k(T(t)z) \in [0,1]$ is nonincreasing for each $z \in X$ note that, if $0 \leqslant t_1 \leqslant t_2$ we have

$$\begin{aligned} k(T(t_1)z) &= \sup_{t\geqslant 0} l(T(t)T(t_1)z) = \sup_{t\geqslant 0} l(T(t+t_1)z) = \sup_{t\geqslant t_1} l(T(t)z) \\ &\geqslant \sup_{t\geqslant t_2} l(T(t)z) = \sup_{t\geqslant 0} l(T(t+t_2)z) = k(T(t_2)z). \end{aligned}$$

It is clear from the definition of k and from the $\mathcal{T}$-invariance of Ξ and Ξ^* that $k(\Xi) = \{0\}$ and $k(\Xi^*) = \{1\}$. Now, if $z \in X$ is such that $k(z) = 0$, then $l(T(t)z) = 0$ for all $t \geqslant 0$. In particular, $0 = l(T(0)z) = l(z)$, and so, $z \in \Xi$, that is, $k^{-1}(0) \subset \Xi$ which shows that $k^{-1}(0) = \Xi$. On the other hand, if $z \in \mathcal{A}$ is such that $k(z) = 1$ and $z \notin \Xi^*$, then $\omega(z) \subset \Xi$. From the continuity of l and the fact that $\omega(z)$ $\mathcal{T}$-attracts z, we obtain $\lim_{t\to\infty} l(T(t)z) = 0$. So there exists $t_0 > 0$ such that $1 = k(z) = \sup_{0\leqslant t\leqslant t_0} l(T(t)z)$. This implies the existence of $t' \in [0,t_0]$ such that $l(T(t')z) = 1$, that is, $T(t')z \in \Xi^*$. Consequently $\omega(z) = \omega(T(t')z) \subset \Xi^*$, which contradicts the fact that $\omega(z) \subset \Xi$ and so, if $k(z) = 1$ for some $z \in \mathcal{A}$ we must have that $z \in \Xi^*$. From this we conclude that $k^{-1}(1) \cap \mathcal{A} \subset \Xi^*$ and so $k^{-1}(1) \cap \mathcal{A} = \Xi^*$.

We now prove that, if $z \in \mathcal{A}$ and $k(T(t)z) = k(z)$ for all $t \geqslant 0$, then $z \in \Xi \cup \Xi^*$. If $z \in \mathcal{A}\backslash(\Xi \cup \Xi^*)$, $\omega(z) \subset \Xi$ and from the definition of k and the fact that $\omega(z)$ $\mathcal{T}$-attracts z we have $k(z) = \lim_{t\to\infty} k(T(t)z) = 0$. Since $k^{-1}(0) = \Xi$, z must belong to Ξ, which is a contradiction.

Next we prove the continuity of k. We split the proof into three cases:

Case 1: Continuity of k in Ξ^*.

Since $l(z) \leqslant k(z) \leqslant 1$, for all $z \in X$, given $z_0 \in \Xi^*$ and $z \in X$ we have

$$|k(z) - k(z_0)| = 1 - k(z) \leqslant 1 - l(z).$$

This and the continuity of l imply the continuity of k in z_0.

Case 2: Continuity of k in Ξ.

From the continuity of l in Ξ, given $\varepsilon > 0$, there is a $\delta > 0$ such that $l(\mathcal{O}_\delta(\Xi)) \subset [0,\varepsilon)$. Hence Lemma 3.30 implies that there exists $\delta' \in (0,\delta)$ such that $\gamma^+(\mathcal{O}_{\delta'}(\Xi)) \subset \mathcal{O}_\delta(\Xi)$, from which we conclude that $k(\mathcal{O}_{\delta'}(\Xi)) \subset [0,\varepsilon]$.

Case 3: Continuity of k in $X \setminus (\Xi \cup \Xi^*)$.

Given $z_0 \in X \setminus (\Xi \cup \Xi^*)$, from Lemma 3.33, we have either

$$\lim_{t\to\infty} \mathrm{dist}_H(T(t)z_0, \Xi) = 0 \quad \text{or} \quad \lim_{t\to\infty} \mathrm{dist}_H(T(t)z_0, \Xi^*) = 0.$$

Assume that $\lim_{t\to\infty} \mathrm{dist}_H(T(t)z_0, \Xi^*) = 0$ and note firstly that $k(z_0) = 1$. Now, given $\varepsilon > 0$, from the continuity of l in Ξ^* there is a neighborhood V of Ξ^* in X such that $l(V) \subset (1-\varepsilon, 1]$. If $t_0 > 0$ is such that $T(t_0)z_0 \in V$, from the continuity of $T(t_0)\colon X \to X$, let U be a neighborhood of z_0 such that $T(t_0)U \subset V$, from which it follows that $k(z) > 1 - \varepsilon$ for all $z \in U$ (for $T(t_0)z \in V$ and then

$1-\varepsilon < l(T(t_0)z) \leqslant k(z))$. This proves the continuity of k in points z_0 of $X\setminus(\Xi\cup\Xi^*)$ for which $\lim_{t\to\infty} \mathrm{dist}_H(T(t)z_0, \Xi^*) = 0$.

Now, if $z_0 \in X\setminus(\Xi\cup\Xi^*)$ and $\lim_{t\to\infty} \mathrm{dist}_H(T(t)z_0, \Xi) = 0$, it holds that $l(z_0) > 0$. Choose $\delta > 0$ such that $l(\mathcal{O}_\delta(\Xi)) \subset [0, \frac{l(z_0)}{2})$ and, from Lemma 3.30, there is a $\delta' \in (0, \delta)$ such that $\gamma^+(\mathcal{O}_{\delta'}(\Xi)) \subset \mathcal{O}_\delta(\Xi)$. From this, there is a $t_0 > 0$ with the property that $T(t_0)z_0 \in \mathcal{O}_{\delta'}(\Xi)$ and therefore $T(t)z_0 \in \mathcal{O}_\delta(\Xi)$ for all $t \geqslant t_0$. From the continuity of $T(t_0)\colon X \to X$, there is a neighborhood U_1 of z_0 in X such that $T(t_0)U_1 \subset \mathcal{O}_{\delta'}(\Xi)$. Then, for all $z \in U_1$ we have that $T(t)z \in \mathcal{O}_\delta(\Xi)$ for all $t \geqslant t_0$. Finally, from the continuity of l, let U_2 be a neighborhood of z_0 in X such that $l(z) > \frac{l(z_0)}{2}$ for all $z \in U_2$ and write $U := U_1 \cap U_2$, so that for all $z \in U$ it holds that $k(z) = \sup_{0\leqslant t\leqslant t_0} l(T(t)z)$. Reasoning as before we obtain the continuity of k in points z_0 of $X\setminus(\Xi\cup\Xi^*)$ for which $\lim_{t\to\infty} \mathrm{dist}_H(T(t)\Xi_0, \Xi) = 0$.

Let h be the function defined in Lemma 3.39, that is, $h(z) = \sup_{t\geqslant 0} \mathrm{dist}_H(T(t)z, \mathcal{A})$ for $z \in X$, and define $f\colon X \to \mathbb{R}$ by

$$f(z) = k(z) + h(z) \qquad \text{for each } z \in X.$$

The continuity of f follows from the continuity of k (proved above) and h (proved in Lemma 3.39). Since k and h are nonincreasing along solutions of $\mathcal{T}$ (see above and Lemma 3.39), so is f.

Clearly $f(\Xi) = \{0\}$. If $f(z) = 0$ for some $z \in X$, then $h(z) = k(z) = 0$ and we must have $z \in \Xi$. This shows that $f^{-1}(0) = \Xi$. Also, since $f|_{\mathcal{A}} = k|_{\mathcal{A}}$ we have that $f^{-1}(1) \cap \mathcal{A} = k^{-1}(1) \cap \mathcal{A} = \Xi^*$.

Finally, to prove (iv), let $z \in X$ such that $f(T(t)z) = f(z)$ for all $t \geqslant 0$. Since h and k are nonincreasing along solutions, we can see that $h(T(t)z) = h(z)$ and $k(T(t)z) = k(z)$ for all $t \geqslant 0$. Moreover, since $\mathrm{dist}_H(T(t)z, \mathcal{A}) \to 0$ as $t \to \infty$, we have $h(z) = 0$, and hence $z \in \mathcal{A}$. Then the equality $k(T(t)z) = k(z)$ for all $t \geqslant 0$ shows us that $z \in \Xi \cup \Xi^*$ and concludes the proof. □

We are able now to state and prove the main result of this section.

THEOREM 3.41. *Let $\mathcal{T}$ be a semigroup in a metric space X with global attractor $\mathcal{A}$ and a disjoint collection of isolated invariant sets $\mathbf{\Xi} = \{\Xi_1, \cdots, \Xi_n\}$. Then $\mathcal{T}$ is a gradient semigroup with respect to $\mathbf{\Xi}$ if and only if it is a dynamically gradient semigroup with respect to $\mathbf{\Xi}$. In addition, the Lyapunov function $V\colon X \to \mathbb{R}$ of a dynamically gradient semigroup may be chosen in such a way that $V(\Xi_k) = k - 1$, for each $k = 1, \cdots, n$.*

PROOF. It is clear that a gradient semigroup with respect to $\mathbf{\Xi}$ is a dynamically gradient semigroup with respect to $\mathbf{\Xi}$. Suppose now that $\mathcal{T}$ is a dynamically gradient semigroup with respect to $\mathbf{\Xi}$ reordered in such a way that it is a Morse decomposition for $\mathcal{A}$. Let $\varnothing = A_0 \subset A_1 \subset \cdots \subset A_n = \mathcal{A}$ be the sequence of local attractors defined in (3.13) and $\varnothing = A_n^* \subset A_{n-1}^* \subset \cdots \subset A_0^* = \mathcal{A}$ their corresponding repellers such that for each $j = 1, 2, \cdots, n$, we have $\Xi_j = A_j \cap A_{j-1}^*$.

Let h be the function defined in Lemma 3.39 and let k_j be the function constructed in (3.14) for the attractor-repeller pair (A_j, A_j^*), for each $j = 1, \cdots, n$.

Define the continuous function $V\colon X \to \mathbb{R}$ by

$$V(z) = h(z) + \sum_{j=1}^{n} k_j(z) \qquad \text{for each } z \in X.$$

Then V is a Lyapunov function for the dynamically gradient semigroup $\mathcal{T}$ with respect to Ξ. Indeed, since h and each k_j, $1 \leqslant j \leqslant n$, are nonincreasing along solutions of $\mathcal{T}$, so is V.

If $z \in X$ is such that $V(T(t)z) = V(z)$ for all $t \geqslant 0$, then arguing as in the proof of item (iv) of Proposition 3.40 we have $z \in A_j \cup A_j^*$, for each $j = 0, 1, \cdots, n$. Therefore from Lemma 3.38 we have

$$z \in \bigcap_{j=0}^{n} (A_j \cup A_j^*) = \bigcup_{i=1}^{n} \Xi_i.$$

For each $0 \leqslant j \leqslant n$, $k_j\colon \mathcal{A} \to [0,1]$, $k_j^{-1}(0) = A_j$ and $k_j^{-1}(1) \cap \mathcal{A} = A_j^*$. If $k \in \{1, \cdots, n\}$ and $z \in \Xi_k = A_k \cap A_{k-1}^*$, it follows that $z \in A_k \subset A_{k+1} \subset \cdots \subset A_n = \mathcal{A}$ and $z \in A_{k-1}^* \subset A_{k-2}^* \subset \cdots \subset A_0^* = \mathcal{A}$. Hence $k_j(z) = 0$ if $k \leqslant j \leqslant n$ and $k_j(z) = 1$ if $1 \leqslant j \leqslant k-1$, and so

$$V(z) = \sum_{j=1}^{n} k_j(z) = \sum_{j=0}^{k-1} k_j(z) + \sum_{j=k}^{n} k_j(z) = \sum_{j=0}^{k-1} 1 + \sum_{j=k}^{n} 0 = k-1.$$

□

4. Topological structural stability

In this section, we present the results on *topological structural stability* of attractors of gradient semigroups, which is the permanence of the gradient structure of the global attractors for small perturbations of a semigroup. To that end we prove that small perturbations of a dynamically gradient semigroup are dynamically gradient semigroups, that is, for a dynamically gradient semigroup, conditions (G1) and (G2) are stable under perturbations. This result can be found in [**74, 78**].

THEOREM 3.42. *Let $\{\mathcal{T}_\eta\}_{\eta \in [0,1]}$ be a family of semigroups, continuous and collectively asymptotically compact at $\eta = 0$ (see Definition 2.3). Assume that*

(a) *$\mathcal{T}_\eta$ has a global attractor $\mathcal{A}_\eta$, for each $\eta \in [0,1]$, and $\bigcup_{\eta \in [0,1]} \mathcal{A}_\eta$ is bounded;*
(b) *for each $\eta \in [0,1]$, $\mathcal{A}_\eta$ contains a disjoint collection of isolated invariant sets $\Xi_\eta = \{\Xi_{1,\eta}, \cdots, \Xi_{n,\eta}\}$ such that $\lim_{\eta \to 0} \operatorname{dist}_H(\Xi_{i,\eta}, \Xi_{i,0}) = 0$, for all $1 \leqslant i \leqslant n$;*
(c) *there exist $\delta > 0$ and $\eta_0 \in (0,1]$ such that $\Xi_{i,\eta}$ is the maximal invariant set in $\mathcal{O}_\delta(\Xi_{i,\eta})$, $1 \leqslant i \leqslant n$ and $0 \leqslant \eta \leqslant \eta_0$;*
(d) *$\mathcal{T}_0$ is a dynamically gradient semigroup with global attractor $\mathcal{A}_0$ and disjoint collection of isolated invariant sets $\Xi_0 = \{\Xi_{1,0}, \cdots, \Xi_{n,0}\}$.*

Then there exists an $\eta_1 > 0$ such that, for all $\eta \in (0, \eta_1]$, $\mathcal{T}_\eta$ is a dynamically gradient semigroup with respect to Ξ_η.

PROOF. Note firstly that (b) and (c) imply that there exists $\delta > 0$ such that, for suitably small η, if a solution ξ_η satisfies $\operatorname{dist}_H(\xi_\eta(t), \Xi_{i,0}) \leqslant \delta$ for all $t \geqslant t_0$ and for some $t_0 > 0$, then $\xi_\eta(t) \to \Xi_{i,\eta}$ as $t \to \infty$.

We argue by contradiction to prove that for all suitably small η, $\mathcal{T}_\eta$ satisfies (G1). Assume that there is a sequence $\eta_k \to 0$ and corresponding global solutions ξ_k of $\mathcal{T}_{\eta_k}$ in $\mathcal{A}_{\eta_k}$ such that

$$\limsup_{t\to\infty} \operatorname{dist}_H\left(\xi_k(t), \bigcup_{i=1}^{n} \Xi_{i,\eta_k}\right) > \delta. \tag{3.15}$$

From Lemma 2.4 with $\mathbf{s} = \{0\}$, there is a subsequence (which we again denote by $\{\xi_k\}$) and a global solution $\xi_0 \colon \mathbb{R} \to X$ of $\mathcal{T}_0$ such that $\xi_k(t) \to \xi_0(t)$ uniformly in compact subsets of $\mathbb{R}$. From (G1), $\xi_0(t) \to \Xi_{i,0}$, for some $1 \leqslant i \leqslant n$. It follows that, for each $\ell \in \mathbb{N}^*$, there are $t_\ell > 0$ and $k_\ell \in \mathbb{N}$ such that $d(\xi_k(t_\ell), \Xi_{i,0}) < \frac{1}{\ell}$, for each $k \geqslant k_\ell$. From (3.15), there exists $t'_\ell > t_\ell$ such that $d(\xi_{k_\ell}(t), \Xi_{i,0}) < \delta$ for all $t \in [t_\ell, t'_\ell)$ and $d(\xi_{k_\ell}(t_\ell), \Xi_{i,0}) = \delta$. From the continuity of $\mathcal{T}_0$ and the $\mathcal{T}_0$-invariance of $\Xi_{i,0}$ we have $t_{\ell'} - t_\ell \overset{\ell\to\infty}{\longrightarrow} \infty$. From Lemma 2.4 with $\mathbf{s} = \{t_\ell\}_{\ell=1}^{\infty}$, taking subsequences if necessary, there is a global solution $\xi_1 \colon \mathbb{R} \to X$ of $\mathcal{T}_0$ such that $\xi_1(t) = \lim_{\ell\to\infty} \xi_{k_\ell}(t + t_\ell)$ uniformly in compact subsets of $\mathbb{R}$. Then $\operatorname{dist}_H(\xi_1(t), \Xi_{i,0}) \leqslant \delta$ for all $t \leqslant 0$ and, consequently, $\xi_1(t) \to \Xi_{i,0}$ as $t \to -\infty$. From (G1) and (G2), $\xi_1(t) \to \Xi_{j,0} \in \boldsymbol{\Xi}_0$ as $t \to \infty$ with $i \neq j$. From the fact that $\xi_{k_\ell} \to \xi_1(t)$ uniformly in compact subsets of $\mathbb{R}$ we have, for each $m \in \mathbb{N}$, a time $t_m > 0$ and $k_m \in \mathbb{N}$ such that $\operatorname{dist}_H(\xi_k(t_m), \Xi_j) < \frac{1}{m}$ for all $k \geqslant k_m$. Again, from (3.15), there exists $t'_m > t_m$ such that $\operatorname{dist}_H(\xi_{k_m}(t), \Xi_j) < \delta$ for all $t \in [t_m, t'_m)$ and $\operatorname{dist}_H(\xi_{k_m}(t'_m), \Xi_j) = \delta$. Proceeding exactly as before we obtain a global solution $\xi_2 \colon \mathbb{R} \to X$ of $\mathcal{T}_0$ such that $\xi_2(t) \to \Xi_j$ as $t \to -\infty$ and $\xi_2(t) \to \Xi_{r,0} \in \boldsymbol{\Xi}_0$ as $t \to \infty$ with $r \notin \{i, j\}$. In a finite number of steps we construct a homoclinic structure and arrive at a contradiction with (G2). This proves that there is an $\eta_1 > 0$ such that, for all global solution ξ_η in $\mathcal{A}_\eta$ with $\eta \leqslant \eta_1$, we have

$$\lim_{t\to\infty} \operatorname{dist}_H(\xi_\eta(t), \Xi_{i,0}) = 0.$$

The proof that there is an $\eta_2 > 0$ such that for all global solutions ξ_η of $\mathcal{T}_\eta$ in $\mathcal{A}_\eta$ with $\eta \leqslant \eta_2$ we have

$$\lim_{t\to-\infty} \operatorname{dist}_H(\xi_\eta(t), \Xi_{j,0}) = 0$$

is similar. This completes the proof that, for all suitably small η, $\mathcal{T}_\eta$ satisfies (G1).

Let us prove that, for all suitably small η, $\mathcal{T}_\eta$ satisfies (G2). Again we argue by contradiction. Assume that there is a collection $\Xi_{1,0}, \cdots, \Xi_{p,0}$ in $\boldsymbol{\Xi}_0$, a sequence $\eta_k \to 0$, global solutions ξ_k of $\mathcal{T}_{\eta_k}$ in $\mathcal{A}_{\eta_k}$, and times $t_1^k, \cdots, t_p^k$ such that

$$\operatorname{dist}_H(\xi_k(0), \Xi_{i,0}) < \frac{1}{k}, \quad \operatorname{dist}_H(\xi_k(t_i^k), \Xi_{i+1,0}) < \frac{1}{k}, \quad 1 \leqslant i \leqslant p,$$

where $\Xi_{1,0} = \Xi_{p+1,0}$. One can choose a fixed sequence $\Xi_{1,0}, \cdots, \Xi_{p,0}$ since there are only a finite number of possible sequences from $\boldsymbol{\Xi}_0$. Proceeding as in the proof of (G1) we construct a homoclinic structure for $\mathcal{T}_0$ and arrive at a contradiction. $\square$

In the pictorial example of Figure 3.6 we may consider the stable node, the unstable focus and the closure of the homoclinic orbit (adding the saddle to it) as the three isolated invariant sets and we will have an attractor for a dynamically gradient semigroup relative to these isolated invariant sets.

A possible perturbation is described in Figure 3.7, with the isolated invariant sets being the stable node, the unstable focus and the closure of the intersection

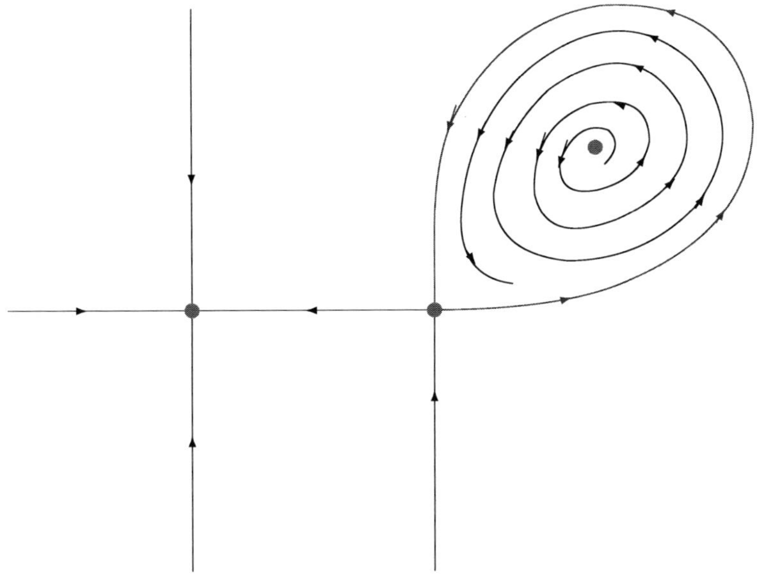

FIGURE 3.6. Unperturbed pictorial example.

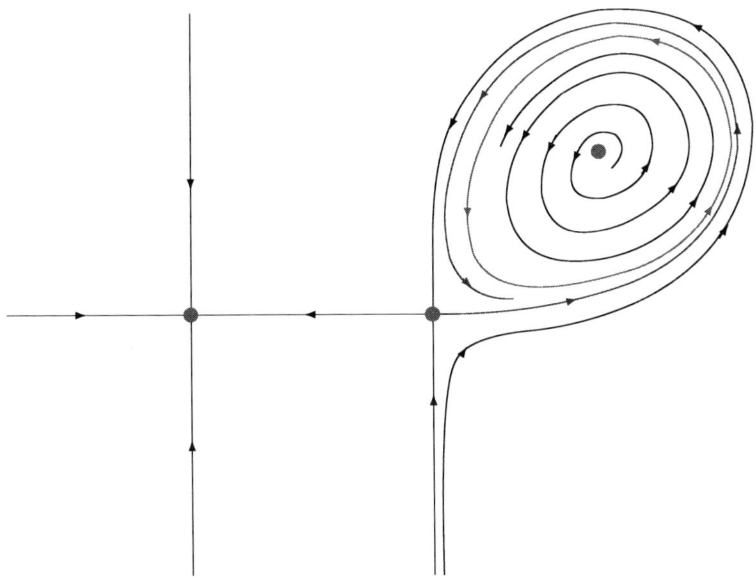

FIGURE 3.7. Perturbed pictorial example.

between the stable manifold of the periodic orbit and the unstable manifold of the saddle.

REMARK 3.43. As examples we may consider each one of the examples considered in Section 2, as well as several of those considered in Section 3, and consider a small C^1 (in the phase space) perturbation of the nonlinearity. The resulting problem (under some mild assumptions) will give rise to associated gradient semigroups.

It is an interesting exercise to consider the several examples and the appropriate conditions which ensure the applicability of Theorem 3.42, and we leave this task to the reader.

Our next result establishes that, for suitably small values of the parameter, if there are connections between two isolated invariant sets of the perturbed problem, then there will be an *indirect* connection between these isolated invariant sets of the unperturbed problem. This feature is essential for the structural stability of Morse-Smale Semigroups in Section 2.

LEMMA 3.44. *With the hypothesis of Theorem 3.42, if $\eta_k \to 0$ and $\xi_k : \mathbb{R} \to X$ is a global solution of $\mathcal{T}_{\eta_k}$ with*

$$\lim_{t\to-\infty} \mathrm{dist}_H(\xi_k(t), \Xi_{i,\eta_k}) = 0 \quad \text{and} \quad \lim_{t\to\infty} \mathrm{dist}_H(\xi_k(t), \Xi_{j,\eta_k}) = 0,$$

then there exist global solutions $\xi^\ell : \mathbb{R} \to X$ of $\mathcal{T}_0$, for $1 \leqslant \ell \leqslant m-1$, $m \leqslant n$, and isolated invariant sets $\{\Xi_{n_\ell}\}_{1\leqslant \ell\leqslant m}$ in $\mathbf{\Xi}_0$ such that $\Xi_{n_1} = \Xi_{i,0}$, $\Xi_{n_{m-1}} = \Xi_{j,0}$ and

$$\lim_{t\to-\infty} \mathrm{dist}_H(\xi^\ell(t), \Xi_{n_\ell}) = 0 \quad \text{and} \quad \lim_{t\to\infty} \mathrm{dist}_H(\xi^\ell(t), \Xi_{n_{\ell+1}}) = 0,$$

for $1 \leqslant \ell \leqslant m-1$.

PROOF. Fix $0 < \delta < \epsilon_0(\mathbf{\Xi_0})$. Since

$$\lim_{t\to-\infty} \mathrm{dist}_H(\xi_k(t), \Xi_{i,\eta_k}) = 0 \quad \text{and} \quad \lim_{\eta\to 0} \mathrm{dist}_H(\Xi_{i,\eta}, \Xi_{i,0}) = 0,$$

taking subsequences, if necessary, we may assume that $\xi_k(t) \to \xi^1(t)$ uniformly in compact subsets of $\mathbb{R}$ with $\xi^1(t) \overset{t\to-\infty}{\longrightarrow} \Xi_{n_1} := \Xi_{i,0}$. Also, (G1) implies that there exists an isolated invariant set $\Xi_{n_2} \in \mathbf{\Xi}_0$ (different from Ξ_{n_1}) such that $\xi^1(t) \overset{t\to\infty}{\longrightarrow} \Xi_{n_2}$. If $\Xi_{n_2} = \Xi_{j,0}$ the proof is done, otherwise there are $t'_\ell > t_\ell$ such that $\mathrm{dist}_H(\xi_{k_\ell}(t_\ell), \Xi_{n_2}) < \frac{\delta}{\ell}$, $\mathrm{dist}_H(\xi_{k_\ell}(t), \Xi_{n_2}) < \delta$ for all $t \in [t_\ell, t'_\ell)$ and $\mathrm{dist}_H(\xi_{k_\ell}(t'_\ell), \Xi_{n_2}) = \delta$. Taking subsequences, if necessary, let $\xi^2(t) = \lim_{\ell\to\infty} \xi_{k_\ell}(t+t_{\ell'})$. Then, since $t'_\ell - t_\ell \to \infty$ as $\ell \to \infty$, $\mathrm{dist}_H(\xi^2(t), \Xi_{n_2}) \leqslant \delta$ for all $t \leqslant 0$ and consequently $\mathrm{dist}_H(\xi^2(t), \Xi_{n_2}) \overset{t\to-\infty}{\longrightarrow} 0$. Furthermore, $\mathrm{dist}_H(\xi^2(t), \Xi_{n_3}) \overset{t\to\infty}{\longrightarrow} 0$ for some $\Xi_{n_3} \in \mathbf{\Xi_0}$ (different from Ξ_{n_1} and Ξ_{n_2}). This reasoning can be done as long as Ξ_{n_ℓ} is different from $\Xi_{j,0}$, and since there are only finitely many isolated invariant sets, the proof is completed in finitely many steps. □

5. Examples

EXAMPLE 3.45. If $f \in C^2(\mathbb{R})$, $\limsup_{|u|\to\infty} \frac{f(u)}{u} < 0$, $\Omega \subset \mathbb{R}^n$ is a bounded smooth domain, $\epsilon \geqslant 0$ and $a \in C(\bar{\Omega}, \mathbb{R}^n)$, consider the initial value problem

$$\begin{cases} u_{tt} + \beta u_t = \Delta u + \epsilon a(x) \cdot \nabla u + f(u), & x \in \Omega,\ t > 0, \\ u(x,t) = 0, & x \in \partial\Omega,\ t > 0, \\ u(\cdot,0) = u_0 \in H^1_0(\Omega), \\ u_t(\cdot,0) = v_0 \in L^2(\Omega), \end{cases}$$

with Dirichlet boundary condition. This equation generates a semigroup $\mathcal{T}_\epsilon$ in $H^1_0(\Omega) \times L^2(\Omega)$, which is dynamically gradient if there exists a finite number of equilibria of $\Delta u = -f(u)$, with all of them hyperbolic, and $\epsilon \geqslant 0$ is suitably small (see [**74**]). Hence, for ϵ sufficiently small, the semigroups $\mathcal{T}_\epsilon$ are gradient as an application of Theorem 3.42.

EXAMPLE 3.46. For each $1 \leqslant m \in \mathbb{N}$, the *time one map* for the problem

$$\begin{cases} \dot{r} = \begin{cases} \pi^{-1}(1 - \frac{1}{2m+1} - r)^3 \sin \frac{\pi}{1-r}, & r < 1 - \frac{1}{2m+1}, \\ -(1 - \frac{1}{2m+1} - r)^2, & r \geqslant 1 - \frac{1}{2m+1}, \end{cases} \\ \dot{\theta} = 2\pi \end{cases} \tag{3.16}$$

has the attractor $\mathcal{A}_m = \{|r| \leqslant 1 - \frac{1}{2m+1}\}$, which is the union of unstable manifolds of Ξ_j, $1 \leqslant j \leqslant 2m+1$, where Ξ_j is the 1-periodic solution corresponding to $r = 1 - \frac{1}{j}$, $1 \leqslant j \leqslant 2m+1$. These periodic solutions are normally hyperbolic (if k is even, the orbit is unstable, if k is odd, the orbit is stable). In this case, it is easy to see that the attractor $\mathcal{A}_m$ is the union of the unstable manifolds of the periodic solutions $\{\Xi_j\}_{1 \leqslant j \leqslant 2m+1}$. Theorem 3.42 implies that any small perturbation of the vector field in (3.16) will lead us to a dynamically gradient semigroup relative to a disjoint collection of isolated invariant sets and the attractor is characterized.

Notes. The *Fundamental Theorem of Dynamical Systems* (as named in [**174**]) states that every dynamical system on a compact metric space (the one defined on a global attractor, for instance) has a geometrical structure described by a (finite or countable) number of sets with an intrinsic recurrent dynamics and a gradient-like dynamics outside them, and has been considered in different frameworks, as in the case of flows ([**105**]) and semiflows on compact spaces ([**192**]), or even compact and noncompact topological spaces ([**146, 181, 182**]). In other words, the attractor can be always described by a (finite or countable) number of invariants and connections between them. This intrinsic dynamics is what is described by dynamically gradient semigroups. Note that, given a system of differential or partial differential equations, the determination of a Lyapunov function is usually a difficult task. Thus, the characterization of gradient systems as dynamically gradient systems for infinite-dimensional spaces enlarges the class of systems where a detailed structure of the attractors is available (see, for instance, [**112**] for this equivalence in the case of multivalued semigroups, or [**48**] for the case of stochastic differential equations). The determination of a Lyapunov function describing all the isolated invariant sets within a global attractor allows a formal description of this gradient attractor from its energy levels ([**4**]), which joins invariants, which are not connected among themselves, with a same value given by the Lyapunov functional. Sometimes, a system can be proved to be gradient by a direct description of the Morse sets and their gradient-like dynamics ([**129**]). The robustness under autonomous ([**5**]) and non-autonomous ([**6**]) perturbation of dynamically gradient semigroups allows us to conclude the robustness of Lyapunov functionals under perturbation.

CHAPTER 4

Neighborhood of a Critical Element

In this chapter we will discuss in detail what happens in the neighborhood of *critical elements.* For us, in what follows, a **critical element** is either an *equilibrium point* or a *periodic orbit.* We will be mostly interested in hyperbolic equilibrium points and on normally hyperbolic periodic orbits, concepts which will be introduced later in this chapter. In most of what follows, we have adapted the results of [**141**]. For additional definitions and results on Banach manifolds, we refer to [**173**].

From now on, unless clearly stated otherwise, X will be a Banach space with norm $\|\cdot\|$.

DEFINITION 4.1. A semigroup $\mathcal{T}$ in X is said to be **continuously differentiable** (or a **C^1-semigroup**) if, for each $t \geqslant 0$, $T(t)\colon X \to X$ is differentiable in X and the map $\mathbb{R}^+ \times X \ni (t,\xi) \mapsto D_xT(t)(\xi) \in \mathcal{L}(X)$ is continuous, where $D_xT(t)(\xi)$ denotes the Frechét derivative of $T(t)$ at ξ.

REMARK 4.2. At this point we call the readers attention to the fact that the continuity of $\mathbb{R}^+ \times X \ni (t,\xi) \mapsto D_xT(t)(\xi) \in \mathcal{L}(X)$ imposes restrictions on the type of semigroup we are considering. For example, if a part of $T(t)$ is only a strongly continuous semigroup, this condition will not be satisfied. When applying the results of this section the reader should be careful.

DEFINITION 4.3. If $L\colon X \to X$ is a bounded linear operator, then the **resolvent set** $\rho(L)$ of L is defined as the set of $\lambda \in \mathbb{C}$ such that $\lambda I - L$ is invertible with bounded linear inverse, where $I\colon X \to X$ denotes the identity operator. The **spectrum** of L is the set $\sigma(L) = \mathbb{C} \setminus \rho(L)$.

The following sections are dedicated to showing that orbits of critical elements are isolated invariant sets and that these isolated invariant sets behave continuously with respect to parameters. We also study the stable and unstable sets of critical elements and their dependence with respect to parameters. As a consequence of the results in this chapter we have, for the case when the isolated invariant sets are critical elements, conditions under which hypotheses (b) and (c) of Theorem 2.14 are satisfied as well as hypotheses (b) and (c) of Theorem 3.42.

1. Invariant manifolds near an equilibrium point

In this section we provide a detailed analysis of the neighborhood of an equilibrium point. In what follows, an invariant manifold is a graph of a Lipschitz map, over a given subspace of a Banach space X, which is $\mathcal{T}$-invariant. We will discuss this in much more detail in the following section. Before stating our next theorem, recall that given a bounded linear operator $V\colon X \to X$, the **spectral radius** of V

is given by

$$r_\sigma(V) = \max_{\lambda\in\sigma(V)} |\lambda|, \tag{4.1}$$

and (see for instance [**113**]) we have

$$r_\sigma(V) = \lim_{n\to\infty} \|V^n\|^{1/n} = \inf_{n\geqslant 1} \|V^n\|^{1/n}.$$

If $V\colon X \to X$ is a bounded linear operator and $r > r_\sigma(V)$, choose $m \geqslant 1$ such that $\|V^j\|^{1/j} \leqslant r$ if $j \geqslant m$ and define a new norm $\|\cdot\|_*$ in X by

$$\|x\|_* = \sup_{0\leqslant i\leqslant m} \frac{\|V^i x\|}{r^i}.$$

Clearly the norm $\|\cdot\|_*$ is equivalent to $\|\cdot\|$ and we have $\|V\|_* \leqslant r$, since

$$\|Vx\|_* = \sup_{0\leqslant i\leqslant m} \frac{\|V^{i+1}x\|}{r^i} = r \sup_{0\leqslant i\leqslant m} \frac{\|V^{i+1}x\|}{r^{i+1}} \leqslant r \sup_{0\leqslant i\leqslant m} \frac{\|V^i x\|}{r^i} = r\|x\|_*.$$

For (X, d_X), (Y, d_Y) metric spaces and a Lipschitz map $N\colon X \to Y$ we define

$$\operatorname{Lip} N = \inf\{C \geqslant 0\colon d_Y(N(x), N(y)) \leqslant C d_X(x, y) \text{ for all } x, y \in X\}.$$

THEOREM 4.4 (Basic properties of invariant manifolds). *Let X be a Banach space, let $L\colon X \to X$ be a bounded linear operator and let $\rho > 0$ such that $\sigma(L) \cap \{\xi \in \mathbb{C}\colon\ |\xi| = \rho\} = \varnothing$. If $\Gamma(t) = \rho e^{it}$, $t \in [0, 2\pi]$, we define the spectral projections*

$$\Pi_2 = \frac{1}{2\pi i}\int_\Gamma (\lambda - L)^{-1} d\lambda, \quad \text{and} \quad \Pi_1 = I - \Pi_s,$$

and the corresponding decomposition of X and L as $X = X_1 \oplus X_2$ and $L = L_1 \oplus L_2$, where $X_j = \Pi_j X$, $L_j = L|_{X_j}$, $j = 1, 2$, and $\sigma(L_2) = \sigma(L) \cap \{\lambda \in \mathbb{C}\colon |\lambda| < \rho\}$, $\sigma(L_1) = \sigma(L) \cap \{\lambda \in \mathbb{C}\colon |\lambda| > \rho\}$.

Then there are $\gamma > 0$ and equivalent norms in X such that if $N\colon X \to X$ is globally Lipschitz continuous with Lip $N \leqslant \gamma$, $N(0) = 0$ *there exist Lipschitz manifolds $S_{(\rho)}$, $U_{(\rho)}$ for $T = L + N$ in X such that:*

(i) *$U_{(\rho)}$ is a graph over X_1, $S_{(\rho)}$ is a graph over X_2;*
(ii) *$TU_{(\rho)} = U_{(\rho)}$, $TS_{(\rho)} = S_{(\rho)} \cap TX$ and the restriction $T|_{U_{(\rho)}}$ is a homeomorphism;*
(iii) *$S_{(\rho)} \cap U_{(\rho)} = \{0\}$;*
(iv) Lip $\left(T|_{S_{(\rho)}}\right) < \rho$ *and* Lip $\left(T|_{U_{(\rho)}}^{-1}\right) < \rho^{-1}$;
(v)

$$S_{(\rho)} = \{x \in X\colon \|T^n x\| = o(\rho^n), \ \text{as } n \to \infty\},$$

$$U_{(\rho)} = \{x \in X\colon \ \text{there is a sequence } \{x_j\}_{j\leqslant 0} \text{ with } x_0 = x, \quad x_{j+1} = T(x_j),\ j < 0,\ \text{and } \|x_j\| = o(\rho^j),\ \text{as } j \to -\infty\}.$$

For ρ^* close to ρ, $S_{(\rho^*)} = S_{(\rho)}$ and $U_{(\rho^*)} = U_{(\rho)}$;
(vi) *There is a continuous map $P\colon X \to U_{(\rho)}$ with*

$$\|Px - x\| \leqslant C \operatorname{dist}_H(x, U_{(\rho)}),$$

for some $C > 0$, $T\circ P = P\circ T$, where $Px = z \in U_{(\rho)}$ if and only if $\|T^n x - T^n z\| = o(\rho^n)$ as $n \to \infty$. Also, $\operatorname{dist}_H(T^n x, U_{(\rho)}) \leqslant C\nu^n \operatorname{dist}_H(x, U_{(\rho)})$ for all $n \geqslant 0$, $x \in X$ and for some constants $\nu < \rho$ and $C > 0$.

If L is an isomorphism, for each γ small, T is a homeomorphism and $TS_{(\rho)} = S_{(\rho)}$.

PROOF. Fix $a, b \in \mathbb{R}$ satisfying $0 < b < \rho < a$, $\sigma(L_1) \subset \{\xi \in \mathbb{C} \colon |\xi| > a\}$ and $\sigma(L_2) \subset \{\xi \in \mathbb{C} \colon |\xi| < b\}$. Choose equivalent norms, as described in the beginning of this section, in X_1 and X_2, respectively, such that $\|L_1^{-1}\| \leqslant a^{-1}$ and $\|L_2\| \leqslant b$. For $X = X_1 \oplus X_2$ consider the norm given by $\|x_1 + x_2\| := \max\{\|x_1\|, \|x_2\|\}$. Note that, with this norm in X, $\|\Pi_j\|_{\mathcal{L}(X)} = 1$, $j = 1, 2$.

Choose γ such that $b + 2\gamma < \rho < a - 2\gamma$. Then T can be written as

$$\begin{aligned} T \colon X_1 \oplus X_2 &\to X_1 \oplus X_2 \\ (x_1, x_2) &\mapsto (L_1 x_1 + N_1(x_1 + x_2), L_2 x_2 + N_2(x_1 + x_2)), \end{aligned}$$

with $\operatorname{Lip} N_1 \leqslant \gamma$ and $\operatorname{Lip} N_2 \leqslant \gamma$, where $N_j = \Pi_j \circ N$, $j = 1, 2$.

Our aim is to find $U_{(\rho)}$ of the form $\{\xi + \theta(\xi) \colon \xi \in X_1\}$, for some $\theta \colon X_1 \to X_2$ with $\theta(0) = 0$ and $\operatorname{Lip} \theta \leqslant 1$ and $TU_{(\rho)} \subset U_{(\rho)}$.

The condition $TU_{(\rho)} \subset U_{(\rho)}$ implies that, for all $\xi \in X_1$, there exists $\hat{\xi} \in X_1$ with $T(\xi + \theta(\xi)) = \hat{\xi} + \theta(\hat{\xi})$, that is,

$$\begin{cases} \hat{\xi} = L_1 \xi + N_1(\xi + \theta(\xi)), \\ \theta_*(\hat{\xi}) = L_2 \theta(\xi) + N_2(\xi + \theta(\xi)), \end{cases} \tag{4.2}$$

with $\theta_* = \theta$, where θ is the fixed point of the map $\theta \mapsto \theta_*$ defined by (4.2).

Let us show that there exists such a fixed point θ and that $U_{(\rho)} = \{\xi + \theta(\xi) \colon \xi \in X_1\}$ satisfies the conditions of the theorem. Let $\theta \colon X_1 \to X_2$ be a map satisfying $\theta(0) = 0$ and $\operatorname{Lip} \theta \leqslant 1$. Then the map $\xi \mapsto -L_1^{-1} N_1(\xi + \theta(\xi))$ is a contraction. In fact,

$$\begin{aligned} \| - L_1^{-1} N_1(\xi + \theta(\xi)) + L_1^{-1} N_1(\check{\xi} + \theta(\check{\xi}))\| &\leqslant a^{-1}\gamma \left[\|\check{\xi} - \xi\| + \|\theta(\check{\xi}) - \theta(\xi)\| \right] \\ &\leqslant 2a^{-1}\gamma \|\check{\xi} - \xi\|, \end{aligned}$$

and $2a^{-1}\gamma < 1$. Hence, for each $\hat{\xi} \in X_1$, there exists a unique $\xi \in X_1$ that satisfies the first equation in (4.2) since, for each $\hat{\xi} \in X_1$, the map

$$\xi \mapsto L_1^{-1}\hat{\xi} - L_1^{-1} N_1(\xi + \theta(\xi)) \tag{4.3}$$

is a contraction and has a unique fixed point $\xi \in X_1$. It follows that (4.2) defines a map $\theta_* \colon X_1 \to X_2$, and clearly $\theta_*(0) = 0$.

Now, given $\hat{\xi}, \hat{\mu} \in X_1$, with $\xi, \mu \in X_1$ the corresponding fixed points of the map (4.3), we have

$$\begin{aligned} \|\hat{\xi} - \hat{\mu}\| &= \|L_1 \xi_1 + N_1(\xi_1 + \theta(\xi)) - L_1 \mu - N_1(\mu + \theta(\mu))\| \\ &\geqslant \|L_1(\xi - \mu)\| - \gamma\Big(\|\xi - \mu\| + \|\theta(\xi) - \theta(\mu)\|\Big) \\ &\geqslant a\|\xi - \mu\| - 2\gamma\|\xi - \mu\| = (a - 2\gamma)\|\xi - \mu\|. \end{aligned}$$

We also have that

$$\begin{aligned} \|\theta_*(\hat{\xi}) - \theta_*(\hat{\mu})\| &= \|L_2 \theta(\xi) + N_2(\xi + \theta(\xi)) - L_2 \theta(\mu) - N_2(\mu + \theta(\mu))\| \\ &\leqslant b\|\xi - \mu\| + \gamma\Big(\|\xi - \mu\| + \|\theta(\xi) - \theta(\mu)\|\Big) \\ &\leqslant (b + 2\gamma)\|\xi_1 - \xi_2\|, \end{aligned}$$

and consequently $\|\theta_*(\hat{\xi}) - \theta_*(\hat{\mu})\| \leqslant \dfrac{b+2\gamma}{a-2\gamma}\|\hat{\xi} - \hat{\mu}\|$, that is, Lip $\theta_* \leqslant \dfrac{b+2\gamma}{a-2\gamma} < 1$.

The set $\mathcal{F}$ of maps $\theta\colon X_1 \to X_2$ with $\theta(0) = 0$, Lip $\theta \leqslant 1$ with the metric

$$d(\theta, \tau) = \sup_{\xi \in X_1,\ \xi \neq 0} \frac{\|\theta(\xi) - \tau(\xi)\|}{\|\xi\|},$$

is a complete metric space, and the map $\theta \mapsto \theta_*$ takes $\mathcal{F}$ into itself. Let us show that $\mathcal{F} \ni \theta \mapsto \theta_* \in \mathcal{F}$ is a contraction. To that end take $\theta, \tau \in \mathcal{F}$, $\hat{\xi} \in X_1$. There exist unique $\xi_\theta, \xi_\tau \in X_1$ such that

$$\hat{\xi} = L_1\xi_\theta + N_1(\xi_\theta + \theta(\xi_\theta)) = L_1\xi_\tau + N_1(\xi_\tau + \tau(\xi_\tau)).$$

Then $\|\hat{\xi}\| \geqslant (a - 2\gamma)\|\xi_\theta\|$ and

$$\begin{aligned} a\|\xi_\theta - \xi_\tau\| \leqslant \|L_1(\xi_\theta - \xi_\tau)\| &= \|N_1(\xi_\theta + \theta(\xi_\theta)) - N_1(\xi_\tau + \tau(\xi_\tau))\| \\ &\leqslant \gamma\Big(\|\xi_\theta - \xi_\tau\| + \|\theta(\xi_\theta) - \tau(\xi_\tau)\|\Big) \\ &\leqslant \gamma\Big(\|\xi_\theta - \xi_\tau\| + \|\theta(\xi_\theta) - \tau(\xi_\theta)\| + \|\tau(\xi_\theta) - \tau(\xi_\tau)\|\Big) \\ &\leqslant \gamma\Big(2\|\xi_\theta - \xi_\tau\| + d(\theta,\tau)\|\xi_\theta\|\Big), \end{aligned}$$

and so

$$\|\xi_\theta - \xi_\tau\| \leqslant \frac{\gamma d(\theta,\tau)}{a - 2\gamma}\|\xi_\theta\| \leqslant \frac{\gamma d(\theta,\tau)}{(a-2\gamma)^2}\|\hat{\xi}\|.$$

We also have

$$\begin{aligned} \|\theta_*(\hat{\xi}) - \tau_*(\hat{\xi})\| &= \|L_2\theta(\xi_\theta) + N_2(\xi_\theta + \theta(\xi_\theta)) - L_2\tau(\xi_\tau) - N_2(\xi_\tau + \tau(\xi_\tau))\| \\ &\leqslant b\|\theta(\xi_\theta) - \tau(\xi_\tau)\| + \gamma\Big(\|\xi_\theta - \xi_\tau\| + \|\theta(\xi_\theta) - \tau(\xi_\tau)\|\Big) \\ &\leqslant (b+\gamma)\|\theta(\xi_\theta) - \tau(\xi_\tau)\| + \gamma\|\xi_\theta - \xi_\tau\| \\ &\leqslant (b+\gamma)\Big(\|\theta(\xi_\theta) - \tau(\xi_\theta)\| + \|\tau(\xi_\theta) - \tau(\xi_\tau)\|\Big) + \gamma\|\xi_\theta - \xi_\tau\| \\ &\leqslant (b+\gamma)d(\theta,\tau)\|\xi_\theta\| + (b+2\gamma)\|\xi_\theta - \xi_\tau\| \\ &\leqslant d(\theta,\tau)\|\hat{\xi}\|\left(\frac{b+\gamma}{a-2\gamma} + \frac{\gamma}{a-2\gamma}\cdot\frac{b+2\gamma}{a-2\gamma}\right) \\ &\leqslant d(\theta,\tau)\|\hat{\xi}\|\frac{b+2\gamma}{a-2\gamma}, \end{aligned}$$

that is,

$$d(\theta_*, \tau_*) \leqslant \frac{b+2\gamma}{a-2\gamma}d(\theta,\tau), \text{ with } \frac{b+2\gamma}{a-2\gamma} < 1,$$

and hence $\mathcal{F} \ni \theta \mapsto \theta_* \in \mathcal{F}$ is a contraction and possesses a unique fixed point which we call θ.

Now we define $U_{(\rho)} := \{\xi + \theta(\xi) \colon \xi \in X_1\}$ and we will prove that $U_{(\rho)}$ satisfies all the claims of this theorem. Of course, (i) is automatically proven for $U_{(\rho)}$. Let us prove that $TU_{(\rho)} = U_{(\rho)}$ and to that, given $x = \xi + \theta(\xi) \in U_{(\rho)}$, we write $Tx = \hat{\xi} + \hat{\eta}$ with $\hat{\xi} \in X_1$ and $\hat{\eta} \in X_2$. It follows that $\hat{\xi} = L_1\xi + N_1(\xi + \theta(\xi))$ and $\hat{\eta} = L_2\theta(\xi) + N_2(\xi + \theta(\xi)) = \theta(\hat{\xi})$ (where the last equality is given by the definition of θ). From this we have $Tx \in U_\rho$ and that $TU_{(\rho)} \subset U_{(\rho)}$.

On the other hand, given $\hat{\xi} + \theta(\hat{\xi}) \in U_\rho$, choose ξ such that $\hat{\xi} = L_1\xi + N_1(\xi + \theta(\xi))$. Then

$$T(\xi + \theta(\xi)) = \hat{\xi} + \theta(\hat{\xi})$$

and $U_{(\rho)} \subset TU_{(\rho)}$. It follows that $TU_{(\rho)} = U_{(\rho)}$ and the claim is proved.

Now, if $x = \xi + \theta(\xi)$ and $z = \zeta + \theta(\zeta)$ are in $U_{(\rho)}$ and $Tx = \hat{\xi} + \theta(\hat{\xi})$ and $Tz = \hat{\zeta} + \theta(\hat{\zeta})$, then $\|x - z\| = \max\{\|\xi - \zeta\|, \|\theta(\xi) - \theta(\zeta)\|\} = \|\xi - \zeta\|$, since $\|\theta(\xi) - \theta(\zeta)\| \leqslant \|\xi - \zeta\|$. Also

$$\|Tx - Tz\| = \|\hat{\xi} - \hat{\zeta}\| \geqslant (a - 2\gamma)\|\xi - \zeta\| = (a - 2\gamma)\|x - z\|,$$

and $a - 2\gamma > \rho$, which proves that

$$\text{Lip}\ (T|_{U_{(\rho)}}^{-1}) \leqslant (a - 2\gamma)^{-1} < \rho^{-1}.$$

Given $x \in U_{(\rho)}$, there exists x_j, $j = -1, -2, \cdots$, in $U_{(\rho)}$ such that $T^{|j|}x_j = x$ and $T(x_j) = x_{j+1}$, for $j < 0$ with $x_0 = x$ (since $TU_{(\rho)} = U_{(\rho)}$). Also

$$\|x_{j+1}\| = \|T(x_j) - T(0)\| \geqslant (a - 2\gamma)\|x_j\|.$$

Hence $\|x_j\| \leqslant \|x\|(a - 2\gamma)^j = o(\rho^j)$ as $j \to -\infty$ which proves that

$$U_{(\rho)} \subset \{x \in X: \text{ there exists } \{x_j\}_{j \leqslant 0} \text{ with } x_0 = x,\ x_{j+1} = T(x_j),\ j < 0 \\ \text{and } \|x_j\| = o(\rho^j) \text{ as } j \to -\infty\}.$$

Now, let $x \in X$, $\{x_j\}_{j \leqslant 0}$, $x_0 = x$, such that $T(x_j) = x_{j+1}$ for $j < 0$ and $\|x_j\| = o(\rho^j)$ as $j \to -\infty$. Let us write

$$x_j = \xi_j + \theta(\xi_j) + \eta_j, \text{ where } \xi_j \in X_1, \eta_j \in X_2,$$

and we have $T(x_j) = x_{j+1}$, hence

$$L_1\xi_j + N_1(\xi_j + \theta(\xi_j) + \eta_j) + L_2[\theta(\xi_j) + \eta_j] + N_2(\xi_j + \theta(\xi_j) + \eta_j) \\ = \xi_{j+1} + \theta(\xi_{j+1}) + \eta_{j+1},$$

that is,

$$\begin{cases} \xi_{j+1} = L_1\xi_j + N_1(\xi_j + \theta(\xi_j) + \eta_j), \\ \theta(\xi_{j+1}) + \eta_{j+1} = L_2(\theta(\xi_j) + \eta_j) + N_2(\xi_j + \theta(\xi_j) + \eta_j), \end{cases}$$

and, if $\hat{\xi}_j := L_1\xi_j + N_1(\xi_j + \theta(\xi_j))$, then $\theta(\hat{\xi}_j) = L_2\theta(\xi_j) + N_2(\xi_j + \theta(\xi_j))$.

Also $\|\hat{\xi}_j - \xi_{j+1}\| \leqslant \gamma\|\eta_j\|$ and

$$\|\eta_{j+1}\| - \|\xi_{j+1} - \hat{\xi}_j\| \leqslant \|\theta(\xi_{j+1}) + \eta_{j+1} - \theta(\hat{\xi}_j)\| \\ \leqslant b\|\eta_j\| + \gamma\|\eta_j\| = (b + \gamma)\|\eta_j\|,$$

that is, $\|\eta_{j+1}\| \leqslant (b + 2\gamma)\|\eta_j\|$. Consequently $\|\eta_0\|(b + 2\gamma)^j \leqslant \|\eta_j\|$ for $j \leqslant 0$.

Now

$$\|\xi_j\| \leqslant \|x_j\| = \max\{\|\xi_j\|, \|\theta(\xi_j) + \eta_j\|\} = o(\rho^j),$$

and $\|\xi_j\| = o(\rho^j)$, so

$$\|\eta_j\| = \|x_j - \xi_j - \theta(\xi_j)\| \leqslant \|x_j\| + 2\|\xi_j\| = o(\rho^j), \text{ as } j \to -\infty.$$

But $\|\eta_0\|(b + 2\gamma)^j \leqslant \|\eta_j\|$ for $j \leqslant 0$ and so $\|\eta_0\| \leqslant \dfrac{\|\eta_j\|}{(b + 2\gamma)^j} \leqslant \dfrac{\|\eta_j\|}{\rho^j} \to 0$ as $j \to -\infty$. Hence $\eta_0 = 0$ and $x = x_0 = \xi_0 + \theta(\xi_0) \in U_{(\rho)}$.

Thus

$$U_{(\rho)} = \{x \in X\colon \text{ there exists } \{x_j\}_{j\leqslant 0} \text{ with } x_0 = x,\ x_{j+1} = T(x_j)\ j < 0 \\ \text{and } \|x_j\| = o(\rho^j) \text{ as } j \to -\infty\}.$$

Let us define $\hat{P}\colon X \to U_{(\rho)}$ by $\hat{P}x = x_1 + \theta(x_1)$, for $x = x_1 + x_2$, and for $n \geqslant 0$ we set

$$P_n = \left(T|_{U_{(\rho)}}\right)^{-n} \circ \hat{P} \circ T^n.$$

We have $\operatorname{dist}_H(x, U_{(\rho)}) \leqslant \|x_2 - \theta(x_1)\| = \|x - \hat{P}x\| \leqslant 2\operatorname{dist}_H(x, U_{(\rho)})$. If $T^n x = x_n = \xi_n + \theta(\xi_n) + \eta_n$, $\xi_n \in X_1$, $\eta_n \in X_2$, using that $T(x_n) = x_{n+1}$, the above computations imply that $\|\eta_{n+1}\| \leqslant (b + 2\gamma)\|\eta_n\|$ and $\|\eta_n\| \leqslant (b + 2\gamma)^n\|\eta_0\|$. But $\|\eta_0\| = \|x - \xi_0 - \theta(\xi_0)\| = \|x - \hat{P}x\|$, so

$$\|\eta_n\| \leqslant (b + 2\gamma)^n\|x - \hat{P}x\|.$$

We also have

$$\begin{cases} \xi_{n+1} + \theta(\xi_{n+1}) + \eta_{n+1} = T(\xi_n + \theta(\xi_n) + \eta_n), \\ \xi_{n+1} = L_1\xi_n + N_1(\xi_n + \theta(\xi_n) + \eta_n), \\ T^{n+1}P_{n+1}x = T^{n+1}\left(T|_{U_{(\rho)}}\right)^{-(n+1)} \hat{P}T^{n+1}x = \hat{P}T^{n+1}x = \xi_{n+1} + \theta(\xi_{n+1}), \\ T^{n+1}P_n x = T^{n+1}\left(T|_{U_{(\rho)}}\right)^{-n} \hat{P}T^n x = T\hat{P}T^n x = T(\xi_n + \theta(\xi_n)) = \hat{\xi}_n + \theta(\hat{\xi}_n), \end{cases}$$

where $\hat{\xi}_n = L_1\xi_n + N_1(\xi_n + \theta(\xi_n))$ and $\theta(\hat{\xi}_n) = L_2\theta(\xi_n) + N_2(\xi_n + \theta(\xi_n))$. Also

$$\|\xi_{n+1} - \hat{\xi}_n\| = \|L_1\xi_n + N_1(\xi_n + \theta(\xi_n) + \eta_n) - L_1\xi_n - N_1(\xi_n + \theta(\xi_n))\| \leqslant \gamma\|\eta_n\|,$$

and since $\operatorname{Lip}\,(T|_{U_{(\rho)}}^{-1}) \leqslant (a - 2\gamma)^{-1}$,

$$\begin{aligned} \|P_{n+1}x - P_n x\| &\leqslant (a - 2\gamma)^{-n-1}\|T^{n+1}P_{n+1}x - T^{n+1}P_n x\| \\ &\leqslant 2(a - 2\gamma)^{-n-1}\|\xi_{n+1} - \hat{\xi}_n\| \leqslant \frac{2\gamma}{(a - 2\gamma)^{n+1}}\|\eta_n\| \\ &\leqslant \frac{2\gamma}{a - 2\gamma}\left(\frac{b + 2\gamma}{a - 2\gamma}\right)^n \|x - \hat{P}x\| \quad \text{ for } n \geqslant 0. \end{aligned}$$

From this and from the fact that $\hat{P}$ is bounded in bounded subsets of X we can prove that $\{P_n\}_{n\geqslant 0}$ is a Cauchy sequence uniformly for x in bounded subsets of X. In fact, for $n \geqslant m$,

$$\begin{aligned} \|P_m x - P_n x\| &\leqslant \frac{2\gamma}{a - 2\gamma}\|x - \hat{P}x\| \sum_{k=n}^{m} \left(\frac{b + 2\gamma}{a - 2\gamma}\right)^k \\ &\leqslant \frac{2\gamma}{a - 2\gamma}\|x - \hat{P}x\| \left(\frac{b + 2\gamma}{a - 2\gamma}\right)^n \left(1 - \frac{b + 2\gamma}{a - 2\gamma}\right)^{-1}. \end{aligned}$$

Define $P\colon X \to U_{(\rho)}$ by $Px := \lim_{m\to\infty} P_m x$ for each $x \in X$. Clearly P is continuous. Making $m \to \infty$ and $n = 0$ in the above inequality and noting that $P_0 = \hat{P}$ we have

$$\|Px - \hat{P}x\| \leqslant K\|x - \hat{P}x\| \quad \text{ and } \quad \|x - Px\| \leqslant (K + 1)\|x - \hat{P}x\|,$$

where $K = \frac{2\gamma}{a-b} > 0$. Since

$$TP_{n+1}x = T\left(T|_{U_{(\rho)}}\right)^{-(n+1)} \hat{P}T^{n+1}x = \left(T|_{U_{(\rho)}}\right)^{-n} \hat{P}T^n(Tx) = P_n Tx,$$

it follows that $TPx = PTx$, for all $x \in X$. Therefore

$$\begin{aligned}\mathrm{dist}_H(T^n x, U_{(\rho)}) &\leqslant \|T^n x - T^n P x\| = \|T^n x - PT^n x\| \leqslant (K+1)\|T^n x - \hat{P}T^n x\| \\ &= (K+1)\|\eta_n\| \leqslant (K+1)(b+2\gamma)^n \|x - \hat{P}x\| \\ &= 2(K+1)(b+2\gamma)^n \mathrm{dist}_H(x, U_{(\rho)}),\end{aligned}$$

and since $b + 2\gamma < \rho$, this shows (vi) and completes the proof for $U_{(\rho)}$.

For $S_{(\rho)} = \{\sigma(\eta) + \eta \colon \eta \in X_2\}$, $\sigma \colon X_2 \to X_1$, the condition $TS_{(\rho)} \subset S_{(\rho)}$ tells us that, for each $\eta \in X_2$ there exists $\hat{\eta} \in X_2$ with

$$\begin{cases} \sigma(\hat{\eta}) = L_1 \sigma_*(\eta) + N_1(\sigma(\eta) + \eta), \\ \hat{\eta} = L_2 \eta + N_2(\sigma(\eta) + \eta) \end{cases} \tag{4.4}$$

with $\sigma_* = \sigma$ and the rest of the proof follows analogously.

Finally, if $x = \xi + \eta \in S_{(\rho)} \cap U_{(\rho)}$, we have $\xi = \sigma(\eta)$, $\eta = \theta(\xi)$ and $\xi = \sigma(\theta(\xi))$. Since Lip $(\sigma \circ \theta) < 1$ we have $\xi = 0$, consequently, $\eta = 0$ and $x = 0$. □

REMARK 4.5. We have Lip $\theta \leqslant \frac{\gamma}{a-b-3\gamma} < 1$. In fact, if

$$\Delta = \sup_{\xi \neq \eta} \frac{\|\theta(\xi) - \theta(\eta)\|}{\|\xi - \eta\|}$$

and $\xi, \eta \in X_1$ we have $\hat{\xi} = L_1\xi + N_1(\xi + \theta(\xi))$, $\hat{\eta} = L_1\eta + N_1(\eta + \theta(\eta))$, $\theta(\hat{\xi}) = L_2\theta(\xi) + N_2(\xi + \theta(\xi))$ and $\theta(\hat{\eta}) = L_2\theta(\eta) + N_2(\eta + \theta(\eta))$, also $\|\hat{\xi} - \hat{\eta}\| \geqslant (a - \gamma(1+\Delta))\|\xi - \eta\|$ and

$$\|\theta(\hat{\xi}) - \theta(\hat{\eta})\| \leqslant [b\Delta + \gamma(1+\Delta)]\|\xi - \eta\| \leqslant \frac{b\Delta + \gamma(1+\Delta)}{a - \gamma(1+\Delta)}\|\hat{\xi} - \hat{\eta}\|.$$

From this we have $\Delta \leqslant \frac{b\Delta + \gamma(1+\Delta)}{a-\gamma(1+\Delta)}$ and since $\Delta \leqslant 1$ we have $\Delta \leqslant \frac{\gamma}{a-b-3\gamma}$.

DEFINITION 4.6. We define the **dimension** of $U_{(\rho)}$ and $S_{(\rho)}$ as $\dim U_{(\rho)} = \dim X_1$ and $\dim S_{(\rho)} = \dim X_2$, respectively.

COROLLARY 4.7 (Stability). *Let $T = L + N$ with $L \in \mathcal{L}(X)$, $\sigma(L) \cap \{\xi \in \mathbb{C} \colon |\xi| = \rho\} = \varnothing$, for some $\rho \leqslant 1$, $N \colon X \to X$, $N(0) = 0$ and* Lip $N \leqslant \gamma$.

If K is a bounded set with $T(K) \supset K$, then $K \subset U_{(\rho)}$. If K is a subset of $U_{(\rho)}$ with $T(K) \subset K$, then K is stable (or asymptotically stable) with respect to T if and only if K is stable (or asymptotically stable) with respect to $T|_{U_{(\rho)}}$.

PROOF. Without loss of generality we may assume that $\rho < 1$. In fact, from Theorem 4.4(iv), $U_{(\rho^*)} = U_{(\rho)}$, for ρ^* close to ρ. From $T(K) \supset K$, given $x \in K$, there is a sequence $\{x_j\}_{j \leqslant 0}$, with $x_0 = x$, $x_{j+1} = T(x_j)$ for $j \leqslant 0$, and $\sup_{j \leqslant 0} \|x_j\| \rho^{-j} < \infty$ (since K is bounded and $\rho < 1$). It follows from Theorem 4.4(iv) that $x \in U_{(\rho)}$, proving that $K \subset U_{(\rho)}$.

Suppose that $K \subset T(K) \subset U_{(\rho)}$ and K is stable with respect to $T|_{U_{(\rho)}}$, that is, given $\epsilon > 0$, there exists $\delta > 0$ such that $\mathrm{dist}_H(x, K) < \delta$ and $x \in U_{(\rho)}$. Then $\mathrm{dist}_H(T^n x, K) < \epsilon$, for all $n \geqslant 0$.

Define

$$V = \left\{ x \in X \colon \mathrm{dist}_H(Px, K) < \delta \ \text{ and } \ \mathrm{dist}_H(x, U_{(\rho)}) < \frac{\epsilon}{C^2} \right\},$$

which is (since, from Theorem 4.4(vi), P is continuous) an open neighborhood of K.

For each $n \geqslant 0$ and $x \in V$ we have

$$\begin{aligned}\operatorname{dist}_H(T^n x, K) &\leqslant \operatorname{dist}_H(T^n Px, K) + \operatorname{dist}_H(T^n x, PT^n x)\\ &\leqslant \operatorname{dist}_H(T^n Px, K) + C\operatorname{dist}_H(T^n x, U_{(\rho)})\\ &< \epsilon + C^2\nu^n \operatorname{dist}_H(x, U_{(\rho)}) < 2\epsilon,\end{aligned}$$

where $0 < \nu < \rho \leqslant 1$, which shows that K is stable with respect to T. The same reasoning applies to asymptotic stability and the converse is trivial. □

2. Differentiability of invariant manifolds

In this section we prove that the invariant manifolds obtained in Section 1 are smooth whenever $T\colon X \to X$ is smooth. We start with a very useful tool which characterizes the continuous differentiability through a condition that does not involve the computation of the derivative.

LEMMA 4.8. *Let X, Y be real Banach spaces, let $Q \subset X$ be an open set and let $f\colon Q \to Y$ be a locally Lipschitz map. Then, f is continuously differentiable in Q if and only if for each $x_0 \in Q$ we have*

$$\|f(x+h) - f(x) - f(x_0+h) + f(x_0)\| = o(\|h\|), \tag{4.5}$$

as $(x, h) \to (x_0, 0)$.

PROOF. If $f \in C^1(Q, Y)$, then for each $x_0 \in Q$ we have

$$\begin{aligned}&\|f(x+h) - f(x) - f(x_0+h) + f(x_0)\|\\ &\qquad\leqslant \left\| \int_0^1 [f'(x+th) - f'(x_0+th) + f'(x_0) - f'(x)]dt\ h \right\|\\ &\qquad+ \|f'(x)h - f'(x_0)h\| \leqslant o(\|h\|) + \|f'(x) - f'(x_0)\|\|h\|\\ &\qquad= o(\|h\|) \quad \text{as } (x, h) \to (x_0, 0).\end{aligned}$$

For the converse, we may assume without loss of generality that Q is a ball and that f is Lipschitz. If the derivative exists on each point of Q, from (4.5) we have

$$\left\| \frac{1}{t}[f(x+th) - f(x)] - \frac{1}{t}[f(x_0+th) - f(x_0)] \right\| = o(\|h\|),$$

as $(x, h) \to (x_0, 0)$. Hence $\|(f'(x) - f'(x_0))h\| = o(\|h\|)$, as $(x, h) \to (x_0, 0)$; that is, given $\epsilon > 0$, there exists $\delta > 0$ such that, if $\|x - x_0\| < \delta$ and $\|h\| < \delta$, then

$$\frac{\|(f'(x) - f'(x_0))h\|}{\|h\|} < \frac{\epsilon}{2},$$

so $\|f'(x) - f'(x_0)\| < \epsilon$ whenever $\|x - x_0\| < \delta$, that is, f' is continuous.

Let us show that if (4.5) holds, then the derivative exists in all points of Q.

Case 1: $X = Y = \mathbb{R}$.

Since f is Lipschitz continuous, it is differentiable almost everywhere. We will show that it is differentiable in all points of Q. Given $x_0 \in Q$ and $\epsilon > 0$, there exists $\delta > 0$ such that, if $|x - x_0| + |h| < \delta$, then

$$|f(x+h) - f(x) - f(x_0+h) + f(x_0)| \leqslant \epsilon|h|.$$

There is x^* with $\|x^* - x_0\| < \delta$ such that $f'(x^*)$ exists. Then, for each $h \neq 0$ small

$$\left|\frac{f(x_0+h)-f(x_0)}{h} - f'(x^*)\right| \leqslant \left|\frac{f(x_0+h)-f(x_0)-f(x^*+h)+f(x^*)}{h}\right| + \left|\frac{f(x^*+h)-f(x^*)}{h} - f'(x^*)\right| \leqslant 2\epsilon$$

and, if

$$\alpha := \limsup_{h\to 0} \frac{f(x_0+h)-f(x_0)}{h} \quad \text{and} \quad \beta := \liminf_{h\to 0} \frac{f(x_0+h)-f(x_0)}{h},$$

the above inequality shows that $\alpha, \beta \in [f'(x^*) - 2\epsilon, f'(x^*) + 2\epsilon]$, that is,

$$0 \leqslant \alpha - \beta \leqslant 4\epsilon.$$

Since $\epsilon > 0$ is arbitrary, it follows that $\alpha = \beta$. Hence $f'(x_0)$ exists.

Case 2: $X = \mathbb{R}$.

Let $\eta \in Y^*$. Then $\mathbb{R} \ni x \mapsto \langle f(x), \eta\rangle \in \mathbb{R}$ is C^1, since

$$\begin{aligned} &|\langle f(x+h) - f(x) - f(x_0+h) + f(x_0), \eta\rangle| \\ &\quad \leqslant \|\eta\|_{Y^*}\|f(x+h) - f(x) - f(x_0+h) + f(x_0)\| = o(|h|), \end{aligned}$$

as $(x, h) \to (x_0, 0)$ and the claim follows from Case 1. We have

$$\begin{aligned} \left|\frac{d}{dx}\langle f(x), \eta\rangle\right| &= \left|\lim_{h\to 0}\frac{\langle f(x+h) - f(x), \eta\rangle}{h}\right| \leqslant \lim_{h\to 0}\frac{|\langle f(x+h) - f(x), \eta\rangle|}{|h|} \\ &\leqslant \|\eta\| \limsup_{h\to 0}\frac{\|f(x+h) - f(x)\|}{|h|} \leqslant \|\eta\| \operatorname{Lip} f, \end{aligned}$$

and therefore $D(x)\colon \eta \mapsto \frac{d}{dx}\langle f(x), \eta\rangle$ belongs to Y^{**}.

We claim that $x \mapsto D(x) \in Y^{**}$ is continuous. In fact, given $\epsilon > 0$, there exists $\delta > 0$ such that, if $\|x - x_0\| < \delta$ and $\|h\| < \delta$, then

$$|\langle f(x+h) - f(x) - f(x_0+h) + f(x_0), \eta\rangle| \leqslant \epsilon\|\eta\|\|h\|,$$

and so

$$\left|\frac{d}{dx}\langle f(x), \eta\rangle - \frac{d}{dx}\langle f(x_0), \eta\rangle\right| \leqslant \epsilon\|\eta\|,$$

that is, $|\langle \eta, D(x) - D(x_0)\rangle| \leqslant \epsilon\|\eta\|$ and therefore D is continuous.

Let $\tau\colon Y \hookrightarrow Y^{**}$ be the canonical inclusion, that is, for each $y \in Y$

$$\tau(y) = \hat{y}, \text{ where } \hat{y} \in Y^{**}, \ \langle \eta, \hat{y}\rangle = \langle y, \eta\rangle, \ \forall \eta \in Y^*.$$

For $x_0 \in Q$, $\eta \in Y^*$ with $\|\eta\| \leqslant 1$ and h small we have

$$\left\langle \eta, \tau\left(\frac{f(x_0+h) - f(x_0)}{h}\right) - D(x_0)\right\rangle = \frac{1}{h}\int_{x_0}^{x_0+h} \langle \eta, D(x) - D(x_0)\rangle dx \overset{h\to 0}{\longrightarrow} 0,$$

uniformly for $\|\eta\| \leqslant 1$. Here we have used that $\langle \eta, D(x)\rangle = \frac{d}{dx}\langle f(x), \eta\rangle$ and that

$$\int_{x_0}^{x_0+h} \langle \eta, D(x)\rangle dx = \int_{x_0}^{x_0+h} \frac{d}{dx}\langle f(x), \eta\rangle dx = \langle f(x_0+h) - f(x_0), \eta\rangle.$$

It follows that

$$\tau\left(\frac{f(x_0+h) - f(x_0)}{h}\right) \to D(x_0), \text{ as } h \to 0.$$

But τ is an isometry, hence $\frac{f(x_0+h)-f(x_0)}{h}$ has limit in Y as $h \to 0$.

Now, lastly, we prove the general case. From Case 2, $t \mapsto f(x+th)\colon \mathbb{R} \to Y$ ($x \in Q, h \in X$) is C^1 for t suitably small, hence the Gâteaux derivative

$$df(x;h) = \frac{d}{dt} f(x+th)|_{t=0}$$

exists, for each $x \in Q$, $h \in X$. The condition (4.5) gives, for fixed $h \in X$,

$$\|f(x+th) - f(x) - f(x_0+th) + f(x_0)\| = o(|t|),$$

for t small and x near x_0. So

$$\|df(x;h) - df(x_0;h)\| \to 0 \text{ as } x \to x_0 \text{ uniformly for } \|h\| = 1,$$

and this implies that $h \mapsto df(x_0;h)$ is a bounded linear map, that f is Fréchet differentiable and that the derivative is continuous. This completes the proof. □

THEOREM 4.9 (C^1 invariant manifolds). *Suppose that X, L, ρ, N, $T = L + N$ satisfy the conditions of Theorem 4.4, and also $N\colon X \to X$ is continuously differentiable with bounded derivative, at least in a neighborhood of $U_{(\rho)}$ (or $S_{(\rho)}$). Then $U_{(\rho)}$ (or $S_{(\rho)}$) is C^1.*

PROOF. Let us prove the result for $U_{(\rho)}$. From Theorem 4.4 $U_{(\rho)} = \{\xi + \theta(\xi)\colon \xi \in X_1\}$, where $\theta\colon X_1 \to X_2$, $\theta(0) = 0$, Lip $\theta \leqslant 1$ is given by

$$\begin{cases} \hat{\xi} = L_1\xi + N_1(\xi + \theta(\xi)), \\ \theta(\hat{\xi}) = L_2\theta(\xi) + N_2(\xi + \theta(\xi)). \end{cases}$$

Since θ is Lipschitz, from Lemma 4.8, it is sufficient to prove that $\lambda_{\xi_0}(\theta) = 0$, for all $\xi_0 \in X_1$, where

$$\lambda_{\xi_0}(\theta) = \limsup_{(\xi,h)\to(\xi_0,0)} \frac{1}{\|h\|} \|\theta(\xi+h) - \theta(\xi) - \theta(\xi_0+h) + \theta(\xi_0)\|,$$

and we already know that $\lambda_{\xi_0}(\theta) \leqslant 2\text{Lip}\,\theta \leqslant 2$. We have

$$\begin{aligned} \widehat{\xi+h} - \hat{\xi} &= L_1 h + N_1(\xi+h+\theta(\xi+h)) - N_1(\xi+\theta(\xi)) \\ &= \Big(L_1 + \frac{\partial N_1}{\partial x_1}(\xi_0+\theta(\xi_0))\Big)h + \frac{\partial N_1}{\partial x_2}(\xi_0+\theta(\xi_0))(\theta(\xi+h)-\theta(\xi)) \\ &\quad + N_1(\xi+h+\theta(\xi+h)) - N_1(\xi+\theta(\xi)) - \frac{\partial N_1}{\partial x_1}(\xi_0+\theta(\xi_0))h \\ &\quad - \frac{\partial N_1}{\partial x_2}(\xi_0+\theta(\xi_0))(\theta(\xi+h)-\theta(\xi)) \\ &= \Big(L_1 + \frac{\partial N_1}{\partial x_1}(\xi_0+\theta(\xi_0))\Big)h + \frac{\partial N_1}{\partial x_2}(\xi_0+\theta(\xi_0))(\theta(\xi+h)-\theta(\xi)) + o(\|h\|), \end{aligned}$$

as $(\xi,h)\to(\xi_0,0)$ in $X_1\times X_1$. Also

$$\begin{aligned}\theta(\widehat{\xi+h})-\theta(\hat{\xi}) &= L_2\theta(\xi+h)-L_2\theta(\xi)+N_2(\xi+h+\theta(\xi+h))-N_2(\xi+\theta(\xi))\\ &=\Big(L_2+\frac{\partial N_2}{\partial x_2}(\xi_0+\theta(\xi_0))\Big)(\theta(\xi+h)-\theta(\xi))+\frac{\partial N_2}{\partial x_1}(\xi_0+\theta(\xi_0))h\\ &\quad+N_2(\xi+h+\theta(\xi+h))-N_2(\xi+\theta(\xi))-\frac{\partial N_2}{\partial x_1}(\xi_0+\theta(\xi_0))h\\ &\quad-\frac{\partial N_2}{\partial x_2}(\xi_0+\theta(\xi_0))(\theta(\xi+h)-\theta(\xi))\\ &=\Big(L_2+\frac{\partial N_2}{\partial x_2}(\xi_0+\theta(\xi_0))\Big)(\theta(\xi+h)-\theta(\xi))+\frac{\partial N_2}{\partial x_1}(\xi_0+\theta(\xi_0))h+o(\|h\|)\end{aligned}$$

as $(\xi,h)\to(\xi_0,0)$ in $X_1\times X_1$. Hence

$$\begin{aligned}&\|\theta(\widehat{\xi+h})-\theta(\hat{\xi})-\theta(\widehat{\xi_0+h})+\theta(\hat{\xi}_0)\|\\ &\quad=\Big\|\Big(L_2+\frac{\partial N_2}{\partial x_2}(\xi_0+\theta(\xi_0))\Big)(\theta(\xi+h)-\theta(\xi)-\theta(\xi_0+h)+\theta(\xi_0))\Big\|\\ &\qquad+o(\|h\|)\\ &\quad\leqslant(b+\gamma)\lambda_{\xi_0}(\theta)\|h\|+o(\|h\|)=[(b+\gamma)\lambda_{\xi_0}(\theta)+o(1)]\|h\|.\end{aligned}\tag{4.6}$$

If $k=\widehat{\xi_0+h}-\hat{\xi}_0$, we have $\|k\|\geqslant(a-2\gamma)\|h\|$ and

$$\begin{aligned}\|\widehat{\xi+h}-(\hat{\xi}+k)\| &= \|\widehat{\xi+h}-\hat{\xi}-\widehat{\xi_0+h}+\hat{\xi}_0\|\\ &=\Big\|\frac{\partial N_1}{\partial x_2}(\xi_0+\theta(\xi_0))\Big(\theta(\xi+h)-\theta(\xi)-\theta(\xi_0+h)+\theta(\xi_0)\Big)\Big\|\\ &\quad+o(\|h\|)\\ &\leqslant\Big(\gamma\lambda_{\xi_0}(\theta)+o(1)\Big)\|h\|.\end{aligned}\tag{4.7}$$

From (4.6) and (4.7), and recalling that $\theta(\hat{\xi}_0+k)=\theta(\widehat{\xi_0+h})$, we have

$$\begin{aligned}&\|k\|^{-1}\|\theta(\hat{\xi}+k)-\theta(\hat{\xi})-\theta(\hat{\xi}_0+k)+\theta(\hat{\xi}_0)\|\\ &\quad=\|k\|^{-1}\|\theta(\widehat{\xi+h})-\theta(\hat{\xi})-\theta(\widehat{\xi_0+h})+\theta(\hat{\xi}_0)-\theta(\widehat{\xi+h})+\theta(\hat{\xi}+k)\|\\ &\quad\leqslant\|k\|^{-1}[\|\theta(\widehat{\xi+h})-\theta(\hat{\xi})-\theta(\widehat{\xi_0+h})+\theta(\hat{\xi}_0)\|+\|\theta(\widehat{\xi+h})-\theta(\hat{\xi}+k)\|]\\ &\quad\leqslant\|k\|^{-1}\{[(b+\gamma)\lambda_{\xi_0}(\theta)+o(1)]\|h\|+\|\widehat{\xi+h}-(\hat{\xi}+k)\|\}\\ &\quad\leqslant\|k\|^{-1}[(b+2\gamma)\lambda_{\xi_0}(\theta)+o(1)]\|h\|\\ &\quad\leqslant\frac{b+2\gamma}{a-2\gamma}\lambda_{\xi_0}(\theta)+o(1),\end{aligned}$$

when $(\xi,h)\to(\xi_0,0)$ and $(\hat{\xi},k)\to(\hat{\xi}_0,0)$. These convergences are equivalent, that is, $(\xi,h)\to(\xi_0,0)$ if and only if $(\hat{\xi},k)\to(\hat{\xi}_0,0)$, since $\hat{\xi}-\hat{\xi}_0=L_1(\xi-\xi_0)+N_1(\xi+\theta(\xi))-N_1(\xi_0+\theta(\xi_0))$ and $k=\widehat{\xi_0+h}-\hat{\xi}_0=L_1h+N_1(\xi_0+h+\theta(\xi_0+h))-N_1(\xi_0+\theta(\xi_0))$. Therefore

$$\lambda_{\hat{\xi}_0}(\theta)\leqslant\frac{b+2\gamma}{a-2\gamma}\lambda_{\xi_0}(\theta),\ \text{for all } \xi_0\in X_1.$$

If $\lambda^* = \sup_{\xi_0 \in X_1} \lambda_{\xi_0}(\theta)$, then $0 \leqslant \lambda^* \leqslant 2$ and

$$\lambda^* = \sup_{\hat{\xi_0} \in X_1} \lambda_{\hat{\xi_0}}(\theta) = \sup_{\xi_0 \in X_1} \lambda_{\hat{\xi_0}}(\theta) \leqslant \frac{b+2\gamma}{a-2\gamma} \sup_{\xi_0 \in X_1} \lambda_{\xi_0}(\theta) = \frac{b+2\gamma}{a-2\gamma}\lambda^*,$$

and since $\dfrac{b+2\gamma}{a-2\gamma} < 1$ we have $\lambda^* = 0$ and $U_{(\rho)}$ is C^1. Similar reasonings show that $S_{(\rho)}$ is C^1. □

3. Local invariant manifolds

In this section we present the local versions of the results presented in the two previous sections, concerning existence and differentiability of invariant manifolds of a critical element. We begin with the definition of local manifolds.

DEFINITION 4.10. Consider the setting and notations of Theorem 4.4. Given a neighborhood V of 0 in X, we define:

$$W_V^u = \{x \in V\colon \text{ there is a sequence } \{x_j\}_{j \leqslant 0} \subset V \text{ with } x_0 = x, \\ x_{j+1} = T(x_j),\ j < 0 \text{ and } \|x_j\| = o(\rho^j), \text{ as } j \to -\infty\}.$$

and

$$W_V^s = \{x \in V\colon \{T^n x\}_{n \in \mathbb{N}} \subset V \text{ and } \|T^n x\| = o(\rho^n) \text{ as } n \to \infty\}.$$

They are called V**-unstable** and V**-stable manifolds** of T in 0, respectively. When the neighborhood V is understood, we denote W_V^u by W_{loc}^u and W_V^s by W_{loc}^s, and call them **local unstable** and **local stable manifolds** of T in 0, respectively.

We recall that for Banach spaces X, Y, we define $C^0(X,Y)$ as the set of continuous and bounded functions from X to Y, and $C^{0+}(X,Y)$ the functions in $C^0(X,Y)$ that are uniformly continuous. Also we set

$$C^{1-}(X,Y) = \left\{\varphi \in C^0(X,Y) : \frac{\|\varphi(x)-\varphi(y)\|}{\|x-y\|} \text{ is bounded for } x \neq y \text{ in } X\right\},$$

$$C^1(X,Y) = \{\varphi \in C^0(X,Y)\colon \varphi \text{ is differentiable and } \varphi' \in C^0(X, \mathcal{L}(X,Y))\},$$

and

$$C^{1+}(X,Y) = \{\varphi \in C^1(X,Y) : \varphi' \in C^{0+}(X, \mathcal{L}(X,Y))\}.$$

DEFINITION 4.11. We say that a Banach space X is C^1 (or C^{1+} or C^{1-}) if there exists a function $\varphi \in C^1(X,\mathbb{R})$ (or $C^{1+}(X,\mathbb{R})$ or $C^{1-}(X,\mathbb{R})$) such that $\varphi(x) = 1$ for $\|x\| \leqslant 1$ and $\varphi(x) = 0$ for $\|x\| \geqslant R$ for some $R \geqslant 0$.

To obtain a C^1 (or $C^{\pm 1}$) Banach space, it suffices that the norm in X (or any equivalent norm in X) is C^1 (or $C^{\pm 1}$). Every Banach space is at least C^{1-}, but there are some (for instance, $C([0,1],\mathbb{R})$) which are not C^1.

THEOREM 4.12. *If $X = X_1 \oplus X_2$, $F\colon X \to X$ is such that $F(0) = 0$, $F'(0) = 0$ and F is C^1 (or C^{1+}) in a neighborhood of 0 and X_1 is a C^1 (or C^{1+}) Banach space, then given $\gamma > 0$, there exist $\delta > 0$ and $\tilde{F}\colon X \to X$ with $\tilde{F} = F$ in a neighborhood of 0, $\operatorname{Lip} \tilde{F} \leqslant \gamma$, $\sup_{x \in X} \|\tilde{F}(x)\| \leqslant \delta\gamma$ and such that $\tilde{F}$ is C^1 (or C^{1+}) in $V_\delta = \{x_1 + x_2 \in X_1 \oplus X_2\colon \|x_2\| < \delta\}$.*

PROOF. There exists $\varphi\colon X_1 \to [0,1] \subset \mathbb{R}$ which is $C^1(X_1,\mathbb{R})$ (or C^{1+}) with $\varphi(x) = 1$ for $\|x\| \leqslant r_0$ (for a fixed $0 < r_0 < 1$) and $\varphi(x) = 0$ for $\|x\| \geqslant 1$. Thus for $R > 0$ the map $X_1 \ni x_1 \mapsto \varphi(x_1/R) \in [0,1]$ is in $C^1(X_1, X_1)$ (or C^{1+}) supported in $\{x_1 \in X_1 \colon \|x_1\| \leqslant R\}$. Therefore

$$\sup_{x_1 \in X_1} \|x_1\varphi(x_1/R)\| \leqslant R,$$

also

$$\sup_{x_1 \in X_1} \left\| \frac{\partial}{\partial x_1}(x_1\varphi(x_1/R)) \right\| = \sup_{x_1 \in X_1} \left\| \varphi(x_1/R)I + \frac{x_1}{R}\varphi'(x_1/R) \right\| \leqslant 1 + \mathrm{Lip}\ \varphi.$$

and $x_1\varphi(x_1/R) = x_1$ for $\|x_1\| \leqslant r_0 R$. Moreover, the same conclusions hold if φ is Lipschitz instead of C^1 (or C^{1+}), with the exception that $X_1 \ni x_1 \mapsto x_1\varphi(x_1/R) \in X_1$ is Lipschitz with Lipschitz constant smaller than or equal to $1 + \mathrm{Lip}\ \varphi$.

Now X_2 is at least C^{1-} and there exists $\psi\colon X_2\colon [0,1]$ Lipschitz with $\psi(x_2) = 1$ for $\|x_2\| \leqslant 1/2$, $\psi(x_2) = 0$ for $\|x_2\| \geqslant 1$ and Lip $\psi \leqslant 2$. For instance the map

$$\psi(x_0) = \begin{cases} 1, & \|x_2\| \leqslant 1/2, \\ 2 - 2\|x_2\|, & 1/2 < \|x_2\| \leqslant 1, \\ 0, & \|x_2\| > 1 \end{cases}$$

has the required properties. Choose $R_1, R_2 \in (0,1]$ small such that F is C^1 (or C^{1+}) in $\{x_1 + x_2 \in X_1 \oplus X_2 \colon \|x_1\| \leqslant R_1,\ \|x_2\| \leqslant R_2\}$. Then

$$\tilde{F}(x_1 + x_2) = F(x_1\varphi(x_1/R_1) + x_2\psi(x_2/R_2))$$

satisfies the required conditions. □

THEOREM 4.13. *Assume that X is a Banach space, V a neighborhood of 0 in X, $L \in \mathcal{L}(X)$, $T \in C^1(V, X)$ (or $C^{1+}(V,X)$) with $\|Tx - Lx\| = o(\|x\|)$ and $T'(x) \to L$ as $x \to 0$, $X = X_1 \oplus X_2$, $LX_j \subset X_j$, $L_j = L|_{X_j}$, $j = 1, 2$, satisfies $r_\sigma(L_2) < 1$, $r_\sigma(L_2)r_\sigma(L_1^{-1}) < 1$ and X_1 is C^1 (or C^{1+}).*

Then there exists $\theta\colon X_1 \to X_2$ C^1 (or C^{1+}) in a neighborhood of 0 $\{\xi \in X_1 \colon \|\xi\| < r\}$ such that $\theta(0) = 0$, $\theta'(0) = 0$ and

$$W^{cu}_{loc} = \{\xi + \theta(\xi) \colon \|\xi\| < r,\ \xi \in X_1\}$$

satisfies:

(1) if $x = x_1 + x_2 \in W^{cu}_{loc}$ and $\bar{x} = Tx$ has $\|\bar{x_1}\| < r$, then $\bar{x} \in W^{cu}_{loc}$, and for some $r_0 > 0$, $x \in W^{cu}_{loc}$ with $\|x_1\| < r_0$ implies that there exists $\hat{x} \in W^{cu}_{loc}$ such that $T\hat{x} = x$;

(2) $T|_{W^{cu}_{loc}}$ is a C^1 (or C^{1+}) diffeomorphism from a neighborhood of $0 \in W^{cu}_{loc}$ onto another neighborhood of 0 in W^{cu}_{loc}. The derivative of $T|_{W^{cu}_{loc}}$ in 0 is L_1;

(3) there exist a neighborhood V_0 of 0 in X and a continuous map $P\colon V_0 \to W^{cu}_{loc}$ such that $P \circ T = T \circ P$ in V_0, $P(x) = x$ for $x \in V_0 \cap W^{cu}_{loc}$,

$$\|x - P(x)\| \leqslant C\mathrm{dist}(x, W^{cu}_{loc}) \quad \text{in } V_0,$$

and if $x_0, x_1 = T(x_0), \cdots, x_n = T(x_{n-1}) = T^n(x_0)$ are all in V_0, then

$$\|T^n(x_0) - T^nP(x_0)\| \leqslant C\rho^n \mathrm{dist}(x_0, W^{cu}_{loc}),$$

where $C > 0$ and $\rho \in (0,1)$ are constants (in fact, given ρ with $r_\sigma(L_2) < \rho < 1$, there exists $C > 0$, V_0 and P such that (3) holds).

PROOF. We choose, as in Theorem 4.4, equivalent norms in X_1 and X_2 such that $\|L_2\| = b < 1$, $\|L_1^{-1}\| = a^{-1}$ and $b < a$. Therefore, given $b < \rho < 1$, there exists $\gamma > 0$ so that $b + 2\gamma < \rho < a - 2\gamma$. For any $\delta > 0$ in $0 < \delta < \delta_0$ there exists $F_\delta \colon X \to X$ with $F_\delta = T$ in a neighborhood of 0, $F_\delta \in C^1(V_\delta, X)$ (or C^{1+}) where $V_\delta = \{x_1 + x_2 \colon \|x_2\| < \delta\}$, $\sup_X \|F_\delta\| \leqslant \gamma\delta$ and Lip $F_\delta \leqslant \gamma$.

From Theorem 4.4, we obtain $\theta_\delta \colon X_1 \to X_2$ and $P_\delta \colon X \to U_{\rho,\delta} = \{\xi + \theta_\delta(\xi) \colon \xi \in X_1\}$ that satisfy properties (1)–(3), except the smoothness property.

The map θ_δ is defined, with $N = F_\delta - L$, by

$$\begin{cases} \hat{\xi} = L_1\xi + N_1(\xi + \theta_\delta(\xi)), \\ \theta_\delta(\hat{\xi}) = L_2\theta_\delta(\xi) + N_2(\xi + \theta_\delta(\xi)). \end{cases}$$

Setting $m(R) = \sup_{\|\xi\| \leqslant R} \|\theta_\delta(\xi)\|$, if $\|\hat{\xi}\| \leqslant R$ we have

$$R \geqslant \|\hat{\xi}\| \geqslant (a - 2\gamma)\|\xi\| \quad \text{and} \quad m(R) = \sup_{\|\hat{\xi}\| \leqslant R} \|\theta_\delta(\hat{\xi})\| \leqslant b\, m\left(\frac{R}{a - 2\gamma}\right) + \gamma\delta,$$

for all $R > 0$. Since $m(R) \leqslant R$ (recall that Lip $\theta_\delta \leqslant 1$) we have $m(R) \leqslant \frac{\gamma\delta}{1-b} < \delta$ for all $R > 0$. Thus Theorem 4.9 says that θ_δ is $C^1(X_1, X_2)$ (or C^{1+}) and if θ is the restriction of θ_δ to $\{x_1 \in X_1 \colon \|x_1\| < r\}$ for some small $r > 0$, then θ satisfies conditions (1)–(3). □

REMARK 4.14. This theorem works, almost with this exact same proof, for the Lipschitz case. Also, with analogous statement and proof, this theorem holds for the center stable manifold.

THEOREM 4.15. *Let X be a Banach space, V a neighborhood of 0 in X, $T \colon V \to X$ a C^1 (or $C^{1\pm}$) map with $T(0) = 0$, $T'(0) = L$ and $\sigma(L) \setminus \{\lambda \in \mathbb{C} \colon |\lambda| = 1\}$ closed. Hence $\sigma(L) \cap \{\lambda \in \mathbb{C} \colon |\lambda| = 1\}$ is a spectral set. Consider the spectral decomposition $X = X_0 \oplus X_1 \oplus X_2$ and $L = L_0 + L_1 + L_2$ associated, respectively, with the spectral sets $\sigma_0 = \sigma(L) \cap \{\lambda \in \mathbb{C} \colon |\lambda| = 1\}$, $\sigma_1 = \sigma(L) \cap \{\lambda \in \mathbb{C} \colon |\lambda| > 1\}$ and $\sigma_2 = \sigma(L) \cap \{\lambda \in \mathbb{C} \colon |\lambda| < 1\}$, and assume that X_0 is C^1 (or $C^{1\pm}$). Then the* **local manifolds** *(which are defined within the proof) W^{ss}_{loc}, W^{cs}_{loc}, W^{c}_{loc}, W^{cu}_{loc} and W^{uu}_{loc} are C^1 (or $C^{1\pm}$) and the tangent spaces at 0 of these manifolds are:*

(i) *$\{0\} \oplus \{0\} \oplus X_2$ for W^{ss}_{loc} and the derivative of $T|_{W^{ss}_{loc}}$ at 0 is L_2;*
(ii) *$X_0 \oplus \{0\} \oplus X_2$ for W^{cs}_{loc} and the derivative of $T|_{W^{cs}_{loc}}$ at 0 is $L_0 \oplus L_2$;*
(iii) *$X_0 \oplus \{0\} \oplus \{0\}$ for W^{c}_{loc} and the derivative of $T|_{W^{c}_{loc}}$ at 0 is L_0;*
(iv) *$X_0 \oplus X_1 \oplus \{0\}$ for W^{cu}_{loc} and the derivative of $T|_{W^{cu}_{loc}}$ at 0 is $L_0 \oplus L_1$;*
(v) *$\{0\} \oplus X_1 \oplus \{0\}$ for W^{uu}_{loc} and the derivative of $T|_{W^{uu}_{loc}}$ at 0 is L_1.*

PROOF. We choose $\psi_j \colon X_j \to [0,1]$ Lipschitz maps with $\psi_j \equiv 1$ near 0 and $\psi_j \equiv 0$ for $\|x\| \geqslant C > 0$, and we can choose ψ_0 C^1 (or $C^{1\pm}$). With $N = T - L$, for $r > 0$ small we have

$$\tilde{N}(x) := N\Big(x_0\psi_0(x_0/r) + x_1\psi_1(x_1/r) + x_2\psi_2(x_2/r)\Big),$$

with $x = x_0 + x_1 + x_2 \in X_0 \oplus X_1 \oplus X_2$, is a C^1 (or $C^{1\pm}$) map in $X_0 \oplus B^{X_1}_{r_1} \oplus B^{X_2}_{r_2}$, for some $r_1, r_2 > 0$, $\tilde{N} = N$ in a neighborhood of 0, and $\sup_{x \in X} \|\tilde{N}(x)\|$ and Lip $\tilde{N}$ are small.

With $\tilde{T} = L + \tilde{N}$ instead of T, we define the local manifolds $W^{uu}_{loc} = U_{(1+\delta)}$, $W^{cs}_{loc} = S_{(1+\delta)}$, $W^{cu}_{loc} = U_{(1-\delta)}$, $W^{ss}_{loc} = S_{(1-\delta)}$ and $W^{c}_{loc} = W^{cu}_{loc} \cap W^{cs}_{loc}$, which are at least C^{1-}.

Now $W^{cu}_{loc} = \{\xi + \theta(\xi) \colon \xi \in X_0 \oplus X_1\}$, where $\theta \colon X_0 \oplus X_1 \to X_2$, $\xi = \xi_0 + \xi_1$ with

$$\begin{cases} \hat{\xi}_0 = L_0\xi_0 + \tilde{N}_0(\xi_0 + \xi_1 + \theta(\xi_0 + \xi_1)), \\ \hat{\xi}_1 = L_1\xi_1 + \tilde{N}_1(\xi_0 + \xi_1 + \theta(\xi_0 + \xi_1)), \\ \theta(\hat{\xi}_0 + \hat{\xi}_1) = L_2\theta(\xi_0 + \xi_1) + \tilde{N}_2(\xi_0 + \xi_1 + \theta(\xi_0 + \xi_1)), \end{cases}$$

and Lip $\tilde{N}_j \leqslant \gamma$ small. In the proof of Theorem 4.9, it is possible to work in

$$\{\xi_0 + \xi_1 \colon \xi_0 \in X_0,\ \xi_1 \in X_1,\ \|\xi_1\| < r\},$$

for any $r > 0$ for which $\tilde{N}$ is (at least) C^1 in $X_0 \oplus B^{X_1}_r \oplus B^{X_2}_r$ and $\sup_{x\in X} \|\tilde{N}(x)\| \leqslant r\gamma$ $(b + 2\gamma < 1)$, since $\|\hat{\xi}_1\| \leqslant r$ implies

$$r \geqslant \|\hat{\xi}_1\| \geqslant b^{-1}\|\xi_1\| - 2\gamma r \quad \text{ and } \quad \|\xi_1\| \leqslant b(1-\gamma)r < r,$$

and $\sup \|\theta\| \leqslant \frac{\gamma r}{1-b} < r$. Therefore $\xi_0 + \xi_1 + \theta(\xi_0 + \xi_1)$ is in the region where $\tilde{N}$ is C^1 (or $C^{1\pm}$) and (as in the proof of Theorem 4.9) for some constant $k < 1$ we obtain

$$\lambda_{\hat{\xi}}(\theta) \leqslant k\lambda_{\xi}(\theta) \leqslant k \sup_{\|\xi_1\|<r} \lambda_{\xi}(\theta),$$

for $\hat{\xi} \in X_0 \oplus X_1$ with $\|\hat{\xi}_1\| < r$, which implies $\lambda_{\xi}(\theta) \equiv 0$ for $\xi \in X_0 \oplus X_1$ with $\|\xi_1\| < r$ and hence W^{cu}_{loc} is C^1 in a neighborhood of 0.

An analogous argument works for W^{uu}_{loc}, W^{cs}_{loc} and W^{ss}_{loc}, and then W^{c}_{loc} is also C^1 (or $C^{1\pm}$) from the Implicit Function Theorem. □

DEFINITION 4.16. The set W^c_{loc} is called the **local center manifold** and we define $\dim W^c_{loc} = \dim X_0$.

4. Continuous dependence of invariant manifolds in C^1 topology

Now we study maps with parameters. We prove that the invariant manifolds vary continuously (in C^1 topology) with respect to perturbations. This study, combined with the λ-lemma of Section 5, will be essential to the proof of structural stability of Morse-Smale semigroups.

Let us start describing the setting at which we will employ the results of this section. Assume that $X \times \mathbb{R} \ni (x,\eta) \mapsto F(x,\eta) \in X$ is continuously differentiable in both variables. Assume that e is a fixed point for $F(\cdot,0)$, that is, $F(e,0) = e$. Then we write

$$\begin{aligned} T\left(\left[\begin{smallmatrix} x \\ r \end{smallmatrix}\right]\right) &= F(x+e,r) - F(e,0) \\ &= D_xF(e,0)x + D_\eta F(e,0)r + F(x+e,r) - F(e,0) \\ &\qquad - D_xF(e,0)x - D_\eta F(e,0)r \\ &= D_xT\left(\left[\begin{smallmatrix} 0 \\ 0 \end{smallmatrix}\right]\right)x + D_\eta T\left(\left[\begin{smallmatrix} 0 \\ 0 \end{smallmatrix}\right]\right)r + T\left(\left[\begin{smallmatrix} x \\ r \end{smallmatrix}\right]\right) - D_xT\left(\left[\begin{smallmatrix} 0 \\ 0 \end{smallmatrix}\right]\right)x - D_\eta T\left(\left[\begin{smallmatrix} 0 \\ 0 \end{smallmatrix}\right]\right)r. \end{aligned} \tag{4.8}$$

From this we can write $\mathscr{T} \colon X \times \mathbb{R} \to X \times \mathbb{R}$ as

$$\mathscr{T}\left(\begin{bmatrix} x \\ r \end{bmatrix}\right) = \begin{bmatrix} T\left(\left[\begin{smallmatrix} x \\ r \end{smallmatrix}\right]\right) \\ r \end{bmatrix} = \begin{bmatrix} L & M \\ 0 & 1 \end{bmatrix}\begin{bmatrix} x \\ r \end{bmatrix} + \mathscr{N}\left(\begin{bmatrix} x \\ r \end{bmatrix}\right), \quad \mathscr{N}\left(\begin{bmatrix} x \\ r \end{bmatrix}\right) = \begin{bmatrix} N\left(\left[\begin{smallmatrix} x \\ r \end{smallmatrix}\right]\right) \\ 0 \end{bmatrix},$$

where

$$L = D_xT\left(\left[\begin{smallmatrix} 0 \\ 0 \end{smallmatrix}\right]\right) \in \mathcal{L}(X), M = D_\eta T\left(\left[\begin{smallmatrix} 0 \\ 0 \end{smallmatrix}\right]\right) \in \mathcal{L}(\mathbb{R}, X) \text{ and}$$
$$N\left(\left[\begin{smallmatrix} x \\ r \end{smallmatrix}\right]\right) = T\left(\left[\begin{smallmatrix} x \\ r \end{smallmatrix}\right]\right) - D_xT\left(\left[\begin{smallmatrix} 0 \\ 0 \end{smallmatrix}\right]\right)x - D_\eta T\left(\left[\begin{smallmatrix} 0 \\ 0 \end{smallmatrix}\right]\right)r.$$

Now, the parameter is included as a variable and we would like to prove that the continuous differentiability of the invariant manifold happens with respect to both variables. This will imply that the invariant manifolds for the map $T(\cdot, r)$ will behave in a C^1 manner whenever $X \times \mathbb{R} \ni (x,r) \mapsto T(x,r) \in X$ is C^1 in both variables.

Let X, Y be Banach spaces, $L\colon X \to X$ and $M\colon Y \to X$ bounded linear maps with $\sigma(L) \setminus \{\xi \in \mathbb{C}\colon |\xi| \leqslant 1\}$ closed and $X = X_1 \oplus X_2$ and $L = L_1 \oplus L_2$ the decomposition associated to the spectral projections Π_i^L, $i = 1, 2$, associated to the spectral sets $\sigma_2(L) = \sigma(L) \cap \{\xi \in \mathbb{C}\colon |\xi| \leqslant 1\}$ and $\sigma_1(L) = \sigma(L) \cap \{\xi \in \mathbb{C}\colon |\xi| > 1\} \subset \{\lambda \in \mathbb{C}\colon |\lambda| \geqslant r\}$, for some $r > 1$, and $M = M_1 + M_2$, $M_i := \Pi_i^L M \in \mathcal{L}(Y, X_j)$, $j = 1, 2$.

Let $N\colon X \times Y \to X$ be a Lipschitz function with Lip N small and $N\left(\left[\begin{smallmatrix}0\\0\end{smallmatrix}\right]\right) = 0$ and define

$$T\left(\left[\begin{smallmatrix}x\\y\end{smallmatrix}\right]\right) = Lx + My + N\left(\left[\begin{smallmatrix}x\\y\end{smallmatrix}\right]\right).$$

In what follows, we will find invariant manifolds for $T\left(\left[\begin{smallmatrix}\cdot\\y\end{smallmatrix}\right]\right)$ studying the map $\mathscr{T}\colon X \times Y \to X \times Y$ given by

$$\mathscr{T}\left(\begin{bmatrix}x\\y\end{bmatrix}\right) = \begin{bmatrix}T\left(\left[\begin{smallmatrix}x\\y\end{smallmatrix}\right]\right)\\ r\end{bmatrix} = \begin{bmatrix}L & M\\ 0 & I_Y\end{bmatrix}\begin{bmatrix}x\\y\end{bmatrix} + \mathscr{N}\left(\begin{bmatrix}x\\y\end{bmatrix}\right),$$

where

$$\mathscr{L} = \begin{bmatrix}L & M\\ 0 & I_Y\end{bmatrix} \in \mathcal{L}(X \times Y) \quad \text{and} \quad \mathscr{N}\left(\begin{bmatrix}x\\y\end{bmatrix}\right) = \begin{bmatrix}N\left(\left[\begin{smallmatrix}x\\y\end{smallmatrix}\right]\right)\\ 0\end{bmatrix},$$

and I_Y is the identity operator on Y.

We have Lip $\mathscr{N}$ = Lip N, and we note that $\sigma(\mathscr{L}) = \sigma(L) \cup \{1\}$. In fact, clearly $1 \in \sigma(\mathscr{L})$ and $\lambda \in \rho(L)\setminus\{1\}$ if and only if $\lambda \in \rho(\mathscr{L})$ with

$$(\lambda - \mathscr{L})^{-1} = \begin{bmatrix}(\lambda - L)^{-1} & (\lambda-1)^{-1}(\lambda - L)^{-1}M\\ 0 & (\lambda-1)^{-1}I_Y\end{bmatrix}.$$

From this we define the projection $\Pi_2^{\mathscr{L}}$ associated to the part of the spectrum of $\mathscr{L}$ contained in $D = \{\lambda \in \mathbb{C}\colon |\lambda| \leqslant 1\}$ as

$$\Pi_2^{\mathscr{L}} = \frac{1}{2\pi i}\int_\gamma (\lambda - \mathscr{L})^{-1} d\lambda = \begin{bmatrix}\Pi_2^L & (I - L_1)^{-1}M_1\\ 0 & I_Y\end{bmatrix},$$

where $\gamma(t) = \nu e^{2\pi i t}$, $t \in [0,1]$, $1 < \nu < r$, and

$$\mathscr{X}_2 = \Pi_2^{\mathscr{L}}(X \times Y) = \left\{\begin{pmatrix}\eta + (I - L_1)^{-1}M_1\, y\\ y\end{pmatrix} : \eta \in X_2,\ y \in Y\right\}.$$

From this we obtain

$$\Pi_1^{\mathscr{L}} = I - \Pi_2^{\mathscr{L}} = \begin{bmatrix}\Pi_1^L & -(I - L_1)^{-1}M_1\\ 0 & 0\end{bmatrix},$$

and hence $\mathscr{X}_1 = \Pi_1^{\mathscr{L}}(X \times Y) = X_1 \oplus \{0\} \times \{0\}$.

If $T\colon X \times Y \to X$ is C^1, then $\mathscr{T}$ is C^1 and $\mathscr{T}\left(\left[\begin{smallmatrix}x\\y\end{smallmatrix}\right]\right) = \left(\left[\begin{smallmatrix}\hat{x}\\\hat{y}\end{smallmatrix}\right]\right)$ implies $y = \hat{y}$ and therefore the invariant manifolds associated to $\mathscr{T}$ are C^1.

From the special form of $\mathscr{T}$ these manifolds also have special forms, namely, $\mathscr{U}_\rho$ is the graph of a map $\vartheta\colon \mathscr{X}_1 \to \mathscr{X}_2$ and define $\tilde{\vartheta}$ by

$$\vartheta\begin{pmatrix}\Pi_1^L x + \alpha y\\ 0\end{pmatrix} = \begin{bmatrix}\tilde{\vartheta}\begin{pmatrix}\Pi_1^L x + \alpha y\\ 0\end{pmatrix}\\ y\end{bmatrix},$$

with $\alpha = -(I - L_1)^{-1}M_1$. Therefore, we obtain

$$\begin{bmatrix} \tilde{\vartheta}\begin{pmatrix} \hat{\xi} + \alpha\hat{y} \\ 0 \end{pmatrix} \\ \hat{y} \end{bmatrix} = \mathscr{L}_2\left(\begin{bmatrix} \xi + \alpha y \\ 0 \end{bmatrix} + \begin{bmatrix} \tilde{\vartheta}\left(\begin{smallmatrix} \xi+\alpha y \\ 0 \end{smallmatrix}\right) \\ y \end{bmatrix}\right) + \mathscr{N}_2\left(\begin{bmatrix} \xi + \alpha y \\ 0 \end{bmatrix} + \begin{bmatrix} \tilde{\vartheta}\left(\begin{smallmatrix} \xi+\alpha y \\ 0 \end{smallmatrix}\right) \\ y \end{bmatrix}\right),$$

which implies

$$\begin{bmatrix} \tilde{\vartheta}\begin{pmatrix} \hat{\xi} + \alpha\hat{y} \\ 0 \end{pmatrix} \\ \hat{y} \end{bmatrix} = \begin{bmatrix} L_2\left(\tilde{\vartheta}\begin{pmatrix} \xi + \alpha y \\ 0 \end{pmatrix} + \alpha y\right) - L_1\alpha y + My \\ y \end{bmatrix}$$
$$+ \begin{bmatrix} N_2\left(\begin{pmatrix} \xi + \alpha y \\ 0 \end{pmatrix} + \begin{pmatrix} \tilde{\vartheta}\begin{pmatrix} \xi + \alpha y \\ 0 \end{pmatrix} \\ y \end{pmatrix}\right) \\ 0 \end{bmatrix}.$$

Therefore $\hat{y} = y$ and

$$\tilde{\vartheta}\left(\begin{smallmatrix} \hat{\xi}+\alpha y \\ 0 \end{smallmatrix}\right) = L_2(\tilde{\vartheta}\left(\begin{smallmatrix} \xi+\alpha y \\ 0 \end{smallmatrix}\right) + \alpha y) - L_1\alpha y + My + N_2\left(\left(\begin{smallmatrix} \xi+\alpha y \\ 0 \end{smallmatrix}\right) + \left(\begin{smallmatrix} \tilde{\vartheta}\left(\begin{smallmatrix} \xi+\alpha y \\ 0 \end{smallmatrix}\right) \\ y \end{smallmatrix}\right)\right),$$

thus we obtain

$$\tilde{\vartheta}\left(\begin{smallmatrix} \hat{\xi}+\alpha y \\ 0 \end{smallmatrix}\right) + \alpha y = L_2(\tilde{\vartheta}\left(\begin{smallmatrix} \xi+\alpha y \\ 0 \end{smallmatrix}\right) + \alpha y) + M_2 y + N_2\left(\left(\begin{smallmatrix} \xi+\alpha y \\ 0 \end{smallmatrix}\right) + \left(\begin{smallmatrix} \tilde{\vartheta}\left(\begin{smallmatrix} \xi+\alpha y \\ 0 \end{smallmatrix}\right) \\ y \end{smallmatrix}\right)\right).$$

Also

$$\left(\begin{smallmatrix} \hat{\xi}+\alpha\hat{y} \\ 0 \end{smallmatrix}\right) = \mathscr{L}_1\left[\begin{smallmatrix} \Pi_1^L & \alpha \\ 0 & 0 \end{smallmatrix}\right]\left(\left(\begin{smallmatrix} \xi+\alpha y \\ 0 \end{smallmatrix}\right) + \left(\begin{smallmatrix} \tilde{\vartheta}\left(\begin{smallmatrix} \xi+\alpha y \\ 0 \end{smallmatrix}\right) \\ y \end{smallmatrix}\right)\right) + \mathscr{N}_1\left(\left(\begin{smallmatrix} \xi+\alpha y \\ 0 \end{smallmatrix}\right) + \left(\begin{smallmatrix} \tilde{\vartheta}\left(\begin{smallmatrix} \xi+\alpha y \\ 0 \end{smallmatrix}\right) \\ y \end{smallmatrix}\right)\right),$$

hence

$$\left(\begin{smallmatrix} \hat{\xi}+\alpha\hat{y} \\ 0 \end{smallmatrix}\right) = \mathscr{L}_1\left(\begin{smallmatrix} \xi+\alpha y \\ 0 \end{smallmatrix}\right) + \mathscr{N}_1\left(\left(\begin{smallmatrix} \xi+\alpha y \\ 0 \end{smallmatrix}\right) + \left(\begin{smallmatrix} \tilde{\vartheta}\left(\begin{smallmatrix} \xi+\alpha y \\ 0 \end{smallmatrix}\right) \\ y \end{smallmatrix}\right)\right)$$

and we obtain

$$\left(\begin{smallmatrix} \hat{\xi}+\alpha\hat{y} \\ 0 \end{smallmatrix}\right) = \left(\begin{smallmatrix} L_1\xi+L_1\alpha y \\ 0 \end{smallmatrix}\right) + \left(\begin{smallmatrix} N_1\left(\left(\begin{smallmatrix} \xi+\alpha y \\ 0 \end{smallmatrix}\right)+\left(\begin{smallmatrix} \tilde{\vartheta}\left(\begin{smallmatrix} \xi+\alpha y \\ 0 \end{smallmatrix}\right) \\ y \end{smallmatrix}\right)\right) \\ 0 \end{smallmatrix}\right)$$

and finally gives us

$$\hat{\xi} = L_1\xi + M_1 y + N_1\left(\left(\begin{smallmatrix} \xi+\alpha y \\ 0 \end{smallmatrix}\right) + \left(\begin{smallmatrix} \tilde{\vartheta}\left(\begin{smallmatrix} \xi+\alpha y \\ 0 \end{smallmatrix}\right) \\ y \end{smallmatrix}\right)\right).$$

Then, making $\theta_y(\xi) = \tilde{\vartheta}\left(\begin{smallmatrix} \xi+\alpha y \\ 0 \end{smallmatrix}\right) + \alpha y$, we have

$$\begin{cases} \hat{\xi} = L_1\xi + M_1 y + N_1\left(\begin{smallmatrix} \xi+\theta_y(\xi) \\ y \end{smallmatrix}\right), \\ \theta_y(\hat{\xi}) = L_2\theta_y(\xi) + M_2 y + N_2\left(\begin{smallmatrix} \xi+\theta_y(\xi) \\ y \end{smallmatrix}\right). \end{cases}$$

In particular, note that $\left[\begin{smallmatrix} 0 \\ 0 \end{smallmatrix}\right]$ is no longer a point in the graph of θ_y. If we define ζ_y as the unique solution of

$$\zeta = L_1\zeta + M_1 y + N_1\left(\begin{smallmatrix} \zeta+\theta_y(\zeta) \\ y \end{smallmatrix}\right)$$

we have $\hat{\zeta_y} = \zeta_y$ and $\theta_y(\zeta_y) = \theta_y(\hat{\zeta_y})$. It follows that

$$\xi_y + \theta_y(\xi) = T\left(\left[\begin{smallmatrix} \xi_y+\theta_y(\xi) \\ y \end{smallmatrix}\right]\right).$$

Now, if $\zeta = \xi - \zeta_y$ and $\hat{\zeta} = \hat{\xi} - \zeta_y$, $\Theta_y(\hat{\zeta}) = \theta_y(\hat{\zeta} + \zeta_y) - \theta_y(\zeta_y)$ we have

$$\begin{cases} \hat{\zeta} = L_1\zeta + N_1\left(\substack{\zeta+\zeta_y+\theta_y(\zeta+\zeta_y)\\ y}\right) - N_1\left(\substack{\zeta_y+\theta_y(\zeta_y)\\ y}\right), \\ \Theta_y(\hat{\zeta}) = L_2\Theta_y(\zeta) + N_2\left(\substack{\zeta+\zeta_y+\theta_y(\zeta+\zeta_y)\\ y}\right) - N_2\left(\substack{\zeta_y+\theta_y(\zeta_y)\\ y}\right). \end{cases}$$

From all these it becomes clear that the differentiability of ϑ with respect to y implies the differentiability of Θ_y with respect to y and the C^1 dependence of the unstable manifold with respect to the parameter y.

5. The λ-lemma

In this section, we present a complete proof of a general version of the λ-lemma, of Palis [**178**], done by D. Henry in [**141**]. For a geometric interpretation of this result, see Figure 4.1. For detailed definitions and results on Banach manifolds see, for instance, [**173**].

THEOREM 4.17 (λ-lemma). *Let X be a Banach space, let $T\colon X \to X$ be a C^1 map in a neighborhood of 0 with $T(0) = 0$ and let $L = T'(0) \in \mathcal{L}(X)$ be such that $\sigma(L)\backslash\{\xi \in \mathbb{C}\colon |\xi| \leqslant 1\}$ is closed.*

Consider the decomposition $X = X_1 \oplus X_2$ of X and $L = L_1 \oplus L_2$ of L given by the spectral projections Π_i^L, $i = 1, 2$, associated to the spectral sets $\sigma_2(L) = \sigma(L) \cap \{\xi \in \mathbb{C}\colon |\xi| \leqslant 1\}$ and $\sigma_1(L) = \sigma(L) \cap \{\xi \in \mathbb{C}\colon |\xi| > 1\} \subset \{\lambda \in \mathbb{C}\colon |\lambda| \geqslant r\}$, for some $r > 1$. Assume that either $\dim X_1 < \infty$ *or $T \in C^{1+}$ in a neighborhood of zero (that is, $x \mapsto T'(x)$ is uniformly continuous near zero), and let W^u_{loc} and W^{cs}_{loc} be the local unstable and center stable manifolds of T in 0, respectively.*

Then there exists a neighborhood V of 0 in X such that if $x = x^0 \in V$, $x^{n+1} = T(x^n) \in V$ for $n \geqslant 0$ and $x^n \to 0$ when $n \to \infty$ and if D^0 is a C^1 embedded manifold in X with $D^0 \cap W^{cs}_{loc} = \{x^0\}$, $T_{x^0}D^0 \oplus T_{x^0}W^{cs}_{loc} = X$ and

$$\begin{aligned} D^n = \{p \in V\colon\ &\text{there exist } p_0, p_1, \cdots, p_n \in V \text{ with } p_0 \in D^0, \\ &p_n = p \text{ and } p_{j+1} = T(p_j), \text{ for } 0 \leqslant j < n\}, \end{aligned}$$

then D^n is a C^1 embedded manifold in X (for n sufficiently large) and $D^n \to D^*$, as $n \to \infty$, in the C^1 topology, where D^* is the connected component of $W^u_{loc} \cap V$ containing 0.

Indeed, for n sufficiently large, D^n and W^u_{loc} have (in convenient coordinates) the form :

$$D^n = \{x^n + \xi + \psi_n(\xi)\colon \xi \in X_1,\ \|\xi\| < r\}$$

and

$$W^u_{loc} \cap V = D^* = \{\xi + \theta(\xi)\colon \xi \in X_1,\ \|\xi\| < r\},$$

where $\psi_n, \theta\colon X_1 \to X_2$ are C^1 and $\|\psi_n - \theta\|_{C^1(B_r(X_1), X_2)} \to 0$ when $n \to \infty$.

PROOF. We split the proof into seven steps which are presented next. Throughout the proof we will use sometimes $X_1 \times X_2$ instead of $X = X_1 \oplus X_2$ to write

$$\begin{aligned} T\colon X_1 \times X_2 &\to X_1 \times X_2 \\ (x_1, x_2) &\mapsto (L_1x_1 + N_1(x_1, x_2), L_2x_2 + N_2(x_1, x_2)). \end{aligned}$$

Step 1. We know, by Theorem 4.12, that there exist $r_0 > 0$ and C^1 (or C^{1+}) maps in a neighborhood of the origin in X, $\theta\colon X_1 \to X_2$ and $\sigma\colon X_2 \to X_1$, such that

$$W^u_{loc} = \{\xi + \theta(\xi)\colon \xi \in X_1,\ \|\xi\| < r_0\} \quad \text{and} \quad W^{cs}_{loc} = \{\sigma(\eta) + \eta\colon \eta \in X_2,\ \|\eta\| < r_0\}.$$

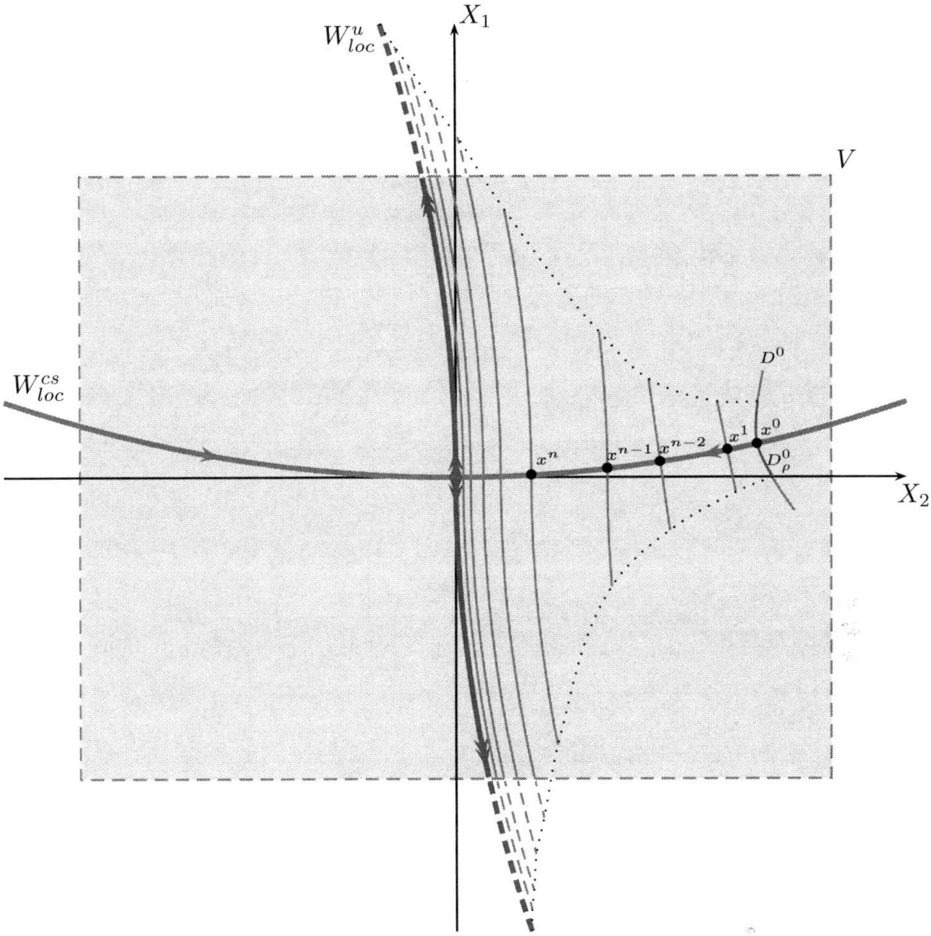

FIGURE 4.1. The local λ-lemma.

Let us define a change of coordinates $h\colon X_1 \times X_2 \to X$ by $h(\xi,\eta) = \xi + \sigma(\eta) + \eta$. Then if $T(h(\xi,\eta)) = h(\hat{\xi},\hat{\eta})$, for $T = L + N$, we have

$$T(\xi+\sigma(\eta)+\eta) = L_1\xi + L_1\sigma(\eta) + L_2\eta + N_1(\xi+\sigma(\eta)+\eta) + N_2(\xi+\sigma(\eta)+\eta) = \hat{\xi} + \sigma(\hat{\eta}) + \hat{\eta},$$

and thus

$$\begin{cases} \hat{\xi} = L_1\xi + L_1\sigma(\eta) + N_1(\xi + \sigma(\eta) + \eta) - \sigma(\hat{\eta}), \\ \hat{\eta} = L_2\eta + N_2(\xi + \sigma(\eta) + \eta). \end{cases}$$

Now, for $\sigma\colon X_2 \to X_1$, we have

$$\begin{cases} \sigma(\tilde{\eta}) = L_1\sigma(\eta) + N_1(\sigma(\eta) + \eta), \\ \tilde{\eta} = L_2\eta + N_2(\sigma(\eta) + \eta), \end{cases}$$

and we can rewrite the previous equation as

$$\begin{cases} \hat{\xi} = L_1\xi + N_1(\xi + \sigma(\eta) + \eta) - N_1(\sigma(\eta) + \eta) + \sigma(\tilde{\eta}) - \sigma(\hat{\eta}), \\ \hat{\eta} = L_2\eta + N_2(\xi + \sigma(\eta) + \eta). \end{cases}$$

Hence, defining $\tilde{N}_1(\xi, \eta) = N_1(\xi + \sigma(\eta) + \eta) - N_1(\sigma(\eta) + \eta) + \sigma(\tilde{\eta}) - \sigma(\hat{\eta})$ and $\tilde{N}_2(\xi, \eta) = N_2(\xi + \sigma(\eta) + \eta)$, we have

$$\begin{aligned} h^{-1} \circ T \circ h \colon X_1 \times X_2 &\to X_1 \times X_2 \\ (\xi, \eta) &\mapsto (L_1\xi + \tilde{N}_1(\xi, \eta), L_2\eta + \tilde{N}_2(\xi, \eta)), \end{aligned}$$

with $\tilde{N}^1, \tilde{N}^2$ C^1 (or C^{1+}) maps near zero and also $\tilde{N}_1(0, \eta) = 0$, since if $\xi = 0$, we have $\hat{\eta} = L_2\eta + N_2(\sigma(\eta) + \eta) = \tilde{\eta}$ and hence $\hat{\xi} = 0$.

For $\tilde{T} = h^{-1} \circ T \circ h$, the local center stable manifold is $\{\xi = 0\}$. We can then assume, without loss of generality (or differentiability), that $W^{cs}_{loc} = X_2$, that is, $N_1(0, \eta) = 0$, for $\eta \in X_2$, $\|\eta\| < r_0$.

In the same manner, the C^1 (or C^{1+}) change of coordinates $x \mapsto (\xi, \eta)$, $x = \xi + \theta(\xi) + \eta$, guarantees that we can assume that $W^u_{loc} = X_1$ near zero, that is, $N_2(\xi, 0) = 0$, for $\|\xi\| < r_0$.

One change of coordinates does not affect the other, in the sense that if $N_1(0, \eta) = 0$, then

$$\tilde{N}_1(0, \eta) = N_1(\xi + \theta(\xi) + \eta)_{|\xi=0} = N_1(0, \eta) = 0,$$

and therefore we can assume, without loss of generality (or differentiability), that $W^u_{loc} = X_1$ and $W^{cs}_{loc} = X_2$.

Step 2. By the previous step, we are able to write

$$\begin{aligned} T \colon X_1 \times X_2 &\to X_1 \times X_2 \\ (x_1, x_2) &\mapsto (L_1x_1 + N_1(x_1, x_2), L_2x_2 + N_2(x_1, x_2)) \end{aligned}$$

as a C^1 (or C^{1+}) map in $\|x_1\| < r_0$, $\|x_2\| < r_0$, with $N_j(0,0) = 0$ and $\left\|\frac{\partial N_i}{\partial x_j}(x_1, x_2)\right\| \to 0$, when $(x_1, x_2) \to (0, 0)$, $N_1(0, x_2) \equiv 0$, $N_2(x_1, 0) \equiv 0$ and

$$\frac{\partial N_2}{\partial x_1}(x_1, x_2) \to 0, \text{ as } x_2 \to 0, \text{ uniformly for } \|x_1\| < r_0, \tag{4.9}$$

if r_0 is small. This condition is the reason behind the hypothesis $\dim X_1 < \infty$ or $T \in C^{1+}$. Using (4.9) directly, the argument that follows requires only that T is a C^1 map in a neighborhood of zero in X. From now on, we denote $N_{i,j} = \frac{\partial N_i}{\partial x_j}$, for $i, j = 1, 2$.

We can choose equivalent norms in X_1, X_2 such that $\|L_1^{-1}\| = a^{-1} < 1$, $\|L_2\| = b$, with $1 \leqslant b < a$. Now we choose

$$0 < \gamma < \min\left\{a - 1, \frac{a - b}{4}\right\},$$

and we can assume that $r_0 > 0$ is small enough such that $\|N_{i,j}\| = \|\frac{\partial N_i}{\partial x_j}\| \leqslant \gamma$ in $V := \{(x_1, x_2) \colon \|x_1\| < r_0, \|x_2\| < r_0\}$. In particular $\frac{a-b}{4} < \frac{a}{2}$, and hence $\gamma < \frac{a}{2}$.

Step 3. Let $x^0 \in V$ such that $x^n = T^n(x^0) \in V$ for $n \geqslant 0$ and $x^n \to 0$ as $n \to \infty$. In this way, we can see that $x^n \in W^{cs}_{loc} \cap V \subset X_2$ (since $W^{cs}_{loc} = X_2$) and also $\|x^n\| < r_0$, since they all belong to V.

The tangent space of $W^{cs}_{loc} = X_2$ at x^0 (and at any x^n) is X_2, and the projection $\pi_1 \colon X_1 \oplus X_2 \to X_1$, when restricted to the tangent space $T_{x^0}D^0$ of D^0 at x^0, is an isomorphism over X_1. Indeed, let $\tilde{\pi}_1$ denote the restriction of π_1 to $T_{x^0}D^0$. If $v \in T_{x^0}D^0$ is such that $\tilde{\pi}_1 v = 0$, then $v \in X_2$ and since $T_{x^0}D^0 \cap X_2 = \{0\}$ we have $v = 0$ which implies that $\tilde{\pi}_1$ is injective. Now, given $w \in X_1$, since $X = T_{x^0}D^0 \oplus X_2$ there exist $w_1 \in T_{x^0}D^0$ and $w_2 \in X_2$ such that $w = w_1 + w_2$ and hence $w = \pi_1 w = \pi_1 w_1 = \tilde{\pi}_1 w_1$, which implies that $\tilde{\pi}_1$ is surjective and therefore $\tilde{\pi}_1 \colon T_{x^0}D^0 \to X_1$ is an isomorphism.

The Inverse Function Theorem applied to $\pi_1 \colon D^0 \to X_1$ guarantees that there exist $\rho_0 > 0$ and a neighborhood $D^0_{\rho_0}$ of x^0 in D^0 such that $\pi_1 \colon D^0_{\rho_0} \to U_{\rho_0}$ is invertible with C^1 inverse $\pi_1^{-1} \colon U_{\rho_0} \to D^0_{\rho_0}$, where $U_{\rho_0} = \{\xi \in X_1 \colon \|\xi\| < \rho_0\}$.

Define now, for each $\xi \in U_{\rho_0}$, $\psi_0(\xi) = \pi_1^{-1}\xi - \xi - x^0$. Since $\pi_1\psi_0(\xi) = 0$, this implies that ψ_0 is a C^1 map from X_1 into X_2, $\psi_0(0) = 0$ and clearly

$$D^0_{\rho_0} = \{x^0 + \xi + \psi_0(\xi) \colon \xi \in U_{\rho_0}\}.$$

We will now turn our attention to the sets $E^n = T^n(D^0_\rho) \cap V$, for each $n \geqslant 0$. Consider the map $U_{\rho_0} \ni \xi \mapsto L_1\xi + N_1(x^0 + \xi + \psi_0(\xi)) \in X_1$. The linear operator $L_1 + N_{1,1}(0, x^0) \colon X_1 \to X_1$ is an isomorphism, since $\|L_1^{-1}N_{1,1}(0,x^0)\| \leqslant a^{-1}\gamma < 1$, and $N_{1,2}(0, x_n) = 0$. Then the Inverse Function Theorem implies that there exists a neighborhood U_{ρ_1} of 0 in X_1 such that for any $\hat{\xi} \in U_{\rho_1}$ there exists only one $\xi \in U_{\rho_0}$ such that $\hat{\xi} = L_1\xi + N_1(x^0 + \xi + \psi_0(\xi))$. Define then $\psi_1 \colon U_{\rho_1} \to X_2$ by

$$\psi_1(\hat{\xi}) = L_2\psi_0(\xi) + N_2(x^0 + \xi + \psi_0(\xi)) - N_2(x^0).$$

It is clear that ψ_1 is a C^1 map, $\psi_1(0) = 0$ and we have

$$T(D^0_{\rho_0}) \supset \{x^1 + \xi + \psi_1(\xi),\ \xi \in U_{\rho_1}\}.$$

Assume that we already have defined a C^1 map $\psi_n \colon U_{\rho_n} \to X_2$, with $\psi_n(0) = 0$ and $T^n(D^0_{\rho_0}) \supset \{x^n + \xi + \psi_n(\xi),\ \xi \in U_{\rho_n}\}$. Again, we consider the map $U_{\rho_n} \ni \xi \mapsto L_1\xi + N_1(x^n + \xi + \psi_n(\xi)) \in X_1$, and the linear operator $L_1 + N_{1,1}(0, x^n) \colon X_1 \to X_1$ is also an isomorphism, since $\|L_1^{-1}N_{1,1}(0,x^n)\| \leqslant a^{-1}\gamma < 1$. Then the Inverse Function Theorem implies the existence of a $\rho_{n+1} > 0$ such that for any $\hat{\xi} \in U_{\rho_{n+1}}$ there exists only one $\xi \in U_{\rho_n}$ such that $\hat{\xi} = L_1\xi + N_1(x^n + \xi + \psi_n(\xi))$.

Define then the C^1 map $\psi_{n+1} \colon U_{\rho_{n+1}} \to X_2$ by

$$\psi_{n+1}(\hat{\xi}) = L_2\psi_n(\xi) + N_2(x^n + \xi + \psi_n(\xi)) - N_2(0, x^n).$$

We have $\psi_{n+1}(0) = 0$ and

$$T^{n+1}(D^0_{\rho_0}) \supset \{x^{n+1} + \xi + \psi_{n+1}(\xi),\ \xi \in U_{\rho_{n+1}}\}.$$

Therefore we have defined $\rho_n > 0$ and C^1 maps $\psi_n \colon U_{\rho_n} \to X_2$, with $\psi_n(0) = 0$ and such that

$$T^n(D^0_{\rho_0}) \supset \{x^n + \xi + \psi_n(\xi),\ \xi \in U_{\rho_n}\}.$$

Step 4. Let us define $\Lambda_n \in \mathcal{L}(X_1, X_2)$, for $n \geqslant 0$, by $\Lambda_n = \psi_n'(0)$. Then, by definition of ψ_n, we have

$$\Lambda_{n+1} = [(L_2 + N_{2,2}(0,x^n))\Lambda_n + N_{2,1}(0,x^n)] \circ [L_1 + N_{1,1}(0,x^n)]^{-1},$$

since $\|L_1^{-1}N_{1,1}(0,x^n)\| \leqslant a^{-1}\gamma < 1$ for all $n \geqslant 0$. Moreover,

$$\|\Lambda_{n+1}\| \leqslant \Big((b+\gamma)\|\Lambda_n\| + \gamma\Big)(a-\gamma)^{-1} = \frac{b+\gamma}{a-\gamma}\|\Lambda_n\| + \frac{\gamma}{a-\gamma}, \tag{4.10}$$

for each $n \geqslant 0$.

We now prove that the sequence $\{\|\Lambda_n\|\}_{n\geqslant 0}$ is bounded. Indeed, if $\gamma^* \doteq \frac{\gamma}{a-b-2\gamma}$, then:

(i) if $\|\Lambda_n\| \leqslant \gamma^*$,

$$\|\Lambda_{n+1}\| \leqslant \frac{b+\gamma}{a-\gamma}\gamma^* + \frac{\gamma}{a-\gamma} = \gamma^*;$$

(ii) if $\|\Lambda_n\| > \gamma^*$, assume that $\|\Lambda_{n+1}\| > \|\Lambda_n\|$. Thus

$$\|\Lambda_n\| < \frac{b+\gamma}{a-\gamma}\|\Lambda_n\| + \frac{\gamma}{a-\gamma},$$

which implies that $\|\Lambda_n\| < \gamma^*$ and contradicts our hypothesis. Thus $\|\Lambda_{n+1}\| \leqslant \|\Lambda_n\|$.

Items (i) and (ii) combined guarantee that $\{\|\Lambda_n\|\}_{n\geqslant 0}$ is bounded and if $\alpha \doteq \limsup_{n\to\infty} \|\Lambda_n\|$, then by (4.10) we obtain

$$\alpha \leqslant \frac{b+\gamma}{a-\gamma}\alpha + \frac{\gamma}{a-\gamma},$$

and hence $\alpha \leqslant \gamma^* < \frac{1}{2}$. This implies that there exists $n_0 \geqslant 0$ such that $\|\Lambda_n\| \leqslant \frac{1}{2}$ for all $n \geqslant n_0$.

Let $0 < \rho < \min_{0\leqslant n\leqslant n_0}\{\rho_n\}$ such that ψ_n is a C^1 map defined in $\|\xi\| \leqslant \rho$ and

$$\sup_{\|\xi\|\leqslant\rho} \|\psi_n'(\xi)\| \leqslant \|\Lambda_n\| + \frac{1}{2}, \quad \text{for each } 0 \leqslant n \leqslant n_0, \tag{4.11}$$

and in particular we have $\sup_{\|\xi\|\leqslant\rho} \|\psi_{n_0}'(\xi)\| \leqslant \|\Lambda_{n_0}\| + \frac{1}{2} \leqslant 1$.

Step 5. Now, we prove that for $n \geqslant n_0$ the functions $\psi_n \colon U_{\rho_n} \to X_2$ can be extended to U_{r_n}, where $r_n = \min\{\rho(a-\gamma)^{n-n_0}, r_0\}$. Equation (4.11) implies that $\operatorname{Lip}\psi_{n_0} \leqslant 1$ in U_ρ. Then given $\hat{\xi} \in U_{r_{n_0+1}}$ the map

$$U_{r_{n_0}} \ni \xi \mapsto L_1^{-1}\hat{\xi} - L_1^{-1}N_1(x^{n_0} + \xi + \psi_{n_0}(\xi)) \in U_{r_{n_0}}$$

is a well defined contraction. Indeed, since $N_1(x^{n_0} + \psi_{n_0}(\xi)) = 0$, we have

(4.12)
$$\begin{aligned}\|L_1^{-1}\hat{\xi} - L_1^{-1}N_1(x^{n_0} + \xi + \psi_{n_0}(\xi))\| &\leqslant a^{-1}\|\hat{\xi}\| + a^{-1}\gamma\|\xi\| \leqslant a^{-1}\Big(1 + \frac{\gamma}{a-\gamma}\Big)\|\hat{\xi}\| \\ &= \frac{1}{a-\gamma}\|\hat{\xi}\| \leqslant \frac{1}{a-\gamma} r_{n_0+1} \leqslant r_{n_0},\end{aligned}$$

and since $\operatorname{Lip}\psi_{n_0} \leqslant 1$ we obtain

$$\|L_1^{-1}N_1(x^{n_0} + \xi + \psi_{n_0}(\xi)) - L_1^{-1}N_1(x^{n_0} + \tilde{\xi} + \psi_{n_0}(\tilde{\xi}))\| \leqslant 2a^{-1}\gamma\|\xi - \tilde{\xi}\|.$$

Then, for each $\hat{\xi} \in U_{r_{n_0+1}}$ there exists only one $\xi \in U_{r_{n_0}} = U_\rho$ such that

$$\hat{\xi} = L_1\xi + N_1(x^{n_0} + \xi + \psi_{n_0}(\xi)).$$

Hence, we can extend ψ_{n_0+1} to $U_{r_{n_0+1}}$ by

$$\psi_{n_0+1}(\hat{\xi}) = L_2\psi_{n_0}(\xi) + N_2(x^{n_0} + \xi + \psi_{n_0}(\xi)) - N_2(0, x^{n_0}).$$

Now, since $\|L_1^{-1}N_{1,1}(x^{n_0} + \xi + \psi_{n_0}(\xi)) + L_1^{-1}N_{1,2}(x^{n_0} + \xi + \psi_{n_0}(\xi))\psi_{n_0}'(\xi)\| \leqslant 2a^{-1}\gamma < 1$ for all $\xi \in U_{r_{n_0}}$, we can see that

(4.13)
$$\begin{aligned}\psi_{n_0+1}'(\hat{\xi}) =&[(L_2 + N_{2,2}(x^{n_0} + \xi + \psi_{n_0}(\xi)))\psi_{n_0}'(\xi) + N_{2,1}(x^{n_0} + \xi + \psi_{n_0}(\xi))] \\ &\circ [L_1 + N_{1,1}(x^{n_0} + \xi + \psi_{n_0}(\xi)) + N_{1,2}(x^{n_0} + \xi + \psi_{n_0}(\xi))\psi_{n_0}'(\xi)]^{-1},\end{aligned}$$

for all $\hat{\xi} \in U_{r_{n_0+1}}$. Then for all $\hat{\xi} \in U_{r_{n_0+1}}$ we have

$$\|\psi'_{n_0+1}(\hat{\xi})\| \leqslant \frac{(b+\gamma)}{a-2\gamma}\|\psi'_{n_0}(\xi)\| + \frac{\gamma}{a-2\gamma} \leqslant \frac{(b+\gamma)}{a-2\gamma} + \frac{\gamma}{a-2\gamma} = \frac{b+2\gamma}{a-2\gamma},$$

and therefore

$$\sup_{\hat{\xi}\in U_{r_{n_0+1}}} \|\psi'_{n_0+1}(\hat{\xi})\| \leqslant \frac{b+2\gamma}{a-2\gamma} < 1,$$

and hence Lip $\psi_{n_0+1} \leqslant 1$. Inductively, the extension of ψ_n to U_{r_n} can be done analogously and for any $\hat{\xi} \in U_{r_n}$ there exists $\xi \in U_{r_{n-1}}$ such that

$$(4.14) \qquad \begin{cases} \hat{\xi} = L_1\xi + N_1(x^{n-1} + \xi + \psi_{n-1}(\xi)), \\ \psi_n(\hat{\xi}) = L_2\psi_{n-1}(\xi) + N_2(x^{n-1} + \xi + \psi_{n-1}(\xi)) - N_2(0, x^{n-1}). \end{cases}$$

Since $a - \gamma > 1$ we can see that $\rho \leqslant r_n \leqslant r_{n+1}$ for all $n \geqslant n_0$, also, there exists $n_1 \geqslant n_0$ such that $r_n = r_0$ for all $n \geqslant n_1$, and defining $\lambda_n = \sup_{\xi\in U_{r_0}} \|\psi'_n(\xi)\|$, we have $\lambda_n \leqslant 1$, for all $n \geqslant n_1$.

Step 6. Equation (4.14) implies that

$$\frac{\|\psi_{n+1}(\hat{\xi})\|}{\|\hat{\xi}\|} \leqslant \frac{b+\gamma}{a-\gamma}\frac{\|\psi_n(\xi)\|}{\|\xi\|} + \omega(\|x^n\|),$$

for $\|\hat{\xi}\| \leqslant r_0$, where

$$\omega(\|x^n\|) := \frac{\|N_2(\xi, x^n) - N_2(0, x^n)\|}{\|\xi\|}.$$

Since $\omega(\|x^n\|) \to 0$ as $n \to \infty$, we have $\sup_{\xi\in U_{r_0}} \frac{\|\psi_n(\xi)\|}{\|\xi\|} \to 0$ as $n \to \infty$, and defining $|\!|\!|\psi_n|\!|\!| = \sup_{\xi\in U_{r_0}} \|\psi_n(\xi)\|$, we know that $|\!|\!|\psi_n|\!|\!| \to 0$ as $n \to \infty$. Thus

$$\sup_{\xi\in U_{r_0}} \|N_{2,1}(x^n + \xi + \psi_n(\xi))\| \to 0 \qquad \text{as } n \to \infty.$$

Furthermore from (4.13) we have

$$\lambda_{n+1} \leqslant \frac{b+\gamma}{a-2\gamma}\lambda_n + \frac{1}{a-2\gamma}\sup_{\xi\in U_{r_0}} \|N_{2,1}(x^{n_0} + \xi + \psi_{n_0}(\xi))\|,$$

and hence[1] $\lambda_n \to 0$ as $n \to \infty$. Thus, we have proven that if

$$S_n = \{x^n + \xi + \psi_n(\xi) \colon \xi \in U_{r_0}\},$$

then $S_n \to D^*$ as $n \to \infty$ in the C^1 topology.

Step 7. It just remains to show that $S_n = D^n$, for $n \geqslant n_1$.

(i) Given $(\xi, x^n + \psi_n(\xi)) \in S_n$, then the sequence $p_i = (\xi_i, x^i + \psi_i(\xi_i))$, $i = 0, \cdots, n$, with $\xi_n = \xi$ and $\xi_{i+1} = \hat{\xi}_i$, for $0 \leqslant i < n$, satisfies $p_i \in V$, for all $i = 0, \cdots, n$, $p_0 \in D^0$ and

$$T(p_i) = p_{i+1}, \quad i = 0, \cdots, n-1.$$

Hence $S_n \subset D^n$.

[1] It is simple to show that if $\{a_n\}$ is bounded, $b_n \to 0$ as $n \to \infty$, $\alpha \in (0,1)$, and $a_{n+1} \leqslant \alpha a_n + b_n$ for all $n \in \mathbb{N}$, then $a_n \to 0$.

(ii) If $p = p_n \in D^n$, we have $p_0 \in D^0$, $p_1, \cdots, p_n \in V$, $p_{j+1} = T(p_j)$, for $0 \leqslant j < n$. If $p_j = x^j + \xi_j + \eta_j$ ($\xi_j \in X_1$, $\eta_j \in X_2$) we have

$$\begin{aligned} p_{j+1} = x^{j+1} + \xi_{j+1} + \eta_{j+1} &= T(p_j) = T(x^j + \xi_j + \eta_j) \\ &= Lx^j + L_1\xi_j + L_2\eta_j + N_1(\xi_j, x^j + \eta_j) + N_2(\xi_j, x^j + \eta_j) \\ &= Tx^j + L_1\xi_j + L_2\eta_j - N_1(0, x^j) - N_2(0, x^j) \\ &\quad + N_1(\xi_j, x^j + \eta_j) + N_2(\xi_j, x^j + \eta_j), \end{aligned}$$

and thus $\xi_{j+1} = L_1\xi_j + N_1(\xi_j, x^j + \eta_j) - N_1(0, x^j + \eta_j)$, with $N_1(0, x^j + \eta_j) = 0$ and hence $\|\xi_{j+1}\| \geqslant (a - \gamma)\|\xi_j\|$. Therefore

$$\|\xi_0\| \leqslant \frac{\|\xi_n\|}{(a-\gamma)^n} < \frac{r_0}{(a-\gamma)^n} \leqslant \rho, \text{ for } n \geqslant n_1.$$

Since $p_0 \in D^0$ and $\|\xi_0\| < \rho$, we have $p_0 \in D^0_\rho$ and thus

$$p_n = T^n(p_0) \in S_n,$$

which proves that $D^n \subset S_n$. □

To prove the global version of the λ-lemma, we will need the reversibility of the map T, which is defined below.

Definition 4.18. Let $T\colon X \to X$ be a map in a Banach space X such that the discrete semigroup $\mathcal{T} = \{T^n \colon n \in \mathbb{N}\}$ has a global attractor $\mathcal{A}$. We say that T is **reversible** if $T|_{\mathcal{A}}$ is injective, T is differentiable in $\mathcal{A}$ and $T'(z)\colon X \to X$ is an injective linear operator, for each $z \in \mathcal{A}$.

We note that if T is a map such that $\mathcal{T} = \{T^n \colon n \in \mathbb{N}\}$ has a global attractor $\mathcal{A}$, then the unstable set of each critical element is contained in $\mathcal{A}$. Using reversibility we can prove that the unstable set W^u is a C^1 injectively immersed submanifold of X.

Proposition 4.19. *Let x be a hyperbolic fixed point of a reversible map T. The set $W^u(x) = \bigcup_{n\geqslant 0} T^n(W^u_{loc}(x))$ is an injectively immersed C^1 submanifold of X.*

Proof. According to Theorem 4.13, there exists a C^1 map $\theta\colon X_1 \to X_2$ such that

$$W^u_{loc}(x) = \{x + \xi + \theta(\xi) \colon \xi \in X_1,\ \|\xi\| < r_0\},$$

for some $r_0 > 0$. Thus, given $z \in W^u(x)$, there exists the smallest nonnegative integer n_z such that $z \in T^{n_z}(W^u_{loc}(x))$. Since T is reversible, $T^{n_z}(W^u_{loc}(x))$ is open and thus let U_z be a neighborhood of z in $T^{n_z}(W^u_{loc}(x))$, and let $W_z = T^{-n_z}U_z$ be a neighborhood in $W^u_{loc}(x)$. If $\pi_u\colon W^u_{loc}(x) \to X_1$ is the projection of $W^u_{loc}(x)$ in X_1, that is, $\pi_u(x + \xi + \theta(\xi)) = \xi$, for $\xi \in X_1$ with $\|\xi\| < r_0$, define $\varphi_z\colon \pi_u(W_z) \to U_z$ by $\varphi(\xi) = T^{n_z}(x + \xi + \theta(\xi))$. It is clear that $\{(\pi_u(W_z), \varphi_z)\}$ is a C^1 chart for $W^u(x)$, and since the inclusion $i\colon W^u(x) \to X$ has injective derivative with this chart, $W^u(x)$ is an injectively immersed C^1 submanifold of X. □

Now we are able to prove the global λ-lemma.

Theorem 4.20 (Global λ-lemma). *Let X be a Banach space, $T\colon X \to X$ a map such that the discrete semigroup $\{T^n \colon n \in \mathbb{N}\}$ has a global attractor $\mathcal{A}$. Assume that $T\colon X \to X$ is a C^1 map in a neighborhood of $\mathcal{A}$ and that it is* *reversible. Assume also that* $T(0) = 0$ *and* $L = T'(0) \in \mathcal{L}(X)$ *is such that* $\sigma(L) \setminus \{\xi \in \mathbb{C}\colon |\xi| \leqslant 1\}$ *is closed. Consider the decompositions* $X = X_1 \oplus X_2$ *of* X *and* $L = L_1 \oplus L_2$ *of* L *given by the spectral projections* Π_i^L, $i = 1, 2$, *associated to the spectral sets* $\sigma_2(L) = \sigma(L) \cap \{\xi \in \mathbb{C}\colon |\xi| \leqslant 1\}$ *and* $\sigma_1(L) = \sigma(L) \cap \{\xi \in \mathbb{C}\colon |\xi| > 1\} \subset \{\lambda \in \mathbb{C}\colon |\lambda| \geqslant r\}$, *for some* $r > 1$. *Denote by* W^u_{loc} *and* W^{cs}_{loc} *the local manifolds (unstable and center-stable, respectively) of* 0 *and* $W^u = \bigcup_{n \geqslant 0} T^n(W^u_{loc})$. *In addition, assume that* $\dim X_1 < \infty$ *or that* $T \in C^{1+}$ *in* $\mathcal{A}$ ($\mathcal{A} \ni z \mapsto T'(z)$ *is uniformly continuous).*

Then, given $n_0 \in \mathbb{N}$, *and a submanifold* D^0 *of* X *such that* $D^0 \cap W^{cs}_{loc} = \{x^0\}$, $T_{x^0} D^0 \oplus T_{x^0} W^{cs}_{loc} = X$, *there exist a neighborhood* D^0_ρ *of* x^0 *in* D^0 *and (for* n *sufficiently large) sets* $S^n \subset T^n(D^0_\rho)$ *such that*

$$S^n \to T^{n_0}(W^u_{loc}) \text{ as } n \to \infty,$$

in the C^1 *norm.*

Proof. By the local λ-lemma we know that there exist neighborhoods D^0_ρ of x^0 in D^0, V of 0 in X and sets (for n sufficiently large)

$$R^n = \{x^n + \xi + \psi_n(\xi)\colon \xi \in X_1,\ \|\xi\| < r_0\} \subset T^n(D^0_\rho),$$

where $x^n = T^n(x^0)$, $\psi_n\colon X_1 \to X_2$ are C^1 maps, such that $R^n \to W^u_{loc}$, as $n \to \infty$, in the C^1 norm. More precisely, if $W^u_{loc} = \{\xi + \theta(\xi)\colon \xi \in X_1,\ \|\xi\| < r_0\}$, then

$$\|\psi_n - \theta\|_{C^1(B^{X_1}_{r_0}(0), X_2)} \to 0, \text{ as } n \to \infty.$$

Let $U_r = \{\xi \in X_1 : \|\xi\| < r\}$ for $r > 0$ and define $\varphi, \mu_n\colon U_{r_0} \to T^{n_0}(W^u_{loc})$ by

$$\varphi(\xi) = T^{n_0}(\xi + \theta(\xi)) \text{ and } \mu_n(\xi) = T^{n_0}(x^n + \xi + \psi_n(\xi)),$$

for n sufficiently large. Since $\dim X_1 < \infty$ or $T \in C^{1+}$ in a neighborhood of $\mathcal{A}$,

$$\sup_{\xi \in U_{r_0}} \|\mu_n(\xi) - \varphi(\xi)\| \to 0 \text{ as } n \to \infty,$$

and also

$$\sup_{\xi \in U_{r_0}} \|\mu_n'(\xi) - \varphi'(\xi)\| \to 0, \text{ as } n \to \infty.$$

The proof is complete taking $S^n = T^{n_0}(R^{n-n_0})$, for n sufficiently large. □

6. Permanence and continuity of critical elements

In this section we prove robustness under perturbation of the two elementary invariants for a given dynamical system: equilibria and periodic orbits, which, in many applications, are the basic blocks that describe the internal structure of a global attractor for a semigroup.

6.1. Permanence and continuity of equilibria. Let $\eta \in [0, 1]$ be a parameter. For each $\eta \in [0, 1]$, let T_η be a map in $\mathcal{C}(X)$ and let $\mathcal{E}_\eta$ be its set of fixed points (note that $\mathcal{E}_\eta$ could possibly be empty). In this section, our aim is to study the upper and lower semicontinuity (see Definition 2.1) of the family $\{\mathcal{E}_\eta\}_{\eta \in [0,1]}$ at $\eta = 0$.

We prove upper semicontinuity of the family $\{\mathcal{E}_\eta\}_{\eta \in [0,1]}$ under natural assumptions, as remarked before, this means that the set $\mathcal{E}_\eta$ does not *explode* (becomes suddenly *larger* than $\mathcal{E}_0$) as we move away from $\eta = 0$. We will use Lemma 2.2 to establish the upper and lower semicontinuity, and we remark that, from the continuity of each T_η, each $\mathcal{E}_\eta$ is closed, $\eta \in [0, 1]$.

THEOREM 4.21 (Upper semicontinuity of equilibria). *Let the map* $[0,1] \ni \eta \mapsto T_\eta x \in X$ *be continuous at* $\eta = 0$*, uniformly in compact subsets of* X*, and assume that the family of semigroups* $\{T_\eta^n : n \in \mathbb{N}\}_{\eta\in[0,1]}$ *is collectively asymptotically compact (see Definition 2.3(ii) with* $\mathbb{R}^+$ *replaced by* $\mathbb{N}$*). If* T_η *has a nonempty set of fixed points* $\mathcal{E}_\eta$*, for each* $\eta \in (0,1]$*, and* $\bigcup_{\eta\in(0,1]} \mathcal{E}_\eta$ *is bounded, then* $\mathcal{E}_\eta$ *is compact for each* $\eta \in (0,1]$*,* $\mathcal{E}_0 \neq \varnothing$ *and the family* $\{\mathcal{E}_\eta\}_{\eta\in[0,1]}$ *is upper semicontinuous at* $\eta = 0$*.*

PROOF. For the compactness of $\mathcal{E}_\eta$ simply note that, for each $n \in \mathbb{N}$ and $x_\eta^* \in \mathcal{E}_\eta$, $T_\eta x_\eta^* = x_\eta^*$ and use the asymptotic compactness of $\{T_\eta^n : n \in \mathbb{N}\}$. For the upper semicontinuity of the family $\{\mathcal{E}_\eta\}_{\eta\in[0,1]}$ at $\eta = 0$, simply note that for each sequence $\{\eta_n\} \in (0,1]$ with $\eta_n \to 0$ as $n \to \infty$, the corresponding sequence $\{x_{\eta_n}^*\}$ has a convergent subsequence (from the boundedness of $\bigcup_{\eta\in(0,1]} \mathcal{E}_\eta$ and the collective asymptotic compactness of $\{T_\eta^n : n \in \mathbb{N}\}_{\eta\in[0,1]}$), which we label the same. Now, if $x_{\eta_n}^* \to x^*$ as $n \to \infty$, then since $T_{\eta_n}(x_{\eta_n}^*) = x_{\eta_n}^*$ it follows from the continuity properties of the family $\{T_\eta\}_{\eta\in[0,1]}$ at $\eta = 0$ that

$$x^* = \lim_{n\to\infty} x_{\eta_n}^* = \lim_{n\to\infty} T_{\eta_n}(x_{\eta_n}^*) = T_0(x^*).$$

Hence $T_0 x^* = x^*$ and $x^* \in \mathcal{E}_0 \neq \varnothing$. It follows from Lemma 2.2 that the family $\{\mathcal{E}_\eta\}_{\eta\in[0,1]}$ is upper semicontinuous at $\eta = 0$. □

We have seen already that the lower semicontinuity is a much more complicated matter. It requires additional assumptions to prevent the family $\mathcal{E}_\eta$ collapsing and becoming *smaller* than $\mathcal{E}_0$.

Next we will give conditions under which a point, in the set of fixed points $\mathcal{E}_0$ of a semigroup T_0, is isolated and when it can be approximated by points in $\mathcal{E}_\eta$.

We start with the definition of non-degenerated fixed point. Hereafter in this chapter, unless said otherwise, X is a Banach space, $\mathcal{L}(X)$ is the space of bounded linear operators from X into itself and, for $L \in \mathcal{L}(X)$, $\sigma(L)$ denotes the spectrum of L.

DEFINITION 4.22. We say that $x_0^* \in \mathcal{E}_0$ is **non-degenerated** if T_0 is differentiable at x_0^* and, if $T_0'(x^*) \in \mathcal{L}(X)$ denotes the derivative of T_0 at x_0^*, $1 \notin \sigma(T_0'(x^*))$.

LEMMA 4.23. *Let* $T_0 \in C(X)$ *and let* $x^* \in \mathcal{E}_0$*. If* x^* *is a non-degenerated fixed point of* T_0 *and, for each* $\rho > 0$*, there exists* $\delta > 0$ *such that*

$$\|T_0(y+x^*) - T_0(z+x^*) - T_0'(x^*)(y-z)\|_X \leqslant \rho\|y-z\|_X \text{ for } y,z \in \overline{B_\delta(0)}, \tag{4.15}$$

then x^* *is isolated. Hence, if* $\mathcal{E}_0$ *is compact,* T_0 *satisfies* (4.15) *and* x^* *is non-degenerated for each* $x^* \in \mathcal{E}_0$*, then* $\mathcal{E}_0$ *is finite.*

PROOF. Since x^* is a non-degenerated fixed point of T_0, it follows that $y + x^*$ is a fixed point for T_0 if and only if y is a fixed point for the map

$$\Phi(y) = (I - T_0'(x^*))^{-1}(T_0(y+x^*) - T_0(x^*) - T_0'(x^*)y).$$

Note that $y = 0$ is a fixed point for Φ,

$$\|\Phi(y)\|_X \leqslant \|(I - T_0'(x^*))^{-1}\|_{\mathcal{L}(X)}\|T_0(y+x^*) - T_0(x^*) - T_0'(x^*)y\|_X$$

and also that

$$\begin{aligned}&\|\Phi(y)-\Phi(z)\|_X\\&\quad\leqslant \|(I-T_0'(x^*))^{-1}\|_{\mathcal{L}(X)}\|T_0(y+x^*)-T_0(z+x^*)-T_0'(x^*)(y-z)\|_X.\end{aligned}$$

Since T_0 satisfies (4.15), there exists $\delta>0$ such that

$$\|T_0(y+x^*)-T_0(z+x^*)-T_0'(x^*)(y-z)\|_X\leqslant \frac{1}{2\|(I-T_0'(x^*))^{-1}\|_{\mathcal{L}(X)}}\|y-z\|_X$$

for all $y,z\in\overline{B_\delta(0)}$. It follows that Φ takes $\overline{B_\delta(0)}$ into itself and that it is a contraction in $\overline{B_\delta(0)}$. Hence Φ has a unique fixed point in $\overline{B_\delta(0)}$. As a consequence, x^* is the unique fixed point of T_0 in $\overline{B_\delta(x^*)}$ and hence x^* is an isolated equilibrium point of T_0.

Finally note that, if T_0 satisfies (4.15) for all points of $\mathcal{E}_0$ and all points in $\mathcal{E}_0$ are non-degenerated, then all points in $\mathcal{E}_0$ are isolated. Since $\mathcal{E}_0$ is compact it must be finite. □

We now show that a non-degenerated fixed point x_0^* of T_0 perturbs continuously under a small $\mathcal{C}^1$ perturbation T_η of T_0.

THEOREM 4.24. *Let $\{T_\eta\}_{\eta\in[0,1]}$ be a family of maps in $\mathcal{C}^1(X)$. Assume that $x_0^*\in\mathcal{E}_0$ is non-degenerated, that*

$$\lim_{\eta\to 0}\Big(\|T_\eta(x_0^*)-T_0(x_0^*)\|_X+\|T_\eta'(x_0^*)-T_0'(x_0^*)\|_{\mathcal{L}(X)}\Big)=0,\tag{4.16}$$

and also that, for some $\eta_0\in(0,1]$, given $\rho>0$ there exists $r>0$ such that

$$\|T_\eta(y)-T_\eta(z)-T_\eta'(x_0^*)(y-z)\|_X\leqslant\rho\|y-z\|_X,\tag{4.17}$$

for $y,z\in\overline{B_r(x_0^*)}$ and $\eta\in[0,\eta_0]$.

Then, there exist $\delta>0$ and $\eta_1>0$ such that T_η has a unique fixed point x_η^* in $B_\delta(x_0^*)$ for each $\eta\in[0,\eta_1]$. Furthermore $x_\eta^*\to x_0^*$ as $\eta\to 0$.

PROOF. We first observe that x_η^* is a fixed point for T_η if and only if it is a fixed point for the map

$$\Phi_\eta(y)=(I-T_0'(x_0^*))^{-1}[T_\eta(y)-T_0'(x_0^*)y].$$

Hence,

$$\begin{aligned}\Phi_\eta(y)-x_0^*&=(I-T_0'(x_0^*))^{-1}[T_\eta(y)-T_0(x_0^*)-T_0'(x_0^*)(y-x_0^*)]\\&=(I-T_0'(x_0^*))^{-1}[T_\eta(y)-T_\eta(x_0^*)-T_\eta'(x_0^*)(y-x_0^*)]\\&\quad+(I-T_0'(x_0^*))^{-1}[T_\eta(x_0^*)-T_0(x_0^*)+(T_\eta'(x_0^*)-T_0'(x_0^*))(y-x_0^*)].\end{aligned}$$

Using (4.17) with $\rho=\frac{1}{4\|(I-T_0'(x_0^*))^{-1}\|_{\mathcal{L}(X)}}$, choose $r>0$ such that

$$\|[T_\eta(y)-T_\eta(z)-T_\eta'(x_0^*)(y-z)]\|_X\leqslant\rho\|y-z\|_X$$

for all $y,z\in\overline{B_r(x_0^*)}$ and $\eta\in[0,\eta_0]$. Using (4.16), choose $\eta_1\in(0,\eta_0]$ such that for every $\eta\in[0,\eta_1]$ we have

$$\begin{aligned}&\|T_\eta(x_0^*)-T_0(x_0^*)\|_X\leqslant\frac{r}{4\|(I-T_0'(x_0^*))^{-1}\|_{\mathcal{L}(X)}}\quad\text{and}\\&\|T_\eta'(x_0^*)-T_0'(x_0^*)\|_{\mathcal{L}(X)}\leqslant\frac{1}{4\|(I-T_0'(x_0^*))^{-1}\|_{\mathcal{L}(X)}}.\end{aligned}\tag{4.18}$$

Combining these inequalities, with $z = x_0^*$, we obtain

$$\|\Phi_\eta(y) - x_0^*\|_X \leqslant r \text{ if } y \in \overline{B_r(x_0^*)} \text{ for all } \eta \in [0, \eta_1],$$

that is, $\Phi_\eta(\overline{B_r(x_0^*)}) \subset \overline{B_r(x_0^*)}$ for all $\eta \in [0, \eta_1]$.

Also, for $y, z \in \overline{B_r(x_0^*)}$, we obtain

$$\begin{aligned}\Phi_\eta(y) - \Phi_\eta(z) &= (I - T_0'(x_0^*))^{-1}[T_\eta(y) - T_\eta(z) - T_0'(x^*)(y-z)] \\ &= (I - T_0'(x_0^*))^{-1}[T_\eta(y) - T_\eta(z) - T_\eta'(x_0^*)(y-z)] \\ &+ (I - T_0'(x_0^*))^{-1}[T_\eta'(x_0^*) - T_0'(x^*)](y-z),\end{aligned}$$

and hence

$$\|\Phi_\eta(y) - \Phi_\eta(z)\|_X \leqslant \frac{1}{2}\|y - z\|_X \text{ for each } \eta \in [0, \eta_1].$$

Therefore Φ_η is a contraction in $\overline{B_r(x_0^*)}$ for each $\eta \in [0, \eta_1]$, and has a unique fixed point x_η^* in it.

Finally, note that given $0 < \epsilon < r$, choosing $\eta_2 \in (0, \eta_1]$ such that the first inequality of (4.18) is satisfied with ϵ instead of r, we obtain that Φ_η is a contraction in $\overline{B_\epsilon(x_0^*)}$ for each $\eta \in [0, \eta_2]$ and has a unique fixed point x_η^2 in $\overline{B_\epsilon(x_0^*)}$.

By the uniqueness of x_η^*, we obtain $x_\eta^* = x_\eta^2$ for all $\eta \in [0, \eta_2]$, which implies that

$$\|x_\eta^* - x_0^*\| \leqslant \epsilon \text{ for all } \eta \in [0, \eta_2],$$

and thus $x_\eta^* \to x_0^*$ as $\eta \to 0$ and completes the proof. □

The following result is an immediate consequence of Theorem 4.24 and Lemma 2.2.

THEOREM 4.25 (Lower semicontinuity of non-degenerated equilibria). *Consider a family $\{T_\eta\}_{\eta\in[0,1]}$ of maps in $\mathcal{C}^1(X)$. Assume that $\mathcal{E}_0$ is compact, consists solely of non-degenerated points and that* (4.16) *and* (4.17) *hold for each $x_0^* \in \mathcal{E}_0$. Then the family $\{\mathcal{E}_\eta\}_{\eta\in[0,1]}$ is lower semicontinuous at $\eta = 0$.*

Together, Theorems 4.21 and 4.25 give us the following result.

THEOREM 4.26. *Assume that the conditions of Theorems* 4.21 *and* 4.25 *are satisfied. Then the family $\{\mathcal{E}_\eta\}_{\eta\in[0,1]}$ is continuous at $\eta = 0$. Furthermore, there exists $\tilde{\eta} > 0$ such that $\mathcal{E}_\eta$ is finite and has the same number of elements that $\mathcal{E}_0$, for each $\eta \in [0, \tilde{\eta}]$.*

PROOF. The continuity is clear, so we show the remaining claim. Since $\mathcal{E}_0$ is compact and has only non-degenerated points, it is finite and the points are all isolated. Choose $r_0 > 0$ such that

$$B_{r_0}(x_1) \cap B_{r_0}(x_2) = \varnothing \text{ for each } x_1, x_2 \in \mathcal{E}_0 \text{ with } x_1 \neq x_2.$$

We know, from Theorem 4.25, that there exist $0 < r < r_0$ and $\eta_1 > 0$ such that for each $\eta \in [0, \eta_1]$, T_η has a unique fixed point x_η^* in $\overline{B_r(x_0^*)}$, for each $x_0^* \in \mathcal{E}_0$. Hence $\mathcal{E}_\eta$ has at least as many elements as $\mathcal{E}_0$, for each $\eta \in [0, \eta_1]$. Since $\mathcal{E}_0$ is finite, r and η_1 may be chosen the same for all $x_0^* \in \mathcal{E}_0$.

To complete the claim, assume to the contrary that there exists a sequence $\eta_n \in (0, \eta_1]$, $\eta_n \overset{n\to\infty}{\longrightarrow} 0$, such that T_{η_n} has strictly more fixed points than T_0. From the previous argument, for each n, there must be at least one fixed point y_n^* of T_{η_n} that lies outside of $N = \bigcup_{x_0^*\in\mathcal{E}_0} B_r(x_0^*)$. But then, as in Theorem 4.21, we can extract a convergent subsequence of y_n^* that converges to some $y^* \in \mathcal{E}_0$ which lies outside of N. This contradiction completes the proof. $\square$

DEFINITION 4.27. An equilibrium point e for a differentiable semigroup $\mathcal{T}$ is said to be **hyperbolic** if the spectrum of $D_xT(t)(e)$ is disjoint from the unit circle S^1 in $\mathbb{C}$ for each $t \geqslant 0$, that is, $\sigma(D_xT(t)(e)) \cap S^1 = \varnothing$, for each $t \geqslant 0$.

6.2. Permanence and continuity of periodic solutions. The main purpose of this section is to establish that a normally hyperbolic periodic orbit is an isolated invariant set, that it behaves continuously under perturbations and that the neighborhood where it is a maximal invariant set can be chosen uniformly with respect to perturbations. This is crucial to establish that the nonwandering set of a Morse-Smale semigroup is a disjoint family of isolated invariant sets of a dynamically gradient semigroup.

We will first introduce some general features of periodic solutions of semigroups, and establish some properties of the Poincaré time map and of the spectral projections associated to them. Next we give sufficient conditions for asymptotic stability and instability of periodic solutions. Finally, we establish the permanence of normally hyperbolic periodic solutions and that its orbit is an isolated invariant set and that the neighborhood where it is maximal invariant can be chosen uniformly with respect to perturbations. Our presentation is inspired by that of Henry in [**142**, Chapter 8].

6.3. General features of periodic solutions. Recall that if $\mathcal{T}$ is a semigroup in a Banach space X, a function $\xi\colon \mathbb{R} \to X$ is a p-**periodic solution** if it is a global solution for the semigroup $\mathcal{T}$ and $\xi(t+p) = \xi(t)$ for each $t \in \mathbb{R}$. Any p-periodic solution will be called simply a **periodic solution**. The smallest non-negative p such that $\xi(t+p) = \xi(t)$ for all $t \in \mathbb{R}$ is called the **period** of ξ. Clearly, a 0-periodic solution is a stationary solution. If the period of ξ is greater than zero, we say that ξ is a **nontrivial periodic solution**. For a given p-periodic solution ξ, the set $\xi(\mathbb{R}) = \xi([0,p])$ is called the **orbit** of ξ.

Consider a very simple ordinary differential equation with periodic solutions; that is, the pendulum equation

$$\begin{cases} \dot{x}_1 = x_2, \\ \dot{x}_2 = -\sin x_1, \end{cases} \tag{4.19}$$

which can be written as

$$\ddot{x}_1 + \sin x_1 = 0.$$

In this example there is a continuum of periodic solutions around the zero equilibrium solution. Each two periodic solutions corresponding to two different nearby orbits will have different periods. In this case the period is a function of the amplitude and, as a consequence of that, any periodic solution will be unstable in the usual sense; that is, two periodic solutions, starting very close to one another, will be driven apart if they have very close different amplitude. This shows that any periodic solution of this equation is unstable. Nonetheless, orbits of solutions that start close remain close. This explains why the notion of orbital stability becomes so important when periodic solutions are concerned.

DEFINITION 4.28. Let $\mathcal{T}$ be a semigroup in a Banach space X and let ξ be a periodic solution of it. We say that ξ is:

(a) **orbitally stable** if given a neighborhood U of $\xi(\mathbb{R})$ in X there exists another neighborhood V of $\xi(\mathbb{R})$ in X such that for each $x \in V$, we have $T(t)x \in U$ for all $t \geqslant 0$;

(b) **orbitally asymptotically stable** if it is orbitally stable and given a neighborhood U of $\xi(\mathbb{R})$ in X there exists another neighborhood W of $\xi(\mathbb{R})$ in X such that for each $x \in W$, we have

$$\lim_{t\to\infty} \operatorname{dist}_H(T(t)x, \xi(\mathbb{R})) = 0;$$

(c) **orbitally unstable** if it is not orbitally stable.

We will assume throughout this section that ξ is a nontrivial p-periodic solution of $\mathcal{T}$, that $T(t) \in \mathcal{C}^1(X)$ for each $t \geqslant 0$ and that

(4.20) $$\mathbb{R}^+ \ni t \mapsto D_xT(t)(x)y \in X \text{ is continuous, for each } x, y \in X.$$

Note that a single p-periodic nontrivial solution of a semigroup $\mathcal{T}$ gives us, by translation or *phase change*, infinitely many solutions, that is, for each $\theta \in [0, p)$ the global solution $\xi_\theta(\cdot) = \xi(\cdot + \theta)$ is also a periodic solution. Fix $\theta \in [0, p)$ and define the bounded linear operator $S_\theta(t, s)$ in X by

(4.21) $$S_\theta(t, s) = D_xT(t-s)(\xi_\theta(s)) \text{ for } t \geqslant s.$$

When $\theta = 0$ we define

(4.22) $$S(t, s) = S_0(t, s) = D_xT(t-s)(\xi(s)) \text{ for } t \geqslant s.$$

Before we proceed we recall, from the introduction, the definition of *evolution process.*

DEFINITION 4.29. Let X be a Banach space and let $\mathcal{C}(X)$ be the space of continuous maps from X into itself. A family $\mathcal{S} = \{S(t, s)\colon t \geqslant s\} \subset \mathcal{C}(X)$ is called an **evolution process** if it satisfies

(i) $S(t, t)x = x$ for all $t \in \mathbb{R}$ and for all $x \in X$,
(ii) $S(t, r) \circ S(r, s)x = S(t, s)x$ for all $t \geqslant r \geqslant s$, $x \in X$,
(iii) the map $\{(t, s, x) \in \mathbb{R} \times \mathbb{R} \times X : t \geqslant s\} \ni (t, s, x) \mapsto S(t, s)x \in X$ is continuous.

If $\mathcal{S} = \{S(t, s)\colon t \geqslant s\} \subset \mathcal{L}(X)$ we say that it is a **linear evolution process**.

The results presented next for $\mathcal{S}$ are also true for $\mathcal{S}_\theta = \{S_\theta(t, s)\colon t \geqslant s\}$, for all $\theta \in (0, p)$, with unchanged proofs. Moreover the constants obtained are independent of θ.

PROPOSITION 4.30. *The family $\mathcal{S} = \{S(t, s)\colon t \geqslant s\}$ is a linear evolution process, satisfies*

$$S(t+p, s+p) = S(t, s) \text{ for all } t \geqslant s,$$

and $S(t+np, t) = S(t+p, t)^n$ for all $t \in \mathbb{R}$ and each nonnegative integer n. Also 1 is an eigenvalue of $S(t+p, t)$ for all $t \in \mathbb{R}$.

PROOF. It is clear that $S(t, t)$ is the identity operator in X for each $t \in \mathbb{R}$. Now if $s \leqslant \tau \leqslant t$ we have

$$T(t-s)\xi(s) = T(t-\tau)T(\tau-s)\xi(s),$$

and differentiating both sides with respect to x at $\xi(s)$, we obtain

$$\begin{aligned} S(t,s) &= D_xT(t-s)(\xi(s)) = D_xT(t-\tau)(T(\tau-s)\xi(s))D_xT(\tau-s)(\xi(s)) \\ &= D_xT(t-\tau)(\xi(\tau))D_xT(\tau-s)(\xi(s)) = S(t,\tau)S(\tau,s). \end{aligned}$$

The continuity property of $\mathcal{S}$ follows from (4.20), the fact that $T(t) \in \mathcal{C}^1(X)$ for each $t \geqslant 0$ and that $\mathbb{R}^+ \ni t \mapsto \xi(t) \in X$ is continuous.

Now since ξ is p-periodic, we have

$$S(t+p,s+p) = D_xT(t+p-(s+p))(\xi(s+p)) = D_xT(t-s)(\xi(s)) = S(t,s),$$

and obtain

$$S(t+np,t) = S(t+np,t+(n-1)p)\cdots S(t+2p,t+p)S(t+p,t) = S(t+p,t)^n. \tag{4.23}$$

Also, since $\xi(t+\tau) = T(\tau)\xi(t)$, differentiating both sides with respect to t we have

$$\xi'(t+\tau) = (D_xT(\tau)(\xi(t)))\xi'(t) = S(t+\tau,t)\xi'(t).$$

This implies that if $\xi'(t) = 0$ for some $t \in \mathbb{R}$, then $\xi'(t+\tau) = 0$ for all $\tau \in \mathbb{R}$, and therefore ξ is constant. Since ξ is a nontrivial periodic solution, we must have $\xi'(t) \neq 0$ for all $t \in \mathbb{R}$. Also, in the particular case $\tau = p$ we have

$$\xi'(t) = D_xT(p)(\xi(t))\xi'(t) = S(t+p,t)\xi'(t),$$

and 1 is an eigenvalue of $S(t+p,t)$ for each $t \in \mathbb{R}$. □

From (4.23) and the continuity of $\mathbb{R}^+ \ni t \mapsto D_xT(t)(x) \in \mathcal{L}(X)$ uniformly in compact subsets of X, for all $t \geqslant s$ we have

$$\|S(t,s)\|_{\mathcal{L}(X)} \leqslant Me^{m(t-s)}. \tag{4.24}$$

DEFINITION 4.31. The **period map** is defined as the bounded linear operator

$$U(t) = S(t+p,t) = D_xT(p)(\xi(t)) \quad \text{for each } t \in \mathbb{R}.$$

The nonzero eigenvalues of $U(t)$ are called the **characteristic multipliers**.

REMARK 4.32. Note that $\mathbb{R}^+ \ni t \mapsto U(t) \in \mathcal{L}(X)$ is continuous, since $T(p) \in \mathcal{C}^1(X)$ and $\mathbb{R}^+ \ni t \mapsto \xi(t) \in X$ is continuous.

Clearly U is periodic and $U(t+p) = U(t)$ for each $t \in \mathbb{R}$. We can also define here $U_\theta(t) = S_\theta(t+p,t)$ for each $\theta \in (0,p)$, and again, all the following results remain true for each $\theta \in (0,p)$, and all the obtained results are independent of θ.

Firstly, we have the following.

LEMMA 4.33. *The characteristic multipliers are independent of time. In fact, $\sigma(U(t)) \setminus \{0\}$ is independent of time.*

PROOF. Suppose that $\lambda \neq 0$ and $x \neq 0$ are such that $U(s)x = S(s+p,s)x = \lambda x$ and $t \in [s,s+p]$. Let $y = S(t,s)x$, so $S(s+p,t)y = S(s+p,t)S(t,s)x = S(s+p,s)x = U(s)x = \lambda x$. Hence $y \neq 0$ and $U(t)y = S(t+p,t)y = S(t+p,s)x = S(t+p,s+p)S(s+p,s)x = \lambda S(t+p,s+p)x = \lambda S(t,s)x = \lambda y$. This and the fact that U is p-periodic proves that the characteristic multipliers are independent of time.

If $\lambda \in \rho(U(s))\backslash\{0\}$, $t \in [s,s+p]$ and $y \in X$ let

$$x = \frac{1}{\lambda}[I + S(t,s)(\lambda - U(s))^{-1}S(s+p,t)]y = Ly.$$

Then,

$$
\begin{aligned}
U(t)x = S(t+p,t)x &= \frac{1}{\lambda}[S(t+p,t) + S(t+p,s)(\lambda - U(s))^{-1}S(s+p,t)]y \\
&= \frac{1}{\lambda}S(t,s)[I + U(s)(\lambda - U(s))^{-1}]S(s+p,t)y \\
&= S(t,s)(\lambda - U(s))^{-1}S(s+p,t)y
\end{aligned}
\tag{4.25}
$$

and $(\lambda - U(t))x = y$, that is, $(\lambda - U(t))Ly = y$ for all $y \in X$. Similarly $L(\lambda - U(t))y = y$ for all $y \in X$ and $\lambda \in \rho(U(t)) \setminus \{0\}$.

This and the periodicity of U show that $\rho(U(t)) \setminus \{0\}$ is independent of time. □

Before stating our next result, let us emphasize that, as a consequence of Lemma 4.33, $\sigma(U(t)) \setminus \{0\}$ is independent of t.

THEOREM 4.34. *Suppose that σ_1 is a spectral set for $\sigma(U(t))$ for all $t \in \mathbb{R}$. Then for each $t \in \mathbb{R}$ the space X may be decomposed as $X = X_1(t) \oplus X_2(t)$ with*

(i) *$S(t,s)X_i(s) \subset X_i(t)$ for $t \geqslant s$ and $i = 1, 2$, which allow us to define*

$$S_i(t,s) = S(t,s)|_{X_i(s)} \colon X_i(s) \to X_i(t) \text{ and } U_i(t) = S_i(t+p,t),$$

for each $t \geqslant s$ and $i = 1, 2$;

(ii) *$\sigma(U_1(t)) = \sigma_1$ and $\sigma_2 = \sigma(U_2(t)) = \sigma(U(t)) \setminus \sigma_1$;*

(iii) *if $0 \notin \sigma_1$, then $S_1(t,s)$ is bijective for each $t \geqslant s$.*

PROOF. First note that if $t \geqslant s$ we have

$$S(t,s)U(s) = S(t,s)S(s,s-p) = S(t,t-p)S(t-p,s-p) = U(t)S(t,s),$$

therefore we have

$$U(t)S(t,s) = S(t,s)U(s) \quad \text{and} \quad S(t,s)(\lambda - U(s))^{-1} = (\lambda - U(t))^{-1}S(t,s),$$

for each $\lambda \in \rho(U(t)) \setminus \{0\}$. Hence if Γ is a closed simple rectifiable contour, positively oriented, disjoint from $\sigma(U(t))$, enclosing σ_1 and with σ_2 in its exterior we have

$$S(t,s)P_{\sigma_1}(s) = P_{\sigma_1}(t)S(t,s),$$

where $P_{\sigma_1}(\cdot)$ is the p-periodic family of projections given by

$$P_{\sigma_1}(t) = \frac{1}{2\pi i}\int_\Gamma (\lambda - U(t))^{-1} d\lambda \quad \text{for all } t \in \mathbb{R}.$$

Defining $X_1(t) = P_{\sigma_1}(t)X$ and $X_2(t) = (I - P_{\sigma_1}(t))X$ for each $t \in \mathbb{R}$ we have $X = X_1(t) \oplus X_2(t)$ for each $t \in \mathbb{R}$ and $S(t,s)$ maps $X_i(s)$ into $X_i(t)$ for $t \geqslant s$ and $i = 1, 2$, which proves (i) and (ii).

For (iii), if $S_1(t,s)x = 0$ for some $x \in X_1(s)$ with $x \neq 0$, let $n \in \mathbb{N}$ be such that $s \leqslant t \leqslant s + np$. We have then $0 = S_1(s+np,t)S_1(t,s)x = S_1(s+np,s)x = (U_1(s))^n x$. If $x_1 = (U_1(s))^{n-1}x \neq 0$, we obtain $U_1(s)x_1 = 0$ and $0 \in \sigma_1$. Otherwise $(U_1(s))^{n-1}x = 0$ and we proceed by induction, reaching at most $U_1(s)x = 0$, to obtain $0 \in \sigma_1$. Therefore if $0 \notin \sigma_1$, $S_1(t,s) \colon X_1(s) \to X_1(t)$ is injective.

Now, if $t \geqslant s$, $y \in X_1(t)$, and we choose $n \in \mathbb{N}$ such that $t - np \leqslant s$, we can write

$$y = U_1^n(t)U_1^{-n}(t)y = S_1(t,t-np)U_1^{-n}(t)y = S_1(t,s)S_1(s,t-np)U_1^{-n}(t)y,$$

and defining $x = S_1(s,t-np)U_1^{-n}(t)y$ we have $S_1(t,s)x = y$, and $S_1(t,s)$ is surjective. □

THEOREM 4.35. *Under the conditions of Theorem 4.34, if* $e^{\beta p} = \sup\{|\lambda|\colon \lambda \in \sigma_1\} = r_\sigma(U_1(t))$ *(the spectral radius of* $U_1(t)$*), then for any* $\epsilon > 0$ *there exists* $M_\epsilon > 0$ *such that*

$$\|S_1(t,s)x\| \leqslant M_\epsilon e^{(\beta+\epsilon)(t-s)}\|x\| \text{ for all } t \geqslant s \text{ and } x \in X_1(s), \tag{4.26}$$

and if $0 \notin \sigma_1$ *and* $e^{\gamma p} = \inf\{|\lambda|\colon \lambda \in \sigma_1\} > 0$, *then defining* $S_1(s,t)\colon X_1(t) \to X_1(s)$ *as* $S_1(s,t) = S_1(t,s)^{-1}$ *for each* $t \geqslant s$ *we have, given* $\epsilon > 0$, *there exists* $M_\epsilon > 0$ *such that*

$$\|S_1(s,t)x\| \leqslant M_\epsilon e^{(\gamma-\epsilon)(s-t)}\|x\| \text{ for } t \geqslant s \text{ and } x \in X_1(t).$$

If, for some $\theta \in (-\pi,\pi]$, $\{re^{i\theta}\colon r \geqslant 0\} \cap \sigma_1 = \varnothing$, *there exists a family of bounded invertible operators* $P(t)\colon X_1(s_0) \to X_1(t)$ *for all* $t \in \mathbb{R}$ *with* $P(t+p) = P(t)$, $P(s_0) = I$ *and a bounded operator* C *in* $X_1(s_0)$ *with spectrum* $\sigma(C) = \frac{1}{p}\log(\sigma_1)$ *and for* $x \in X_1(s)$ *and all* t, s

$$S_1(t,s) = P(t)e^{C(t-s)}P^{-1}(s).$$

PROOF. We prove (4.26) in the following way. Let $t \geqslant s$ and write $t = \tau + kp + s$ with $\tau \in [0,p)$, $k \in \mathbb{N}$, and write

$$\begin{aligned} S_1(t,s) &= S_1(\tau+kp+s, kp+s)S_1(kp+s,s) \\ &= S_1(\tau+kp+s,kp+s)S_1(p+s,s)^k = S_1(\tau+kp+s,kp+s)U(s)^k. \end{aligned}$$

Now, since $r_\sigma(U(s)) = \lim_{k\to\infty} \|U(s)^k\|^{\frac{1}{k}} = e^{\beta p}$ we have, given $\epsilon > 0$, there is an M_ϵ such that

$$\|U(s)^k\| \leqslant \tilde{M}_\epsilon e^{(\beta+\epsilon)kp}.$$

To see that the above estimate is uniform in $s \in [0,p]$ note that given $r > 0$, from Remark 4.32, there exists $\delta > 0$ such that

$$\|U(s)P_{\sigma_1}(s) - U(\bar{s})P_{\sigma_1}(\bar{s})\| \leqslant \frac{r}{3},$$

whenever $|s - \bar{s}| < \delta$. Also, given $r > 0$, there is $k_0 \in \mathbb{N}$ such that

$$r_\sigma(U(\bar{s})P_{\sigma_1}(\bar{s})) \leqslant \|U(\bar{s})^k P_{\sigma_1}(\bar{s}))\|^{\frac{1}{k}} + \frac{r}{3},$$

for all $k \geqslant k_0$. Lastly, it also holds that given $r > 0$ there is an equivalent norm in X (denoted the same), such that

$$\|U(\bar{s})P_{\sigma_1}(\bar{s})\| \leqslant r_\sigma(U(\bar{s})P_{\sigma_1}(\bar{s})) + \frac{r}{3}.$$

From these we obtain for $k \geqslant k_0$:

$$\begin{aligned} \|U(s)^k P_{\sigma_1}(s)\|^{\frac{1}{k}} &\leqslant \|U(s)P_{\sigma_1}(s)\| \leqslant \|U(\bar{s})P_{\sigma_1}(\bar{s})\| + \frac{r}{3} \\ &\leqslant r_\sigma(U(\bar{s})P_{\sigma_1}(\bar{s})) + \frac{2r}{3} \leqslant \|U(\bar{s})^k P_{\sigma_1}(\bar{s})\|^{\frac{1}{k}} + r. \end{aligned}$$

Thus the choice of M_ϵ can be made uniform in a small neighborhood of $\bar{s}$ for each $\bar{s} \in [0,p]$ and, from the compactness of $[0,p]$, uniform in $[0,p]$.

On the other hand $\|S_1(\tau+r,r)\| \leqslant M_0$ and that is independent of $r \in \mathbb{R}$ and of τ as a consequence of the continuity of $\mathbb{R}^+ \ni t \mapsto D_xT(t)(x) \in \mathcal{L}(X)$ uniformly in compact subsets of X. Thus

$$\|S_1(t,s)\| \leqslant M_0\tilde{M}_\epsilon e^{(\beta+\epsilon)kp} =: M_0\tilde{M}_\epsilon e^{-(\beta+\epsilon)(t-s-kp)}e^{(\beta+\epsilon)(t-s)} \leqslant M_\epsilon e^{(\beta+\epsilon)(t-s)}.$$

If $0 \notin \sigma_1$, $S_1(t,s)$ is invertible and the lower estimate is obtained in a similar way.

Now assume that there is a line $\Gamma = \{re^{i\theta} : r \geqslant 0\}$, $\theta \in (-\pi, \pi]$, disjoint from σ_1, and choose any $s_0 \in \mathbb{R}$. Then $U_1(s_0) = e^{pC}$, where

$$C = \frac{1}{2\pi i p} \int_{\Gamma_1} (\lambda - U_1(s_0))^{-1} \log \lambda \, d\lambda;$$

here Γ_1 is a rectifiable, closed and simple curve surrounding σ_1 but disjoint from Γ and we consider the branch of the logarithm associated with the branch cut Γ. Then $\sigma(pC) = \log(\sigma_1)$ and hence $\sigma(e^{pC}) = \sigma_1$.

For $t \in \mathbb{R}$ define

$$P(t) = S_1(t, s_0)e^{-C(t-s_0)} \colon X_1(s_0) \to X_1(t),$$

and recall that $S_1(t, s_0) = S_1(s_0, t)^{-1}$ is $t < s_0$. Then

$$\begin{aligned} P(t+p) &= S_1(t+p, s_0)e^{-C(t+p-s_0)} \\ &= S_1(t+p, s_0+p)U_1(s_0)e^{-Cp}e^{-C(t-s_0)} = P(t). \end{aligned}$$

If $t \geqslant s$, it follows that

$$S_1(t,s) = P(t)e^{C(t-s)}P(s)^{-1}$$

and we may use this equation to define $S_1(t,s)$ for all t and s, and hence

$$S_1(s,t)S_1(t,s) = I_{X_1(s)}.$$

□

6.4. Permanence of periodic solutions under perturbations. Consider $\{\mathcal{T}_\eta\}_{\eta\in[0,1]} \subset C^1(X)$ to be a continuous and collectively asymptotically compact at $\eta = 0$ family of semigroups and assume that $\mathbb{R}^+ \times [0,1] \ni (t,\eta) \mapsto D_x T_\eta(t)(x) \in \mathcal{L}(X)$ is continuous, uniformly in compact subsets of X.

REMARK 4.36. This condition may not be satisfied in applications coming from semilinear differential equations in infinite-dimensional Banach spaces for which the main part of the equation generates only a strongly continuous semigroup.

The uniform continuity in time will restrict the applications to the case of semigroups which become uniformly continuous for positive times (analytic semigroups for example).

When applying the theory developed in this section the reader must pay special attention to this condition in order to avoid mistakes.

Typically, for semigroups in finite-dimensional spaces or for semigroups coming from parabolic problems, it is not difficult to find perturbations under which the above condition is satisfied.

We also assume that, if $x_\eta \to x_0$ as $\eta \to 0$ and $x_\eta(t, x_\eta) = T_\eta(t)x_\eta$, $t \geqslant 0$, then $\dot{x}_\eta(t, x_\eta) \to \dot{x}_0(t, x_0)$ as $\eta \to 0$, uniformly in compact subsets of $\mathbb{R}^+$.

Assume that $\mathcal{T}_0$ has a nontrivial periodic solution $\xi_0 \colon \mathbb{R} \to X$ with period $p > 0$. Let $S_0(t,s) = D_x T_0(t-s)(\xi_0(s)) \in \mathcal{L}(X)$, $t \geqslant s$, $U_0(t) = S_0(t+p,t)$, $t \geqslant 0$, and assume that 1 is an isolated simple eigenvalue for $U_0(t)$, $t \geqslant 0$.

DEFINITION 4.37. We say that $\xi_0 \colon \mathbb{R} \to X$ is a **normally hyperbolic periodic solution** if $\sigma(U_0(t)) \cap \{\lambda \in \mathbb{C} \colon |\lambda| = 1\} = \{1\}$ and 1 is an isolated simple eigenvalue for $U_0(t)$, $t \geqslant 0$ ($\dim R(P_{\{1\}}) = 1$).

THEOREM 4.38. *Assume that the above conditions are satisfied. Then, for a suitably small neighborhood V of $\{\xi_0(t)\colon t \in [0,p]\}$ in X, there are $\eta_0 > 0$, $\delta_0 > 0$ such that, for each $\eta \in (0,\eta_0)$, there is a unique $p(\eta)$-periodic solution $\xi_\eta(\cdot)$ in V, with $|p(\eta) - p| \leqslant \delta_0$, $p(\eta) \to p$ as $\eta \to 0$ and $\xi_\eta(t) \to \xi_0(t)$ as $\eta \to 0$, uniformly in compact subsets of $\mathbb{R}$.*

Furthermore, $1 = \sigma(U_\eta(t)) \cap \{\lambda \in \mathbb{C}\colon |\lambda| = 1\}$ is an isolated eigenvalue of $U_\eta(t) = D_x T_\eta(p(\eta))(\xi_\eta(t)))$ and the orbit ξ_η is an isolated invariant set in V for all $\eta \in [0,\eta_0]$.

PROOF. Let $\mathcal{S}_0 = \{S_0(t,s)\colon t \geqslant s\}$ be the evolution process defined above and let $U_0(t) = S_0(t+p,t)$ be the associated period map. It is clear that $U_0(0)\dot{\xi}_0(0) = \dot{\xi}_0(0)$. Use the spectral projection P associated to the spectral set $\sigma_0 = \{1\}$, that is,

$$P = \frac{1}{2\pi i}\int_\gamma (\lambda - U_0(0))^{-1} d\lambda$$

with $\gamma(t) = 1 + \delta e^{2\pi i t}$, $t \in [0,1]$, and δ such that $\{\lambda \in \mathbb{C}\colon |\lambda - 1| \leqslant \delta\} \cap \sigma(U_0(0)) = \{1\}$. Consider $X = [\dot{\xi}_0(0)] \oplus Z_1$ with $PX = [\dot{\xi}_0(0)]$ and $Z_1 = \ker P$ and $P\dot{\xi}_0(0) = \dot{\xi}_0(0)$. Define $\langle x_0^*, t\dot{\xi}_0(0) + z\rangle = t$ for all scalar t and $z \in Z_1$. Clearly Z_1 is the kernel of x_0^* such that $\langle x_0^*, \dot{\xi}_0(0)\rangle = 1$. Furthermore,

$$\langle (I^* - U_0(0)^*)x_0^*, t\dot{\xi}_0 + z\rangle = 0, \qquad \text{for all scalar } t \text{ and } z \in Z_1,$$

and consequently $U_0(0)^* x_0^* = x_0^*$.

Define the hyperplane $W = \{x \in X\colon \langle x_0^*, x - \xi_0(0)\rangle = 0\} = \xi_0(0) + Z_1$. Note that $\langle x_0^*, \xi_0(t) - \xi_0(0)\rangle$ is zero at $t = 0$ and strictly increasing for $t \in [-\tau_0, \tau_0]$ with τ_0 suitably small. Hence, $\langle x_0^*, \xi_0(\tau_0) - \xi_0(0)\rangle > 0$ and $\langle x_0^*, \xi_0(-\tau_0) - \xi_0(0)\rangle < 0$. If $x_1 \in W$, $\|x_1 - \xi_0(0)\| + |\eta|$ is sufficiently small, there is a unique $\tau = \tau(x_1,\eta) \in [-\tau_0, \tau_0]$ such that

$$g(\tau, x_1, \eta) = \langle x_0^*, T_\eta(\tau + p)x_1 - \xi_0(0)\rangle = 0$$

(for $\langle x_0^*, T_\eta(t+p)x_1 - \xi_0(0)\rangle$ must change sign in $[-\tau_0, \tau_0]$ if $\|x_1 - \xi_0(0)\| + |\eta|$ is suitably small, for the uniqueness we used that the solution is C^1 in time - at least for a dense subset of x_1's in W - and continuous dependence of the time derivative of the solution $t \mapsto T_\eta(t)x_1$ with respect to η uniformly in compact time intervals).

Define, for each suitably small η, the first recurrence map or **Poincaré map** $G(\cdot,\eta) : W \to W$ by $G(x_1,\eta) = T_\eta(\tau(x_1,\eta) + p)x_1$, where $\tau(x_1,\eta)$ is as before. Now

$$g(\tau, x_1, \eta) = \langle x_0^*, x(p + \tau, x_1, \eta) - \xi_0(0)\rangle = 0$$

is a well defined C^1 function from the Implicit Function Theorem, $\tau(x_1,\eta)$ is a well defined C^1 function since

$$g(0, \xi_0(0), 0) = 0 \quad \text{and} \quad \frac{\partial g}{\partial \tau}(0, \xi_0(0), 0) = \langle x_0^*, \dot{\xi}_0(0)\rangle = 1.$$

Now we prove the existence of a fixed point $x_1 = G(x_1,\eta) \in W$, for each η small. Hence, the existence of a periodic solution of period $p_\eta = p + \tau(x_1(\eta),\eta)$ follows.

We know that $G(\xi_0(0), 0) = \xi_0(0)$, and

$$G_x(\xi_0(0),0)z = \dot{\xi}_0(0)\tau_x(\xi_0(0),0)z + U_0(0)z \quad \text{for } z \in Z_1.$$

Then

$$0 = \langle x_0^*, G_x(\xi_0(0),0)z\rangle = \tau_x(\xi_0(0),0)z + \langle x_0^*, U_0(0)z\rangle$$

and so $\tau_x(\xi_0(0), 0)z = 0$.

Thus, in the space $Z_1 = \{z\colon \langle x_0^*, z\rangle = 0\} = R(U_0(0)\big|_{Z_1})$, $G_x(\xi_0(0), 0) = U_0(0)\big|_{Z_1}$. But $\sigma(U_0(0)\big|_{Z_1}) = \sigma(U_0(0)) \setminus \{1\}$ and so $G_x(\xi_0(0), 0) - I$ has a bounded inverse on Z_1. By the Implicit Function Theorem, there exists a unique $x_1(\eta)$ with $x_1(\eta) - \xi_0(0)$ in Z_1, $\|x_1(\eta) - \xi_0(0)\| < r$, such that

$$G(x_1(\eta), \eta) = x_1(\eta).$$

The solution $\xi_\eta\colon \mathbb{R} \to X$ passing through $x_1(\eta)$ is a p_η periodic solution.

If we define $H_\eta\colon Z_1 \to Z_1$ as $H_\eta(z) = G(z + x_1(\eta), \eta) - x_1(\eta)$, for all $z \in Z_1$, we have $z = 0$ is a hyperbolic fixed point of H_η. Hence, the only global bounded solution $\{z_n\}_{n\in\mathbb{Z}}$ of the discrete semigroup $\{H_\eta^n\colon n \in \mathbb{N}\}$, that is, such that $H_\eta(z_n) = z_{n+1}$ for all $n \in \mathbb{Z}$, which remains in a small neighborhood of zero (independently of η for all suitably small η - here we use that $D_xT_\eta(p_\eta)(\xi_\eta(t)) \to D_xT_0(p)(\xi_0(t))$ in $\mathcal{L}(X)$ as $\eta \to 0$) is the trivial solution $z_n = 0$ for all $n \in \mathbb{Z}$. Hence, the periodic orbit obtained above is isolated and since the neighborhood where we obtain the characterization of the stable and unstable manifold (see Theorem 4.4) is uniform in η the periodic orbits are uniformly isolated. □

The following result is a direct consequence of Theorem 4.38 and is stated here to emphasize the constraint implied by our assumption that $\mathbb{R}^+ \times [0, 1] \ni (t, \eta) \mapsto D_xT_\eta(t)(x) \in \mathcal{L}(X)$ is continuous uniformly in compact subsets of X.

THEOREM 4.39. *For $\eta \in [0, 1]$, consider the abstract semilinear problem*

$$\begin{cases} \dot{x} + Ax = f(x, \eta), \\ x(0) = x_0 \in X, \end{cases} \tag{4.27}$$

where $-A$ is the generator of a strongly continuous semigrup which is uniformly continuous for $t > 0$, $f(\cdot, \eta)\colon X \to X$ is continuously differentiable and such that $f(x, \eta) \to f(x, 0)$ in X as $\eta \to 0$, uniformly in bounded subsets of X, together with $f'(x, \eta) \to f'(x, 0)$ in $\mathcal{L}(X)$ as $\eta \to 0$, uniformly for x in compact subsets of X. Under these conditions (4.27) is locally well posed. Assume that all solutions $x(\cdot, x_0, \eta)$ of (4.27) are defined for all $t \geqslant 0$ and define $T_\eta(t)x_0 = x(t, x_0, \eta)$. Assume that $\xi_0\colon \mathbb{R} \to X$ is a nontrivial p-periodic normally hyperbolic solution of $\mathcal{T}_0$.

Then, the conditions of Theorem 4.38 are satisfied and its conclusions hold, that is, for a suitably small neighborhood V of $\{\xi_0(t)\colon t \in [0, p]\}$ in X, there are $\eta_0 > 0$, $\delta_0 > 0$ such that, for each $\eta \in (0, \eta_0)$, there is a unique $p(\eta)$-periodic solution $\xi_\eta(\cdot)$ in V, with $|p(\eta) - p| \leqslant \delta_0$, $p(\eta) \to p$ as $\eta \to 0$ and $\xi_\eta(t)\xi_0(t)$ as $\eta \to 0$, uniformly in compact subsets of $\mathbb{R}$. Furthermore, $1 = \sigma(U_\eta(t)) \cap \{\lambda \in \mathbb{C}\colon |\lambda| = 1\}$ is an isolated eigenvalue of $U_\eta(t) = D_xT_\eta(p(\eta))(\xi_\eta(t))$ and the orbit ξ_η is an isolated invariant set in V for all $\eta \in [0, \eta_0]$.

The normal hyperbolicity is a consequence of the continuity of the spectrum for the period map. That is, ξ_η is a normally hyperbolic p_η-periodic orbit if and only if the corresponding time map $U_\eta(t) = S_\eta(t+p_\eta, t) = D_xT_\eta(p_\eta)(\xi_\eta(t))$ has $\{1\}$ as a simple eigenvalue and the rest of the spectrum is off the unit circle. In this case

$$S_0(t+p,t) = e^{-Ap} + \int_t^{p+t} e^{A(p+t-s)} f_x(\xi_0(s))S_0(s,t)ds,$$

$$S_\eta(t+p_\eta,t) = e^{-Ap_\eta} + \int_t^{p_\eta+t} e^{A(p_\eta+t-s)} f_x(\xi_\eta(s))S_\eta(s,t)ds.$$

Note that

$$\|e^{-Ap_\eta} - e^{-Ap}\|_{\mathcal{L}(X)} \to 0 \quad \text{as } \eta \to 0,$$

and the uniform continuity of e^{-At} at $t = p$ is crucial to the proof of the continuity of the time map. Also, uniformly for s in compact intervals of $\mathbb{R}$,

$$\|f_x(\xi_\eta(s)) - f_x(\xi_0(s))\|_{\mathcal{L}(X)} \to 0 \quad \text{as } \eta \to 0.$$

For parabolic problems one may need to deal with f acting between fractional powers of the space X but the result is similar. It follows that

$$\|S_\eta(t+p_\eta,t) - S_0(t+p,t)\|_{\mathcal{L}(X)} \to 0 \quad \text{as } \eta \to 0.$$

This ensures the continuity of the spectrum of $U_\eta(t)$ as η goes to zero and that normal hyperbolicity holds uniformly for all η suitably small. That is the key ingredient to ensure the existence of a fixed open set where $\xi_\eta(\mathbb{R})$ is the maximal invariant set for all η suitably small.

Notes. The study of critical elements, especially equilibria and periodic orbits, is crucial to apply perturbation results for semigroups that arise from differential (or integral) equations. By understanding the invariant manifolds around a critical element, one can draw results regarding isolation and permanence under perturbations, which among other things, are crucial to apply the results of Chapter 3. Moreover, one can obtain differentiability and C^1 permanence under perturbations, which help us obtain conditions to apply the results of Chapter 5. The results presented in this chapter are as in [**141**], which is a topic among many others treated by D. Henry in his collection of handwritten notes. We included also a generalization of the λ-lemma of Palis ([**178**]) to the infinite-dimensional case, which is also present in Henry's handwritten notes. More results on invariant manifolds can be found in [**27, 156, 157, 177, 184**]. Permanence and continuity of equilibria and periodic orbits are some of the key ingredients to obtain the topological structural stability for global attractors when dealing with applications of the abstract results of Chapter 3 and Chapter 5 to some parabolic partial differential equations. The results presented here follow the ideas in [**142**] for periodic orbits. The reader can see [**25, 137, 144**] for more results on existence and persistence of critical elements and invariant manifolds.

CHAPTER 5

Morse-Smale Semigroups

In this chapter we prove that a suitable small perturbation of a Morse-Smale semigroup has an attractor with the *same phase diagram structure* (isolated invariant sets and connections) as the attractor for the unperturbed semigroup. We show that Morse-Smale semigroups are a special case of dynamically gradient semigroups. Then, we show that the structure of the global attractor is robust under perturbation, that is, not only do the isolated invariant sets behave continuously under perturbations, but they also maintain the same network of connections. This leads to an isomorphism between the phase diagrams of the limit and the perturbed attractors (they give rise to the same oriented graph with the vertices being the isolated invariant sets and there is an oriented edge between two vertices if there is a global solution connecting the corresponding isolated invariant sets).

1. Basic notions

We begin our discussion with the definition of the nonwandering set for a semigroup $\mathcal{T}$.

DEFINITION 5.1. Let $\mathcal{T}$ be a semigroup with a global attractor $\mathcal{A}$. We define the **nonwandering set** $\Omega_{\mathcal{T}}$ of $\mathcal{T}$ as the set of points $x \in X$ such that given $t_0 \geqslant 0$ and a neighborhood V_x of x in X, there exists $t \geqslant t_0$ such that $T(t)V_x \cap V_x \neq \varnothing$. Clearly $\Omega_{\mathcal{T}} \subset \mathcal{A}$.

With this at hand, we can define the concept of a Morse-Smale semigroup (introduced in 1960 by S. Smale in [**195**]).

DEFINITION 5.2. Let $\mathcal{T}$ be a semigroup with a global attractor $\mathcal{A}$. We say that $\mathcal{T}$ is a **Morse-Smale semigroup** if it satisfies the following conditions:

(a) $\mathcal{T}$ is **reversible**, that is, $T(t)|_{\mathcal{A}}$ is injective, $T(t)$ is differentiable in $\mathcal{A}$ and $D_xT(t)(z)\colon X \to X$ is an injective bounded linear operator, for all $t \geqslant 0$ and $z \in \mathcal{A}$;

(b) there exists a finite number of hyperbolic equilibrium points $z_1, \cdots, z_n$ and a finite number of normally hyperbolic periodic orbits $\xi_1, \cdots, \xi_m\colon \mathbb{R} \to X$ of $\mathcal{T}$ such that

$$\Omega_{\mathcal{T}} = \bigcup_{i=1}^{n} \{z_i\} \cup \bigcup_{j=1}^{m} \xi_j(\mathbb{R}).$$

We will often represent and treat $\Omega_{\mathcal{T}}$ as a list of sets

$$\{z_1, \cdots, z_n, \xi_1(\mathbb{R}), \cdots, \xi_m(\mathbb{R})\},$$

and write $\Xi \in \Omega_{\mathcal{T}}$ to mean that either $\Xi = \{z_i\}$, $1 \leqslant i \leqslant n$, or $\Xi = \xi_j(\mathbb{R})$, $1 \leqslant j \leqslant m$;

(c) if $\Xi \in \Omega_{\mathcal{T}}$, then $\dim W^u_{loc}(\Xi) < \infty$;

(d) if $\Xi, \hat{\Xi} \in \Omega_{\mathcal{T}}$, then $W^u(\Xi)$ and $W^s_{loc}(\hat{\Xi})$ intersect transversally, that is, if $z \in W^u(\Xi) \cap W^s_{loc}(\hat{\Xi})$, then $T_z W^u(\Xi) + T_z W^s_{loc}(\hat{\Xi}) = X$, where $T_z W^u(\Xi)$ and $T_z W^s_{loc}(\hat{\Xi})$ are the tangent spaces of $W^u(\Xi), W^s_{loc}(\hat{\Xi})$, respectively, in X at the point z.

REMARK 5.3. Note that in item (d) of Definition 5.2, the manifolds $W^u(\Xi)$ and $W^s_{loc}(\hat{\Xi})$ may not have intersection at all, and (d) is automatically satisfied.

Our first goal will be to relate the Morse-Smale semigroups and the dynamically gradient semigroups, that is, to express both concepts in the same terminology. We begin by showing that the ω-limit and the α_ϕ-limit of points are contained in the nonwandering set.

LEMMA 5.4. *Let $\mathcal{T}$ be a semigroup with a global attractor $\mathcal{A}$ and a nonwandering set $\Omega_{\mathcal{T}}$. Then $\Omega_{\mathcal{T}}$ is compact, $T(t)\Omega_{\mathcal{T}} \subset \Omega_{\mathcal{T}}$ for all $t \geqslant 0$, $\omega(x) \subset \Omega_{\mathcal{T}}$ for all $x \in X$ and $\alpha_\phi(x) \subset \Omega_{\mathcal{T}}$ for all $x \in \mathcal{A}$.*

PROOF. To show the compactness of $\Omega = \Omega_{\mathcal{T}}$ it suffices to show that it is closed, since $\Omega \subset \mathcal{A}$. Let $\{z_n\} \subset \Omega$ such that $z_n \to z \in \mathcal{A}$, $t_0 \geqslant 0$ and V_z a neighborhood of z in X. Since $z_n \to z$, there exists $n_0 \in \mathbb{N}$ such that V_z is a neighborhood of $z_{n_0} \in \Omega$, thus there exists $t \geqslant t_0$ such that $T(t)V_z \cap V_z \neq \varnothing$, and shows that $z \in \Omega$.

Let $z \in \Omega$, $t, t_0 \geqslant 0$ and V a neighborhood of $T(t)z$. From the continuity of $T(t)$, $T(t)^{-1}(V)$ is a neighborhood of z and hence, there exists $\tau \geqslant t_0$ such that $T(\tau)(T(t)^{-1}(V)) \cap T(t)^{-1}(V) \neq \varnothing$, which implies that $T(\tau)V \cap V \neq \varnothing$, $\tau \geqslant t_0$. From this we conclude that $T(t)z \in \Omega$ and that $T(t)\Omega \subset \Omega$, for all $t \geqslant 0$.

For the last statement, let $z \in \omega(x) \subset \mathcal{A}$, for some $x \in X$, $t_0 \geqslant 0$ and V a neighborhood of z in X. Let $\{t_n\} \subset [0, \infty)$ such that $t_n \to \infty$ and $z = \lim_{n\to\infty} T(t_n)x$, thus there exists $n_0 \in \mathbb{N}$ such that $T(t_n)x \in V$ for all $n \geqslant n_0$ and we choose $n_1 \geqslant n_0$ such that $t := t_{n_1} - t_{n_0} \geqslant t_0$. Thus $T(t_{n_1})x \in V$ and $T(t_{n_1})x = T(t_{n_1} - t_{n_0})T(t_{n_0})x = T(t)T(t_{n_0})x \in T(t)V$, which shows us that $T(t)V \cap V \neq \varnothing$ and consequently $z \in \Omega$. The proof for the α_ϕ-limit is analogous. $\square$

COROLLARY 5.5. *With the hypothesis of Lemma 5.4, if $X = A$ and $\mathcal{T}$ is injective, then $\Omega_{\mathcal{T}}$ is invariant.*

PROOF. To see this let $z \in \Omega_{\mathcal{T}}$. Since $T(s)$ is injective and $\mathcal{A}$ is compact, it follows that $T(s)$ is a homeomorphism and thus there exists a unique $x \in \mathcal{A}$ such that $T(s)x = z$. Let $t_0 \geqslant 0$ and V a neighborhood of x in X. Again, since $T(s)$ is a homeomorphism, $T(s)V$ is a neighborhood of z and thus there exists $t \geqslant t_0$ such that $T(t)(T(s)V) \cap T(s)V \neq \varnothing$. Therefore $T(t)V \cap V \neq \varnothing$, which concludes the invariance of $\Omega_{\mathcal{T}}$. $\square$

Now we can show that every Morse-Smale semigroup is a generalized dynamically gradient semigroup. We start with some preparatory results which will be useful throughout the chapter.

PROPOSITION 5.6. *Let $\mathcal{T}$ be a Morse-Smale semigroup with nonwandering set*

$$\Omega_{\mathcal{T}} = \{z_1, \cdots, z_n, \xi_1(\mathbb{R}), \cdots, \xi_m(\mathbb{R})\},$$

where the z_i *is a hyperbolic equilibria and* $\xi_j\colon \mathbb{R} \to X$ *is a normally hyperbolic periodic solution,* $1 \leqslant i \leqslant n$*,* $1 \leqslant j \leqslant m$*. Let* $\Xi, \hat{\Xi} \in \Omega$*. If* $W^u(\Xi) \cap W^s_{loc}(\hat{\Xi}) \neq \varnothing$*, then* $\dim W^u(\Xi) \geqslant \dim W^u(\hat{\Xi})$.

PROOF. Since $\mathcal{T}$ is a Morse-Smale semigroup, if $z \in W^u(\Xi)\cap W^s_{loc}(\hat{\Xi})$ is a point of transversal intersection, then $T_z W^u(\Xi) + T_z W^s_{loc}(\hat{\Xi}) = X$. For $\hat{z} \in \hat{\Xi}$ we have $T_{\hat{z}} W^u_{loc}(\hat{\Xi}) \oplus T_{\hat{z}} W^c_{loc}(\hat{\Xi}) \oplus T_{\hat{z}} W^s_{loc}(\hat{\Xi}) = X$ and

$$\begin{aligned}\dim W^u(\Xi) &= \dim T_z W^u(\Xi) \geqslant \dim T_{\hat{z}} W^u(\hat{\Xi}) + \dim T_{\hat{z}} W^c_{loc}(\hat{\Xi})\\ &= \dim W^u(\hat{\Xi}) + \dim T_{\hat{z}} W^c_{loc}(\hat{\Xi}),\end{aligned}$$

where $\dim T_{\hat{z}} W^c_{loc}(\hat{\Xi}) = 0$ if $\hat{\Xi}$ is a hyperbolic equilibrium and $\dim T_{\hat{z}} W^c_{loc}(\hat{\Xi}) = 1$ if $\hat{\Xi}$ is a normally hyperbolic periodic orbit. □

LEMMA 5.7. *Let* $\mathcal{T}$ *be a Morse-Smale semigroup with nonwandering set*

$$\Omega_{\mathcal{T}} = \{z_1, \cdots, z_n, \xi_1(\mathbb{R}), \cdots, \xi_m(\mathbb{R})\},$$

where the z_i *is a hyperbolic equilibria and* $\xi_j\colon \mathbb{R} \to X$ *is a normally hyperbolic periodic solution,* $1 \leqslant i \leqslant n$*,* $1 \leqslant j \leqslant m$*. If* $W^u(\Xi) \cap W^s_{loc}(\hat{\Xi}) \neq \varnothing$*, then there exists a submanifold* $D^0 \subset W^u(\Xi)$ *such that* $D^0 \cap W^s_{loc}(\hat{\Xi}) = \{x^0\}$ *and* $T_{x^0} D^0 \oplus T_{x^0} W^s_{loc}(\hat{\Xi}) = X$.

PROOF. Let $x^0 \in W^u(\Xi)\cap W^s_{loc}(\hat{\Xi})$ be a point of transversal intersection. There exist a neighborhood V of x^0 in $W^u(\Xi)$, a neighborhood W of 0 in $T_{x^0} W^u(\Xi)$ and a C^1 open map $\psi\colon W \to V$. By Proposition 5.6, $\dim W = \dim W^u(\Xi) \geqslant \dim W^u(\hat{\Xi})$ and hence there exists a vector subspace F of $T_{x^0} W^u(\Xi)$ such that $F \oplus T_{x^0} W^s_{loc}(\hat{\Xi}) = X$.

Then $D^0 = \psi(F)$ has the desired property. □

We now discuss the *openness of transversality.* Such results can be found in [**144**] - where they are referred to as *Thom's Transversality Theorems* - in a very abstract and general framework. We only use the rather particular property of these results which are stated and proved next.

LEMMA 5.8. *Let* X *be a Banach space,* X_1, X_2 *vector subspaces of* X *such that* X_1 *is closed and* $X_1 \oplus X_2 = X$*. Let also* $\theta, \tilde{\theta}\colon X_1 \to X_2$*,* $\sigma, \tilde{\sigma}\colon X_2 \to X_1$ *be continuously differentiable maps such that* $\theta(0) = \sigma(0) = 0$ *and* $\mathrm{Lip}\,\theta$*,* $\mathrm{Lip}\,\tilde{\theta}$*,* $\mathrm{Lip}\,\sigma$*,* $\mathrm{Lip}\,\tilde{\sigma} < 1$*. Consider the following immersed* C^1 *submanifolds*

$$M = \{y + \theta(y)\colon y \in X_1, \|y\| \leqslant r\}, \quad N = \{\sigma(x) + x\colon x \in X_2, \|x\| \leqslant r\},$$

and

$$\tilde{M} = \{y + \tilde{\theta}(y)\colon y \in X_1, \|y\| \leqslant r\}, \quad \tilde{N} = \{\tilde{\sigma}(x) + x\colon x \in X_2, \|x\| \leqslant r\}.$$

If $\tilde{\theta}$ *is* C^1*-close to* θ *and* $\tilde{\sigma}$ *is* C^1*-close to* σ*, then* $\tilde{M}$ *and* $\tilde{N}$ *have a nonempty intersection.*

PROOF. From the hypotheses, there exists a constant $c \in (0, 1)$ such that

$$\mathrm{Lip}\,\theta, \mathrm{Lip}\,\tilde{\theta}, \mathrm{Lip}\,\sigma, \mathrm{Lip}\,\tilde{\sigma} \leqslant c,$$

and also there exists $r > 0$ such that

$$\|\theta - \tilde{\theta}\|_{C^1(B_r^{X_1}(0), X_2)} \leqslant (1 - c)r \text{ and } \|\sigma - \tilde{\sigma}\|_{C^1(B_r^{X_2}(0), X_1)} \leqslant (1 - c)r. \tag{5.1}$$

Note that $y+\tilde{\theta}(y)=\tilde{\sigma}(x)+x$ is equivalent to $\tilde{\sigma}(\tilde{\theta}(y))=y$. Define $g\colon \bar{B}_r^{X_1}(0)\to \bar{B}_r^{X_1}(0)$ by $g(x)=\tilde{\sigma}(\tilde{\theta}(x))$. We claim that g is well defined. Indeed, for all $x\in \bar{B}_r^{X_1}(0)$,

$$\|\tilde{\theta}(x)\|\leqslant \|\tilde{\theta}(x)-\theta(x)\|+\|\theta(x)-\theta(0)\|\leqslant \|\tilde{\theta}(x)-\theta(x)\|+cr\leqslant r.$$

Analogously,

$$\|\tilde{\sigma}(\tilde{\theta}(x))\|\leqslant \|\tilde{\sigma}(\tilde{\theta}(x))-\sigma(\tilde{\theta}(x))\|+\|\sigma(\tilde{\theta}(x))-\sigma(0)\|\leqslant \|\tilde{\sigma}(\tilde{\theta}(x))-\sigma(\tilde{\theta}(x))\|+cr\leqslant r$$

for all $x\in \bar{B}_r^{X_1}(0)$. Note that g is a contraction, since

$$\|g(x)-g(y)\|\leqslant \text{Lip }(\tilde{\sigma})\text{Lip }(\tilde{\theta})\|x-y\|\leqslant c^2\|x-y\|.$$

Hence g has a unique fixed point, which shows that $\tilde{M}$ and $\tilde{N}$ intersect. □

Using this lemma we can prove the *transitive property of connections* (see Figure 5.1).

PROPOSITION 5.9. *Let $\mathcal{T}$ be a Morse-Smale semigroup with nonwandering set $\Omega_{\mathcal{T}}=\{z_1,\cdots,z_n,\xi_1(\mathbb{R}),\cdots,\xi_m(\mathbb{R})\}$ and $\Xi_1,\Xi_2,\Xi_3\in\Omega_{\mathcal{T}}$. If $W^u(\Xi_1)\cap W^s_{loc}(\Xi_2)\neq\varnothing$ and $W^u(\Xi_2)\cap W^s_{loc}(\Xi_3)\neq\varnothing$, then $W^u(\Xi_1)\cap W^s_{loc}(\Xi_3)\neq\varnothing$.*

PROOF. From Lemma 5.7, if x_{12} is the point of transversal intersection between $W^u(\Xi_1)$ and $W^s_{loc}(\Xi_2)$, we can choose a submanifold D^0 of X such that $D^0\cap W^s_{loc}(\Xi_2)=\{x_{12}\}$, $T_{x_{12}}D^0\oplus T_{x_{12}}W^s_{loc}(\Xi_2)=X$ and $D^0\subset W^u(\Xi_1)$. Choosing n_0 such that, if x_{23} is the point of transversal intersection between $W^u(\Xi_2)$ and $W^s_{loc}(\Xi_3)$, $x_{23}\in T^{n_0}(W^u_{loc}(\Xi_2))$, the global λ-lemma implies that there exist a neighborhood D^0_ρ of x_{12} in D^0 and sets $R^n\subset T^n(D^0_\rho)\subset W^u(\Xi_1)$ such that

$$R^n\to T^{n_0}(W^u_{loc}(\Xi_2)),$$

in the C^1 norm. Hence, by the *openness of the transversality* (Lemma 5.8), there exists a point $x_{13}\in W^u(\Xi_1)\cap W^s_{loc}(\Xi_3)$ (in which the intersection is transversal). □

Now we prove that a Morse-Smale semigroup is a dynamically gradient semigroup.

THEOREM 5.10. *Let $\mathcal{T}$ be a Morse-Smale semigroup with global attractor $\mathcal{A}$ and nonwandering set $\Omega_{\mathcal{T}}=\{z_1,\cdots,z_n,\xi_1(\mathbb{R}),\cdots,\xi_m(\mathbb{R})\}$. Then $\mathcal{T}$ is a dynamically gradient semigroup with $\Omega_{\mathcal{T}}$ as its disjoint set of isolated invariant families.*

PROOF. Let us first prove that $\mathcal{T}$ satisfies (G2). Assume that there exists a homoclinic structure in $\Omega_{\mathcal{T}}$. From Proposition 5.9 we obtain that there is a homoclinic solution, that is, there exist $\Xi\in\Omega_{\mathcal{T}}$ and a global solution $\xi\colon\mathbb{R}\to X$ such that $\xi(t_0)\notin\Omega$, for some $t_0\in\mathbb{R}$, and $\xi(t)\stackrel{t\to\pm\infty}{\to}\Xi$. Let $x^0=\xi(t_0)$ (which we can take in $W^s_{loc}(\Xi)$), a neighborhood V of x^0 in X and $D^0\subset V$ a submanifold transversal to $W^s_{loc}(\Xi)$ in x^0. Then, the global λ-lemma implies that there exist a neighborhood D^0_ρ of x^0 in D^0 and subsets S_n of $T^n(D^0_\rho)\subset T^n(V)$ such that $S_n\to T^{n_0}(W^u_{loc}(\Xi))$, where $n_0\in\mathbb{N}$ is such that $x^0\in T^{n_0}(W^u_{loc}(\Xi))$, which proves that $x^0\in\Omega_{\mathcal{T}}$ and gives us a contradiction. Therefore $\mathcal{T}$ satisfies (G2).

Now we prove that $\mathcal{T}$ satisfies (G1). Let $\xi\colon\mathbb{R}\to\mathcal{A}$ be a global solution with $\xi(\mathbb{R})\notin\Omega$ and let $x=\xi(0)$. From Lemma 5.4 we know that $\omega(x),\alpha_\xi(x)\subset\Omega_{\mathcal{T}}$. Since $\omega(x)$ and $\alpha_\xi(x)$ are both connected it follows immediately that (G1) is satisfied. □

Now we prove the converse result.

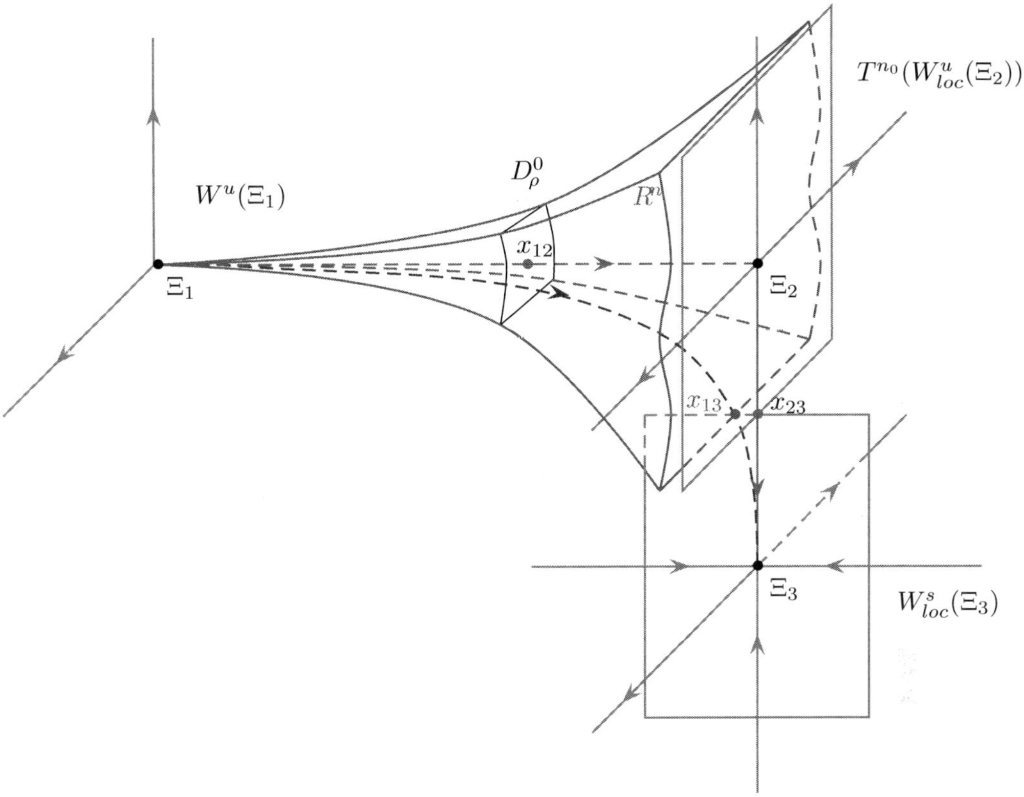

FIGURE 5.1. Indirect connections between Ξ_1 and Ξ_3 generate a direct connection between Ξ_1 and Ξ_3.

THEOREM 5.11. *Let $\mathcal{T}$ be a reversible dynamically gradient semigroup associated with the disjoint set of isolated invariants given by $\boldsymbol{\Xi} = \{z_1, \cdots, z_n, \xi_1(\mathbb{R}), \cdots, \xi_m(\mathbb{R})\}$, where each z_i is a hyperbolic equilibrium point and $\xi_j : \mathbb{R} \to X$ is a normally hyperbolic periodic orbit, for all $1 \leqslant i \leqslant n$ and $1 \leqslant j \leqslant m$. Assume that if $\Xi \in \boldsymbol{\Xi}$, then $\dim W^u_{loc}(\Xi) < \infty$ and that, if $\Xi, \hat{\Xi} \in \boldsymbol{\Xi}$, then $W^u(\Xi)$ and $W^s_{loc}(\hat{\Xi})$ intersect transversally. Then $\mathcal{T}$ is a Morse-Smale semigroup and its nonwandering set $\Omega_{\mathcal{T}}$ coincides with $\boldsymbol{\Xi}$.*

PROOF. It just remains to show the last assertion, that is, that $\Omega_{\mathcal{T}} = \boldsymbol{\Xi}$. It is clear that $\boldsymbol{\Xi} \subset \Omega_{\mathcal{T}}$. Now assume that there exists $x \in \Omega_{\mathcal{T}} \setminus \boldsymbol{\Xi}$ and consider a global solution $\xi_1 : \mathbb{R} \to X$ through x. By (G1), there exist Ξ_1 and Ξ_2 in $\boldsymbol{\Xi}$ such that $\Xi_1 \neq \Xi_2$ and

$$\Xi_1 \overset{t\to-\infty}{\longleftarrow} \xi_1(t) \overset{t\to\infty}{\longrightarrow} \Xi_2.$$

Since $x \in \Omega_{\mathcal{T}}$ and using the already constructed ξ_1, it is possible to construct a global solution $\xi_2 : \mathbb{R} \to X$ such that

$$\Xi_2 \overset{t\to-\infty}{\longleftarrow} \xi_2(t) \overset{t\to\infty}{\longrightarrow} \Xi_3,$$

for some $\Xi_3 \in \boldsymbol{\Xi}$, and by (G2) we must have $\Xi_3 \notin \{\Xi_1, \Xi_2\}$. Using this process inductively, we reach a contradiction since $\boldsymbol{\Xi}$ only has a finite number of isolated invariant families and therefore, $\Omega_{\mathcal{T}} \subset \boldsymbol{\Xi}$ and this completes the proof. $\square$

We can now join our results to obtain the following.

COROLLARY 5.12. *Let $\mathcal{T}$ be a reversible semigroup with finitely many equilibria and finitely many periodic orbits $\boldsymbol{\Xi} = \{z_1, \cdots, z_n, \xi_i(\mathbb{R}), \cdots, \xi_m(\mathbb{R})\}$, where each z_i is a hyperbolic equilibrium point and $\xi_j \colon \mathbb{R} \to X$ is a normally hyperbolic periodic orbit, for all $1 \leqslant i \leqslant n$ and $1 \leqslant j \leqslant m$. Assume also that, if $\Xi \in \boldsymbol{\Xi}$, then $\dim W^u_{loc}(\Xi) < \infty$ and, if $\Xi_1, \Xi_2 \in \boldsymbol{\Xi}$, then $W^u(\Xi_1)$ and $W^s_{loc}(\Xi_2)$ intersect transversally. Then $\mathcal{T}$ is a Morse-Smale semigroup if and only if it is a dynamically gradient semigroup with respect to $\boldsymbol{\Xi}$.*

2. Structural stability of Morse-Smale semigroups

In this subsection we explore the behavior of the structure of the global attractor of a Morse-Smale semigroup under small perturbations. More precisely, we establish that the phase diagrams, for the global attractor of a Morse-Smale semigroup and of a suitably small perturbation of it, are isomorphic, that is, the connections between the isolated invariant sets (equilibria and periodic orbits) are the same for the limiting semigroup and the perturbed semigroup. We start with the definition of phase diagram commutativity.

DEFINITION 5.13. Let $\mathcal{T}$ and $\mathcal{S}$ be two dynamically gradient semigroups. We say that there exists a **phase diagram isomorphism** between the attractors of $\mathcal{T}$ and $\mathcal{S}$ if they satisfy the following condition: if $\boldsymbol{\Xi}_{\mathcal{T}} = \{\Xi_1, \cdots, \Xi_p\}$ and $\boldsymbol{\Xi}_{\mathcal{S}} = \{\hat{\Xi}_1, \cdots, \hat{\Xi}_p\}$ are the disjoint collection of isolated invariant families for $\mathcal{T}$ and $\mathcal{S}$, respectively, then there exists a bijection $\mathcal{D} \colon \boldsymbol{\Xi}_{\mathcal{T}} \to \boldsymbol{\Xi}_{\mathcal{S}}$ such that, for some $1 \leqslant i, j \leqslant p$, there exists a global solution $\xi \colon \mathbb{R} \to X$ of $\mathcal{T}$ satisfying

$$\lim_{t \to -\infty} \operatorname{dist}_H(\xi(t), \Xi_i) = 0 \text{ and } \lim_{t \to \infty} \operatorname{dist}_H(\xi(t), \Xi_j) = 0,$$

if and only if there is a global solution $\psi \colon \mathbb{R} \to X$ of $\mathcal{S}$ such that

$$\lim_{t \to -\infty} \operatorname{dist}_H(\psi(t), \mathcal{D}(\Xi_i)) = 0 \text{ and } \lim_{t \to \infty} \operatorname{dist}_H(\psi(t), \mathcal{D}(\Xi_j)) = 0.$$

We can now state the main result of this chapter.

THEOREM 5.14. *Let $\{\mathcal{T}_\eta\}_{\eta \in [0,1]}$ be a family of semigroups which is continuous and collectively asymptotically compact at $\eta = 0$. Assume that*

(a) *$\mathcal{T}_\eta$ has global attractor $\mathcal{A}_\eta$ for each $\eta \in [0,1]$ and $\bigcup_{\eta \in (0,1]} \mathcal{A}_\eta$ is bounded;*
(b) *$\{\mathcal{T}_\eta\}_{\eta \in [0,1]} \subset C^1(X)$ and $\mathbb{R}^+ \times [0,1] \ni (t,\eta) \mapsto T_\eta(t)'(x) \in \mathcal{L}(X)$ is continuous for each $x \in X$;*
(c) *$\mathcal{T}_0$ is a dynamically gradient semigroup relative to the disjoint collection of isolated invariant sets $\boldsymbol{\Xi}_0 = \{z_{1,0}, \cdots, z_{n,0}, \xi_{1,0}(\mathbb{R}), \cdots, \xi_{m,0}(\mathbb{R})\}$ where $z_{i,0}$ is a hyperbolic equilibria and $\xi_{j,0} \colon \mathbb{R} \to X$ is a normally hyperbolic periodic solution, $1 \leqslant i \leqslant n$, $1 \leqslant j \leqslant m$.*

Then

(i) *there is an* $\eta_0 \in (0,1]$ *such that, for each* $\eta \in (0,\eta_0]$, *there are hyperbolic equilibria* $\{z_{1,\eta},\cdots,z_{n,\eta}\}$ *and normally hyperbolic periodic solutions* $\{\xi_{1,\eta}(\mathbb{R}),\cdots,\xi_{m,\eta}(\mathbb{R})\}$ *such that*

$$\Xi_\eta = \{z_{1,\eta},\cdots,z_{n,\eta},\xi_{1,\eta}(\mathbb{R}),\cdots,\xi_{m,\eta}(\mathbb{R})\}$$

is a disjoint collection of isolated invariant sets for $\mathcal{T}_\eta$, *for all suitably small* $\eta > 0$, *and*

$$\lim_{\eta\to 0} \operatorname{dist}_H(z_{j,\eta}, z_{j,0}) = 0 \quad \text{and} \quad \lim_{\eta\to 0} \operatorname{dist}_H(\xi_{i,\eta}(\mathbb{R}), \xi_{i,0}(\mathbb{R})) = 0; \tag{5.2}$$

(ii) *there exist* $\delta > 0$ *and* $\eta_1 \in (0,\eta_0]$ *such that, if* $\eta \in (0,\eta_1]$ *and* $\Xi_\eta \in \mathbf{\Xi}_\eta$, *then* Ξ_η *is the maximal invariant set in* $\mathcal{O}_\delta(\Xi_\eta)$;

(iii) *there exists an* $\eta_2 \in (0,\eta_1]$ *such that* $\mathcal{T}_\eta$ *is a dynamically gradient semigroup relative to* $\mathbf{\Xi}_\eta$, *for all* $\eta \in (0,\eta_2]$.

In addition, if $\mathcal{T}_0$ *is a Morse-Smale semigroup, then*

(iv) *there exists an* $\eta_3 \in (0,\eta_2]$ *such that* $\mathcal{T}_\eta$ *is a Morse-Smale semigroup and there exists a phase diagram isomorphism between the attractors of* $\mathcal{T}_\eta$ *and* $\mathcal{T}_0$, *for all* $\eta \in (0,\eta_3]$.

PROOF. Items (i) and (ii) follow from Theorems 4.24 and 4.38. From Theorem 3.42, item (iii) follows. It remains to prove (iv). We can chose η_2 in such a way that each equilibrium $z^* \in \{z_{1,\eta},\cdots,z_{n,\eta}\}$ is hyperbolic and each periodic solution $\xi^* \in \{\xi_{1,\eta},\cdots,\xi_{m,\eta}\}$ is normally hyperbolic, $0 < \eta < \eta_2$.

Define $\mathcal{D}\colon \mathbf{\Xi}_0 \to \mathbf{\Xi}_\eta$ by

$$\mathcal{D}(z_i) = z_{i,\eta}, \text{ for } i = 1,\cdots,n, \quad \text{and} \quad \mathcal{D}(\xi_j) = \xi_{j,\eta}, \text{ for } j = 1,\cdots,m,$$

which is clearly a bijection.

Now if $W^u(\Xi_{i,0}) \cap W^s_{loc}(\Xi_{j,0}) \neq \varnothing$, then using the results of Section 2, which ensure that the unstable and stable manifolds behave continuously in the C^1 norm, and Lemma 5.8, there exists $\eta_3 \in (0,\eta_2]$ such that $W^u(\Xi^*_{i,\eta}) \cap W^s_{loc}(\Xi_{j,\eta}) \neq \varnothing$, for all $\eta \in (0,\eta_3]$.

This shows that if there exists a connection between $\Xi_{i,0}$ and $\Xi_{j,0}$, then there exists a connection between $\Xi_{i,\eta}$ and $\Xi_{j,\eta}$. It remains to show the converse.

Assume that we have a sequence $\eta_k \to 0$ and a sequence of global solutions $\xi_k\colon \mathbb{R} \to X$ of $\mathcal{T}_{\eta_k}$ such that

$$\lim_{t\to-\infty} \operatorname{dist}_H(\xi_k(t), \Xi_{i,\eta_k}) = 0 \text{ and } \lim_{t\to\infty} \operatorname{dist}_H(\xi_k(t), \Xi_{j,\eta_k}) = 0.$$

Then we know by Lemma 3.44 that there is a finite collection of global solutions $\varphi_\ell\colon \mathbb{R} \to X$ of $\mathcal{T}_0$, $\ell = 1,\cdots,m-1$ $(m \leqslant n)$, together with isolated invariant sets Ξ_{n_ℓ} in $\mathbf{\Xi}_0$, such that $\Xi_{n_1} = \Xi_{i,0}$, $\Xi_{n_m} = \Xi_{j,0}$ and

$$\Xi_{n_\ell} \overset{t\to-\infty}{\longleftarrow} \varphi_\ell(t) \overset{t\to\infty}{\longrightarrow} \Xi_{n_{\ell+1}}, \text{ for all } \ell = 1,\cdots,m-1.$$

But Proposition 5.9 ensures the existence of a global solution $\varphi\colon \mathbb{R} \to X$ of $\mathcal{T}_0$ such that

$$\Xi_i \overset{t\to-\infty}{\longleftarrow} \varphi(t) \overset{t\to\infty}{\longrightarrow} \Xi_j,$$

and we conclude the proof. □

3. Examples

In this section we will describe a few applications for Theorem 5.14, and also an example where we lose connections when making a perturbation on a semigroup due to the lack of transversality.

3.1. Automatic transversality for a class of ODE's. Following [**125**], let us consider an ordinary differential equation

$$\begin{cases} \dot{x} = f(x), \\ x(0) = x_0 \in \mathbb{R}^n, \end{cases} \tag{5.3}$$

where $f\colon \Omega \subset \mathbb{R}^n \to \mathbb{R}^n$ is a C^1 map, with Ω convex and open. We consider $\mathscr{L}$ the set of $n \times n$ matrices of the form

$$L = \begin{bmatrix} a_1 & b_1 & 0 & c_1 \\ c_2 & \ddots & \ddots & 0 \\ 0 & \ddots & \ddots & b_{n-1} \\ b_n & 0 & c_n & a_n \end{bmatrix}, \tag{5.4}$$

where a_i, b_i and c_i satisfy $b_i, c_i \geqslant 0$, for $i = 1, \cdots, n$, and

$$\prod_{i=1}^{n} b_i + \prod_{i=1}^{n} c_i > 0.$$

If $\{e_j\}_{j=1}^n$ is a basis in E and $\mathcal{L}(E)$ is the set of linear operators on E, we define $\mathscr{L}\{e_j\} \subset \mathcal{L}(E)$ the subset of operators for which the matrix representation with respect to the basis $\{e_j\}_{j=1}^n$ belongs to $\mathscr{L}$. Thus, from [**125**, Theorem 5], if $f\colon \Omega \subset E \to E$ is a C^1 function, with Ω convex and open, and is such that there exists a fixed basis $\{e_j\}_{j=1}^n$ such that for each $x \in \Omega$, the derivative $f'(x) \in \mathscr{L}\{e_j\}$, then given Ξ_1, Ξ_2 two critical elements of (5.3) (either a hyperbolic point or a normally hyperbolic orbit) we have $W^u(\Xi_1)$ and $W^s(\Xi_2)$ intersect transversally, provided that:
(i) Ξ_1 or Ξ_2 is a normally hyperbolic orbit;
(ii) Ξ_1 and Ξ_2 are both hyperbolic points and the spectra $\sigma(S_k)$ of the operators $S_k = e^{f(\Xi_k)}$, $k = 1, 2$, have the property that there is $h \in \{1, \cdots, (\tilde{n} - 1)/2 + 1\}$ such that

$$\nu_h^- > 1 \quad \text{and} \quad \mu_{h+1}^+ < 1,$$

where $\tilde{n} = n$ if n is odd and $\tilde{n} = n + 1$ if n is even, and ν_h^- and μ_h^+ are given in [**125**, Theorem 2].

Thus, if (5.3) generates a reversible semigroup $\mathcal{T}$ with a global attractor $\mathcal{A}$, with a finite number of hyperbolic equilibria and of normally hyperbolic orbits, then $\mathcal{T}$ is a Morse-Smale semigroup.

3.2. Reaction-diffusion equations. Let Ω be a bounded smooth domain in $\mathbb{R}^n$, $n \geqslant 1$, and let f be a C^k-function, $k \geqslant 2$, from $\overline{\Omega} \times \mathbb{R}$ into $\mathbb{R}$. We consider the equation

$$\begin{cases} u_t = \Delta u + f(x, u), & t > 0,\ x \in \Omega, \\ u = 0, & t > 0,\ x \in \partial\Omega, \\ u = u_0, & t = 0,\ x \in \Omega, \end{cases} \tag{5.5}$$

where $u_0 \in W_0^{1,p}(\Omega)$, $p > n$. Since $W_0^{1,p}(\Omega)$, $p > n$, is continuously embedded in $C^0(\overline{\Omega})$, the equation (5.5) defines a *local semiflow* $\mathcal{T}_f$ on $X = W_0^{1,p}(\Omega)$, $p > n$. This semiflow is dynamically gradient, with associated Lyapunov function

$$\Phi(\varphi) = \int_\Omega \Big(\frac{1}{2}|\nabla\varphi(x)|^2 - F(x,\varphi(x))\Big)dx, \tag{5.6}$$

which decreases along nonconstant trajectories of $\mathcal{T}_f$. Here $F(x,y)$ is the primitive $F(x,y) = \int_0^y f(x,s)ds$ of $f(x,\cdot)$. To obtain a semigroup (that is, a global semiflow), we will add the following dissipative condition: we assume that there exist a positive constant C_0 and a constant $\mu < \lambda_1$, where $\lambda_1 > 0$ is the first eigenvalue of the operator $-\Delta$ with homogeneous Dirichlet boundary conditions, such that, for any $y \in \mathbb{R}$ and $x \in \overline{\Omega}$, we have

$$f(x,y)y < C_0 + \mu y^2, \quad F(x,y) < C_0 + \frac{1}{2}\mu y^2 . \tag{5.7}$$

Under this additional assumption (5.7), we obtain the following proposition.

PROPOSITION 5.15. *If condition* (5.7) *holds, then all the solutions of* (5.5) *are global, that is, they exist for all times* $t \geqslant 0$. *In addition, the semigroup* $\mathcal{T}_f$ *admits a global attractor* $\mathcal{A}_f$ *in* X *and there exists a positive constant* K_0, *depending only on* C_0, μ *and* Ω, *such that*

$$\sup_{\varphi\in\mathcal{A}_f} (\|\varphi\|_{W^{2,p}} + \|\varphi\|_{L^\infty}) \leqslant K_0. \tag{5.8}$$

Let $\mathcal{G}^k$, $k \geqslant 1$, be the subset of functions of $C^k(\overline{\Omega}\times\mathbb{R},\mathbb{R})$ satisfying (5.7), endowed with the Whitney C^k topology. It follows from [**34**, Theorem 1.1] that there exists a generic set $\mathcal{G}^k_{MS}$ in $\mathcal{G}^k$, $k \geqslant 2$, such that, for any $f \in \mathcal{G}^k_{MS}$, every equilibrium point φ of (5.5) is hyperbolic and, if φ_- and φ_+ are any two such equilibria, their stable and unstable manifolds intersect transversally, that is,

$$W^u(\varphi_-) \pitchfork W^s(\varphi_+).$$

Thus, from [**34**, Theorem 1.1], we at once deduce the following theorem.

THEOREM 5.16. *For any* f *belonging to the generic set* $\mathcal{G}^k_{MS}$, $k \geqslant 2$, *the semigroup* $\mathcal{T}_f$ *generated by* (5.5) *is a* C^2 *semigroup, which is a Morse-Smale semigroup on* X *with global attractor* $\mathcal{A}_f$ *satisfying the estimates* (5.8). *Moreover, since* $\mathcal{T}_f$ *is dynamically gradient, its nonwandering set* $\Omega_{\mathcal{T}_f}$ *reduces to a finite set of hyperbolic equilibrium points* $\mathcal{Z}_f = \{\varphi_1, \cdots, \varphi_m\}$.

We now fix a function $f_0 \in \mathcal{G}^k_{MS}$, $k \geqslant 2$, and consider perturbations of equation (5.5), that is, we consider the equation

$$\begin{cases} u_t = \Delta u + \eta a(x)\cdot\nabla u + f_0(x,u), & t > 0,\ x \in \Omega, \\ u = 0, & t > 0,\ x \in \partial\Omega, \\ u = u_0, & t = 0,\ x \in \Omega, \end{cases} \tag{5.9}$$

where $u_0 \in X$, $a\colon \bar{\Omega} \to \mathbb{R}^n$. Then, from Theorem 5.14, for all η suitably small the semigroup associated to (5.9) is Morse-Smale and there is a phase diagram isomorphism between the attractors of the perturbed and unperturbed semigroups.

3.3. Application to the damped wave equation. Let Ω be a bounded smooth domain in $\mathbb{R}^n$, $n = 1, 2, 3$. Let f be a C^k-function, $k \geqslant 2$, from $\overline{\Omega} \times \mathbb{R}$ into $\mathbb{R}$. We choose a positive constant $\gamma > 0$ and consider the damped wave equation

$$\begin{cases} u_{tt} + \gamma u_t - \Delta u = f(x, u), & t > 0, \ x \in \Omega, \\ u(x, t) = 0, & t > 0, \ x \in \partial\Omega, \\ (u(x, 0), u_t(x, 0)) = (u_0, v_0), & \end{cases} \tag{5.10}$$

where $(u_0, v_0) \in Y$ with $Y = H_0^1(\Omega) \times L^2(\Omega)$ if $n = 1$ and $Y = (H^2(\Omega) \cap H_0^1(\Omega)) \times H_0^1(\Omega)$ if $n = 2$ or 3. Then (5.10) generates a local semiflow $S_f(t)$ on Y. This semiflow is dynamically gradient with associated Lyapunov function

$$\Phi(\varphi, \psi) = \int_\Omega (\frac{1}{2}|\nabla\varphi(x)|^2 + \frac{1}{2}|\psi(x)|^2 - F(x, \varphi(x)))dx, \tag{5.11}$$

which decreases along nonconstant trajectories of $S_f(\cdot)$. To obtain a global semiflow, as in the parabolic case, we will assume the dissipative condition (5.7). Under this additional assumption (5.7), again as in the parabolic case, we obtain the following proposition.

PROPOSITION 5.17. *If condition* (5.7) *holds, then all the solutions of* (5.10) *are global. In addition, the semiflow $S_f(t)$ admits a compact global attractor $\mathcal{A}_f^D$ in Y and there exists a positive constant K_0^D, depending only on C_0, μ, γ and Ω, such that*

$$\sup_{(\varphi,\psi)\in\mathcal{A}_f^D} (\|(\varphi, \psi)\|_{H^2\times H^1} + \|\varphi\|_{L^\infty}) \leqslant K_0^D. \tag{5.12}$$

For $k \geqslant 3$ (and $k \geqslant 2$ if $n = 1$), we introduce the space $\mathcal{G}_D^k = \{f \in C^k(\overline{\Omega}\times\mathbb{R}, \mathbb{R})\}$ (respectively $\mathcal{G}_D^k = \{f \in C^k(\overline{\Omega} \times \mathbb{R}, \mathbb{R})\colon f(x, 0) = 0,$ for all $x \in \partial\Omega\}$ in the case of Dirichlet boundary conditions if $n \geqslant 2$) which, in addition, satisfy the dissipative condition (5.7). We endow this space with the Whitney C^k topology. It follows from [**36**, Theorem 1.1] that there exists a generic set $\mathcal{G}_{D,MS}^k$ in $\mathcal{G}_D^k$, $k \geqslant 2$, such that, for any $f \in \mathcal{G}_{D,MS}^k$, every equilibrium point $(\varphi, 0)$ of (5.10) is hyperbolic and, if $(\varphi_-, 0)$ and $(\varphi_+, 0)$ are any two such equilibria, their stable and unstable manifolds intersect transversally, that is,

$$W^u((\varphi_-, 0)) \pitchfork W^s((\varphi_+, 0)).$$

Thus, from [**36**, Theorem 1.1], we at once deduce the following theorem.

THEOREM 5.18. *For any f belonging to the generic set $\mathcal{G}_{D,MS}^k$, $k \geqslant 2$, the semigroup S_f, generated by* (5.10), *is C^2, and is a C^1-reversible Morse-Smale semigroup on Y, with global attractor $\mathcal{A}_f^D$ satisfying the estimates* (5.12). *Moreover, since S_f is dynamically gradient, its nonwandering set Ω_{S_f} reduces to a finite set of hyperbolic equilibrium points $\mathcal{Z}_f^D = \{(\varphi_1, 0), \cdots, (\varphi_m, 0)\}$.*

From Theorem 5.14, suitably small perturbations (as in the previous example) of the semigroup associated to (5.10) are Morse-Smale and there is a phase diagram isomorphism between the attractors of the perturbed and unperturbed semigroups.

3.4. An example of nontransversality. In this subsection we recall the brief example done in the introduction in which we *do not* have the structural stability, due to a lack of transversality in the intersection of the unstable manifold and the local stable manifold of two equilibria.

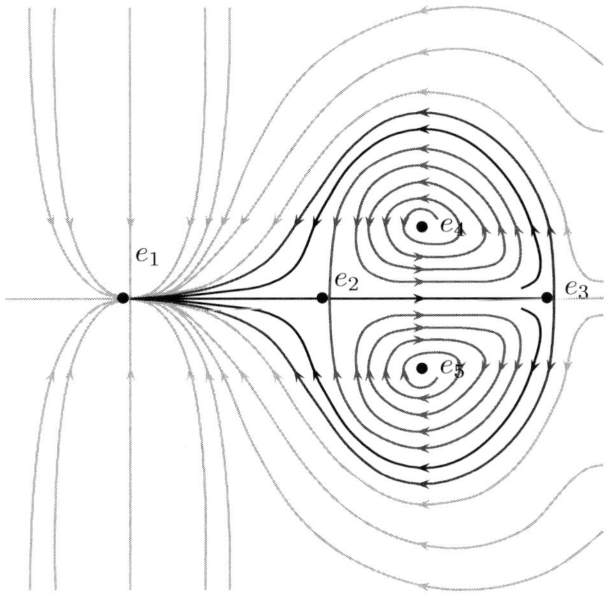

FIGURE 5.2. The *alien head*.

Consider the attractor given in Figure 5.2: note that the unstable manifold $W^u(e_2)$ of e_2 and the local stable manifold $W^s_{loc}(e_3)$ of e_3 do not intersect transversally.

Now, we can make a small autonomous perturbation to this attractor to obtain the attractor given in Figure 5.3, and we can see that there is no phase diagram isomorphism between these objects, since the connection between e_2 and e_3 is lost.

Notes. It is always a very difficult task to show the internal characterization of global attractors. Morse-Smale semigroups have been considered in many works and are the class of examples which shows the most detailed structure of attractors for infinite-dimensional dynamical systems ([**35**,**36**,**133**,**147**,**147**,**175**,**176**,**186**]). The concept was firstly developed in a finite-dimensional setting (see, for instance, [**121**, **178**–**180**]). First results on Morse-Smale semigroups can be found in [**141**–**143**].

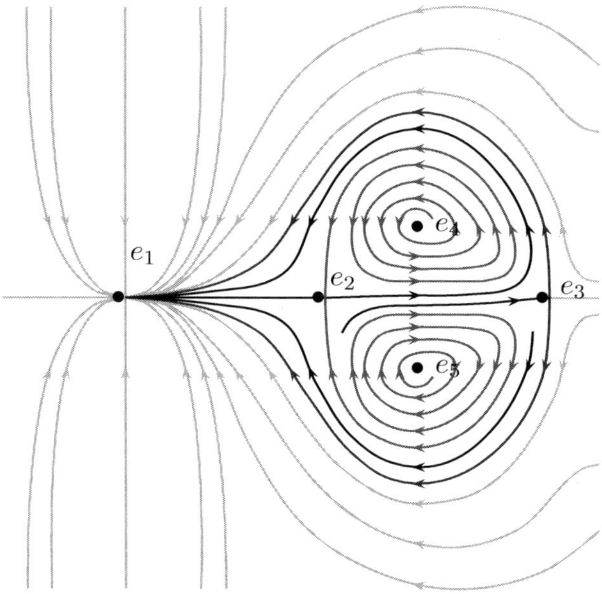

FIGURE 5.3. The perturbed *alien head*.

Part 2

Non-autonomous Theory

CHAPTER 6

Non-autonomous Dynamical Systems and Their Attractors

Recently, the analysis of qualitative properties of *non-autonomous problems* in general phase spaces (infinite-dimensional Banach spaces or general metric spaces) has received a lot of attention (see for instance [**153**], [**140**], [**201**], [**202**], [**78**], [**89**], [**43**], [**98**], [**190**], [**181**], [**116**], [**79**], [**99**], [**149**] and [**150**]). In particular, the study of *pullback*, *skew-product* and *uniform attractors* has developed a wide and deep research area, providing qualitative information for the asymptotic dynamics of an increasing number of non-autonomous models of phenomena from different areas of knowledge such as, among others, Physics, Biology, Economics and Engineering.

To motivate our approach, we consider a simple non-autonomous ordinary differential equation in $\mathbb{R}^n$. Let $\mathbb{K}$ be either $\mathbb{R}^+$ or $\mathbb{R}$ and let $f\colon \mathbb{K}\times\mathbb{R}^n \to \mathbb{R}^n$ be a family of vector fields (one for each $t\in\mathbb{K}$). Assume that f satisfies *good* conditions, in the sense that we have existence and uniqueness of solutions for the initial value problem

$$\begin{cases} \dot{u} = f(t,u), \ t > s, \\ u(s) = u_0 \in \mathbb{R}^n, \end{cases} \tag{6.1}$$

for each $s\in\mathbb{K}$ and $u_0\in\mathbb{R}^n$. Suppose also that the solution $u(\cdot,s,u_0)$ of (6.1) is defined in $[s,\infty)$ for each $s\in\mathbb{K}$ and $u_0\in\mathbb{R}^n$ and that the map $\mathcal{P}\times\mathbb{R}^n \ni (t,s,u_0)\mapsto u(t,s,u_0)\in\mathbb{R}^n$ is continuous, where $\mathcal{P}=\{(t,s)\in\mathbb{K}^2\colon t\geqslant s\}$.

For each t, $f(t,\cdot)$ is the vector field that drives the solution at time t. Hence, the path described by the solution in $\mathbb{R}^n$ between s and $s+\tau$ will depend on the vector fields $f(t,\cdot)$ for $t\in[s,s+\tau]$, and therefore it will depend on both the initial time s and elapsed time τ. When $f(t,\cdot)=f(\cdot)$ is independent of t, that is, (6.1) is autonomous, the path described by the solution will no longer depend on s but only on the elapsed time τ.

The change from a time independent to a time dependent f may easily be underestimated in the study of the asymptotics of (6.1). Hence, we insist on calling the readers' attention to the fact that in the *autonomous case* there is only one vector field driving the solution at all times, whereas in the *non-autonomous case* there are infinitely many vector fields driving the solution (one for each instant of time), and these vector fields evolve with time, in a prescribed way, just as the solution does. Once we are aware of the inherent difficulty of non-autonomous problems we can start to come up with ways to look at its asymptotic dynamics.

When $\mathbb{K}=\mathbb{R}$ we can define the solution operator $S(t,s)u_0=u(t,s,u_0)$ for each $(t,s)\in\mathcal{P}$ and the family $\mathcal{S}=\{S(t,s)\colon (t,s)\in\mathcal{P}\}\subset C(\mathbb{R}^n)$ is an *evolution process* in $\mathbb{R}^n$, that is, a family $\mathcal{S}=\{S(t,s)\colon (t,s)\in\mathcal{P}\}\subset C(\mathbb{R}^n)$ such that $S(t,t)x=x$, for all $t \in \mathbb{R}$ and $x \in \mathbb{R}^n$, $S(t,s) = S(t,r)S(r,s)$, for all $t \geqslant r \geqslant s$, and such that the map $\mathcal{P} \times \mathbb{R}^n \ni (t,s,x) \mapsto S(t,s)x \in \mathbb{R}^n$ is continuous.

When f is independent of time, and hence $u(t,s,f,u_0) = u(t-s,0,f,u_0)$, the asymptotic dynamics can be achieved by either making $t \to \infty$, that is, seeing what happens to the state at the final time t when t is driven further and further to the future, or making $s \to -\infty$, and thus seeing what happens to the state at time t when the initial time s is driven further and further to the past.

Now if f is time dependent these two asymptotic dynamics give rise to completely different scenarios. We may study the asymptotics with respect to the elapsed time $t-s$ (when $t-s \to \infty$) or with respect to s (when $s \to -\infty$ and t is arbitrary but fixed). These are called, respectively, *forward* and *pullback dynamics*, and are in general unrelated. It is only natural that they are unrelated, since the set of vector fields driving the solution depends on time.

For an evolution process $\mathcal{S}$, a *pullback attractor* is a family $\{A(t)\}_{t\in\mathbb{R}}$ such that $\bigcup_{t\in\mathbb{R}} A(t)$ is bounded, $A(t)$ is compact for each $t \in \mathbb{R}$, $S(t,s)A(s) = A(t)$ for all $t \geqslant s$ and $\operatorname{dist}_H(S(t,s)B, A(t)) \to 0$ as $s \to -\infty$ for each bounded subset B of $\mathbb{R}^n$ and for each $t \in \mathbb{R}$.

A *bounded global solution* of $\mathcal{S}$ is a function $\xi\colon \mathbb{R} \to \mathbb{R}^n$ such that $S(t,s)\xi(s) = \xi(t)$ for each $t \geqslant s$, and in this case, assuming that it has a bounded pullback attractor $\{A(t)\}_{t\in\mathbb{R}}$, we can characterize it by

$$A(t) = \{\xi(t)\colon \xi \text{ is a bounded global solution of } \mathcal{S}\}.$$

Therefore this object has much to say about the dynamics of (6.1), since it consists of all bounded global solutions of (6.1). It is also important to know that $\{A(t)\}_{t\in\mathbb{R}}$ may not have any forward attraction property (see, for instance, [**78**, Section 1.3] or [**162**]). In particular, when f is independent of time, it coincides with the global attractor of the associated semigroup.

One must note that the theory of pullback attractors requires that the vector fields are defined for all times in the real line $\mathbb{R}$ (that is, $\mathbb{K} = \mathbb{R}$). In many models the backwards history or evolution is simply not available, and the vector fields are only known after a given fixed initial time, say, for each time in $\mathbb{K} = \mathbb{R}^+$. If we artificially define the vector fields for negative times, the pullback attractors will strongly depend on such extension, since most of the pullback asymptotics will be driven by these artificially introduced vector fields.

When f is time independent and $\mathbb{K} = \mathbb{R}^+$, the following important remark is due: in many situations, there is absolutely no reason to extend f for negative values of time by the same time independent vector field f. Yet, we speak about global solutions for such autonomous problems as we did in Chapter 1. The explanation of that will come together with the natural framework that we will establish for the non-autonomous case. Also, when $\mathbb{K} = \mathbb{R}^+$, the pullback attractor does not play a direct role in the asymptotic dynamics of (6.1). Nonetheless, as we will see later, it appears again in a natural and essential way, in the understanding of the asymptotic behavior of solutions of (6.1) within the appropriate framework. Let us pursue this a little further.

The crucial point here is to understand that, for a non-autonomous differential equation, there are infinitely many vector fields (one for each time) driving the evolution, whereas in the autonomous case there is only one. Therefore the evolution of the vector fields (called *driving semigroup*) must be taken into account as well.

Once one realizes this feature it becomes clear how rich and difficult the subject *dynamics of non-autonomous dynamical systems* really is.

This approach has been followed by Chepyzhov and Vishik in their seminal work on the forward dynamics of non-autonomous problems; see [**98**, Chapter VII]. *Assuming the backwards uniqueness property* for the driving semigroup, they consider the *uniform attractor* for a family of semiprocesses, indexed in the symbol space. They prove that the uniform attractor can be characterized as the union of the kernel sections at zero corresponding to all the elements in the global attractor of the driving semigroup, for which the driving is naturally defined, in a unique way, for negative times. It occurs, very often and in a natural way, that backwards uniqueness does not hold in the attractor of the driving semigroup. Having that in mind, we present the theory *avoiding the requirement of backwards uniqueness* for the driving semigroup. In fact, we show that nonuniqueness is what naturally occurs.

One can already see the difficulties that arise when dealing with non-autonomous problems by looking at simple concepts, such as hyperbolicity for instance. Indeed, in the autonomous case we choose an equilibrium (fixed point of the solution operator), linearize the solution operator around it and compute the spectrum of the linearized operator to decide if it is hyperbolic or not. In the non-autonomous context we have no way to identify which solutions play the special role of equilibria and, even if we were able to single out these solutions, we would not know how to verify if they are hyperbolic or not. Even the simple matter of obtaining a hyperbolic global solution as a small non-autonomous perturbation of a hyperbolic equilibrium involves highly nontrivial results on the robustness of exponential dichotomies (see [**73**] and [**75**]).

In what follows, we introduce a general method that provides a way to study a non-autonomous differential equation, and explain some important features from autonomous equations which are often overlooked. The method consists in considering the family of nonlinearities as a base flow driven by the time shift operator acting on the *closure* of $\{f(t+\cdot,\cdot)\colon t\in\mathbb{R}^+\}$ where nonlinearity $f(\cdot,\cdot)$ is the nonlinearity occurring in the original equation (see [**193**] and [**194**]).

To ensure the proper understanding of the framework, we will continue speaking about the problem (6.1), and leave the more general abstract framework to the next section. We consider $\mathcal{C}$ the set of continuous functions $f\colon\mathbb{R}^+\times\mathbb{R}^n\to\mathbb{R}^n$ such that for each bounded subset $B\subset\mathbb{R}^n$ there exist a constant $L_f(B)>0$ and an increasing continuous function $w_{f,B}\colon\mathbb{R}^+\to\mathbb{R}^+$ with $w_{f,B}(0)=0$ such that

$$\|f(t_1,x_1)-f(t_2,x_2)\|\leqslant w_{f,B}(|t_1-t_2|)+L_f(B)\|x_1-x_2\|, \tag{6.2}$$

for $t_1,t_2\in\mathbb{R}^+$ and $x_1,x_2\in B$.

In $\mathcal{C}$ we consider the metric ρ of the uniform convergence in compact subsets of $\mathbb{R}^+\times\mathbb{R}^n$. Suppose, in addition, that f is *dissipative*, that is, there exist $M>0$ and $\delta>0$ such that

$$\langle f(t,x),x\rangle\leqslant-\delta\quad\text{for } t\in\mathbb{R}^+ \text{ and } x\in\mathbb{R}^n \text{ with } \|x\|>M, \tag{6.3}$$

where $\langle\cdot,\cdot\rangle$ denotes the scalar product in $\mathbb{R}^n$.

Denote by Σ the closure, with respect to ρ, of the set of all forward translates of f, that is,

$$\Sigma(f)=\overline{\{f(s+\cdot,\cdot)\colon s\in\mathbb{R}^+\}}^{\rho},$$

known as the **positive hull** of the function f in the metric space $(\mathcal{C},\rho)$ (see [**98**], [**193**] and [**194**] for more details), and define the *shift operator* $\Theta(t)\colon \Sigma(f)\to\Sigma(f)$ by

$$\Theta(t)f(\cdot,\cdot)=f(t+\cdot,\cdot).$$

It is simple to show that $\Theta=\{\Theta(t)\colon t\in\mathbb{R}^+\}$ defines a semigroup in $\Sigma(f)$, which we usually denote simply by Θ, called the **driving semigroup**. Also, it follows from the Arzelà-Ascoli Theorem that $\Sigma(f)$ is compact and Θ has a global attractor $\mathscr{S}$ in $\Sigma(f)$.

Thus, to treat the asymptotics of a non-autonomous differential equation, we consider a combination of a base flow Θ on $\Sigma(f)$ and, for each $\sigma\in\Sigma(f)$, the semiflow $\mathbb{R}^+\times\mathbb{R}^n\ni(t,u_0)\mapsto\mathcal{K}(t,\sigma)u_0\in\mathbb{R}^n$ where, for each $u_0\in\mathbb{R}^n$, the map $\mathbb{R}^+\ni t\mapsto\mathcal{K}(t,\sigma)u_0\in\mathbb{R}^n$ is the solution of the initial value problem

$$\begin{cases}\dot{u}=\sigma(t,u), & t>0,\\ u(0)=u_0\in\mathbb{R}^n.\end{cases}\tag{6.4}$$

Note that the conditions imposed on f ensure that the solutions of (6.4) are defined for all $t\geqslant 0$. Also the map $\mathbb{R}^+\times\Sigma(f)\ni(t,\sigma)\mapsto\mathcal{K}(t,\sigma)\in\mathcal{C}(\mathbb{R}^n)$ satisfies

(i) $\mathcal{K}(0,\sigma)x=x$ for all $x\in\mathbb{R}^n$ and $\sigma\in\Sigma(f)$;
(ii) $\mathcal{K}(t+s,\sigma)=\mathcal{K}(t,\Theta(s)\sigma)\mathcal{K}(s,\sigma)$ for all $t\geqslant s$ and $\sigma\in\Sigma(f)$;
(iii) the map $\mathbb{R}^+\times\Sigma(f)\times\mathbb{R}^n\ni(t,\sigma,x)\mapsto\mathcal{K}(t,\sigma)x\in\mathbb{R}^n$ is continuous.

The second property is commonly know as the *cocycle property* and, due to this name, the map $\mathcal{K}$ is called *cocycle semiflow*, or simply *cocycle*. These properties define a cocycle and the association with the differential equation (6.4) is no longer needed. In any case, one interprets $\mathcal{K}(t,\sigma)u_0$ as the solution at time t that has started in the state u_0 at time zero subject to the non-autonomous driving term $\sigma\in\Sigma(f)$.

Now we can talk about *evolution processes* and *pullback attractors*. If $\eta\colon\mathbb{R}\to\Sigma(f)$ is a global solution for the semigroup Θ, then the family $\mathcal{S}_\eta=\{S_\eta(t,s)\colon t\geqslant s\}\subset C(\mathbb{R}^n)$ defined by

$$S_\eta(t,s)=\mathcal{K}(t-s,\eta(s))\quad\text{for } t\geqslant s,$$

defines an evolution process. In fact, for $t\geqslant s\geqslant r$, using the cocycle property, we have

$$S_\eta(t,s)S_\eta(s,r)=\mathcal{K}(t-s,\Theta(s-r)\eta(r))\mathcal{K}(s-r,\eta(r))=\mathcal{K}(t-r,\eta(r))=S_\eta(t,r),$$

and all other properties of an evolution process are satisfied, using the properties of the driving semigroup Θ and the cocycle $\mathcal{K}$. It is not difficult to show (we will see this later in the chapter) that, under the conditions imposed on f, $\mathcal{S}_\eta$ has a pullback attractor $\{A_\eta(t)\}_{t\in\mathbb{R}}$.

REMARK 6.1. Assume that $\mathbb{K}=\mathbb{R}^+$ and also that f is independent of time. Then $\Sigma(f)=\{f\}$ and $\mathscr{S}=\{f\}$. From the Θ-invariance of $\mathscr{S}$, there is a global solution $\eta\colon\mathbb{R}\to\{f\}$ and we can define

$$S_\eta(t+s,s)=\mathcal{K}(t,\eta(s))=\mathcal{K}(t,f),$$

for all $t\geqslant 0$ and $s\in\mathbb{R}$. Therefore at ∞ the vector field f can indeed be defined for all $t\in\mathbb{R}$. Readers should know that this remark really deserves deep reflection.

Together, $\mathcal{K}$ and Θ determine the non-autonomous problem (6.1), and the pair $(\mathcal{K},\Theta)_{(\mathbb{R}^n,\Sigma(f))}$ will be called a *non-autonomous dynamical system* on $(\mathbb{R}^n,\Sigma(f))$, and when there is no confusion, it will be simply denoted by the cocycle $\mathcal{K}$. With this object one can also define an associated semigroup (see [**98**], [**194**] and [**193**]) $\Pi = \{\Pi(t)\colon t \geqslant 0\}$ on $\mathbb{X} = \mathbb{R}^n \times \Sigma(f)$, with the metric $d_{\mathbb{X}}((x_1,\sigma_1),(x_2,\sigma_2)) = \|x_1 - x_2\| + \rho(\sigma_1,\sigma_2)$, by setting

$$\Pi(t)(x,\sigma) = (\mathcal{K}(t,\sigma)x, \Theta(t)\sigma), \text{ for } t \geqslant 0 \text{ and } (x,\sigma) \in \mathbb{X},$$

and the properties of Θ and $\mathcal{K}$ ensure that Π satisfies all the properties of a semigroup. This semigroup Π is called the *skew-product semiflow* associated with $\mathcal{K}$. Also, the set $\{(x,\sigma) \in \mathbb{X}\colon \|x\| \leqslant M, \sigma \in \mathscr{S}\}$ is a compact subset of $\mathbb{X}$ which attracts bounded subsets of $\mathbb{X}$. It follows that Π has a global attractor $\mathbb{A}$ in $\mathbb{X}$.

We will see later that in fact we have

$$\mathbb{A} = \bigcup_{\eta} \bigcup_{t\in\mathbb{R}} A_\eta(t) \times \{\eta(t)\} \tag{6.5}$$

where the union is taken over all global solutions η of Θ in $\mathscr{S}$.

The asymptotic set of states, for $t \to \infty$, of the solutions of (6.1) is $\mathcal{A}$ =the projection on the first coordinate of $\mathbb{A}$, and it is called the *uniform attractor* for (6.1), in the terminology of Chepyzhov and Vishik [**98**]. Thus, if we only look at the uniform attractor we may lose perspective of what is indeed the asymptotic dynamics of (6.1). That is, if $\sigma \in \mathscr{S}$, a global solution $\xi\colon \mathbb{R} \to \mathscr{S}$ of Θ with $\xi(0) = \sigma$ corresponds to a natural extension (given by the evolution of (6.1)) of σ for negative times, that is, we can consider the extension $\sigma(s+t,x) = \xi(s)(t,x)$, for all $t \geqslant 0$ and $s \in \mathbb{R}$. Of course if $s \geqslant 0$ we have $\xi(s)(t,x) = \Theta(s)\xi(0)(t,x) = \sigma(t+s,x)$, for all $t \geqslant 0$.

Another important remark here is that in general Θ is not injective in $\mathscr{S}$, and we give a simple example to settle this once and for all.

EXAMPLE 6.2. Consider the function $f(t,x) = \alpha(t)g(x)$ for $t \geqslant 0$ and $x \in \mathbb{R}^n$, where $g \in \mathcal{C}$ satisfies (6.2) and (6.3), and $\alpha\colon \mathbb{R}^+ \to [1,2]$ a globally Lipschitz function with $\alpha(t) = 1$ if $t \in ((2n)^2,(2n+1)^2)$, $\alpha(t) = 2$ if $t \in ((2n+1)^2+1, (2n+2)^2-1)$, $n \in \mathbb{N}$, and linear elsewhere.

It is clear that $f \in \mathcal{C}$ and it also satisfies (6.2) and (6.3). For this such f, the function $\sigma(t,x) = 2g(x)$, for $t \geqslant 0$, is in $\mathscr{S}$ and there are infinitely many global solutions $\eta\colon \mathbb{R} \to \mathscr{S}$ of Θ such that $\eta(0) = \sigma$. In fact, considering $t_n^1 = \frac{(2n+1)^2+(2n+2)^2}{2}$ and $\eta_n^1(\cdot) = \Theta(\cdot + t_n^1)f\colon [-t_n^1,\infty) \to \Sigma(f)$, $\{\eta_n^1\}$ converges in the metric ρ, passing to a subsequence if needed, to $\eta_1(t) = 2g$ for all $t \in \mathbb{R}$, whereas if $t_n^2 = (2n+1)^2+1$ and $\eta_n^2(\cdot) = \Theta(\cdot + t_n^2)f\colon [-t_n^2,\infty) \to \Sigma(f)$, $\{\eta_n^2\}$ converges in the metric ρ, passing to a subsequence if needed, to $\eta_2(t) = 2g$ for $t \geqslant 0$ and $\eta_2(t) = g$ if $t \leqslant -1$.

Clearly $\eta_1(0) = \eta_2(0) = 2g$, which already gives nonuniqueness, but one can repeat this reasoning and construct, for the same example, infinitely many global solutions of Θ through $\sigma = 2g \in \mathscr{S}$.

Having settled this, *we will not assume backwards uniqueness* for Θ in $\Sigma(f)$, and thus, given a non-autonomous differential equation such as (6.1), we need to deal with *four* different, but closely related, evolution operators:

(a) the *semigroup* Θ on $\Sigma(f)$, which is associated to the dynamics of the time-dependent nonlinearities appearing in the equation;

(b) the *non-autonomous dynamical system* $(\mathcal{K}, \Theta)_{(\mathbb{R}^n, \Sigma(f))}$ or the *cocycle* $\mathcal{K}$;
(c) the *skew-product semiflow* Π defined on the product space $\mathbb{X}$;

and lastly, for each global solution $\eta\colon \mathbb{R} \to \Sigma(f)$ of Θ, the *evolution process* $\mathcal{S}_\eta$ given by

(d) $S_\eta(t,s)x = \mathcal{K}(t-s, \eta(s))x$ for all $t \geqslant s$ and $x \in \mathbb{R}^n$.

To each one of the objects described above, we will present a corresponding notion of *attractor*, and each one of them will play its part in itemizing the asymptotic dynamics of (6.1). They are the following:

(a) the *global attractor* $\mathscr{S}$ for Θ in $\Sigma(f)$;
(b) the *uniform attractor* $\mathcal{A}$ for $\mathcal{K}$ in $\mathbb{R}^n$;
(c) the *global attractor* $\mathbb{A}$ for Π in $\mathbb{X}$;
(d) for each global solution $\eta\colon \mathbb{R} \to \mathscr{S}$ of Θ, the *pullback attractor* $\{A_\eta(t)\}_{t\in\mathbb{R}}$ of the evolution process $\mathcal{S}_\eta$.

REMARK 6.3. There is another notion of attractor, widely used in the literature and commonly associated to the non-autonomous dynamical systems $(\mathcal{K}, \Theta)_{X,\Sigma}$, called the **cocycle attractor** (see [**153**], for instance), and given by a family $\{A(\sigma)\}_{\sigma\in\mathscr{S}}$. We have chosen not to present it here, since its introduction would require the assumption that through each σ in the global attractor $\mathscr{S}$ of $\Theta(t)$ *passes a unique global solution* $\eta\colon \mathbb{R} \to \mathscr{S}$.

In fact, if ξ and η are two distinct global solutions through σ, the cocycle attractor $A(\sigma)$ should coincide with the pullback attractor at time zero $A_\xi(0)$ of the evolution process

$$S_\xi(t,s) = \mathcal{K}(t-s, \xi(s)),$$

which also coincides with the pullback attractor at time zero $A_\eta(0)$ of the evolution process

$$S_\eta(t,s) = \mathcal{K}(t-s, \eta(s)),$$

and that will not happen, in general.

In the next section we present the precise definitions of all these objects, as well as conditions ensuring their existence in each case. Moreover, we present the relationship between pullback and uniform attractors, which leads to a detailed description of the uniform attractor and provides further understanding of its dynamical structures (associated to limiting evolution processes) and allows us to talk about upper and lower semicontinuity, and topological and geometrical structural stability, at least in the case of a suitable non-autonomous perturbation of a semigroup.

1. Pullback, uniform and skew-product attractors

Motivated by the discussion carried out in the previous section, we establish an abstract framework to work on. In order to study the asymptotics of a *non-autonomous problem* we need to consider the following families of operators: the driving semigroup and the associated cocycle, the skew-product semiflow, and the evolution processes obtained from the cocycle considering global solutions of the driving semigroup.

DEFINITION 6.4. Let (X, d_X) and (Σ, d_Σ) be metric spaces. Consider $\Theta = \{\Theta(t)\colon t \geqslant 0\} \subset C(\Sigma)$ a semigroup (in the sense of Definition 1.1) in Σ, called the **driving semigroup**. A **cocycle** relative to Θ is a family $\mathcal{K} = \{\mathcal{K}(t,\sigma) \in C(X) \colon t \geqslant 0, \sigma \in \Sigma\}$ which satisfies

(i) $\mathcal{K}(0,\sigma)x = x$, for all $x \in X$ and $\sigma \in \Sigma$;
(ii) $\mathcal{K}(t+s,\sigma) = \mathcal{K}(t,\Theta(s)\sigma)\mathcal{K}(s,\sigma)$, for all $t,s \geqslant 0$ and $\sigma \in \Sigma$;
(iii) the map $\mathbb{R}^+ \times \Sigma \times X \ni (t,\sigma,x) \mapsto \mathcal{K}(t,\sigma)x \in X$ is continuous.

Property (ii) is called the **cocycle property**. Define in $\mathbb{X} = X \times \Sigma$, with the product metric, the **skew-product semiflow** or **skew-product semigroup** $\Pi = \{\Pi(t) \colon t \geqslant 0\} \subset C(\mathbb{X})$ defined by

$$\Pi(t)(x,\sigma) = (\mathcal{K}(t,\sigma)x, \Theta(t)\sigma) \quad \text{for } t \geqslant 0 \text{ and } (x,\sigma) \in \mathbb{X}.$$

Using the semigroup property of Θ and the cocycle property of $\mathcal{K}$ we have

$$\begin{aligned}\Pi(t+s)(x,\sigma) &= (\mathcal{K}(t+s,\sigma)x, \Theta(t+s)\sigma) \\ &= (\mathcal{K}(t,\Theta(s)\sigma)\mathcal{K}(s,\sigma)x, \Theta(t)\Theta(s)\sigma) = \Pi(t)\Pi(s)(x,\sigma),\end{aligned}$$

and since the other properties of Definition 1.1 of a semigroup are clearly satisfied, Π is in fact a semigroup in $\mathbb{X}$.

When Σ is a unitary set, say $\Sigma = \{\sigma_0\}$, we have $\Theta(t)\sigma_0 = \sigma_0$ for all $t \geqslant 0$, and defining

$$T(t)x = \mathcal{K}(t,\sigma_0)x, \quad \text{for all } t \geqslant 0 \text{ and } x \in X, \tag{6.6}$$

the skew-product semigroup Π is reduced to $\Pi(t)(x,\sigma_0) = (T(t)x,\sigma_0)$ for $t \geqslant 0$ and $\mathcal{T} = \{T(t) \colon t \geqslant 0\}$ defines a semigroup in X. This is the situation that arises when we consider an autonomous differential equation. For a non-autonomous differential equation, Σ *will not be a unitary set.*

Let us now derive conditions for the existence of an attractor for the skew-product semigroup Π in $\mathbb{X}$. Recall that a semigroup has a global attractor if and only if it is asymptotically compact and bounded dissipative; see Theorem 1.23.

Define $\pi_X \colon \mathbb{X} \to X$ as the usual projection in the first coordinate, that is, $\pi_X(x,\sigma) = x$ for each $(x,\sigma) \in \mathbb{X}$. Thus we obtain

$$\pi_X\big(\Pi(t)(x,\sigma)\big) = \mathcal{K}(t,\sigma)x, \quad \text{for all } t \geqslant 0 \text{ and } (x,\sigma) \in \mathbb{X}, \tag{6.7}$$

and defining $\pi_\Sigma \colon \mathbb{X} \to \Sigma$ as the usual projection in the second coordinate, that is, $\pi_\Sigma(x,\sigma) = \sigma$ for each $(x,\sigma) \in \mathbb{X}$, we obtain

$$\pi_\Sigma\big(\Pi(t)(x,\sigma)\big) = \Theta(t)\sigma, \quad \text{for all } t \geqslant 0 \text{ and } (x,\sigma) \in \mathbb{X}. \tag{6.8}$$

The bounded dissipativity property of Π is equivalent to saying that there are bounded subsets B_0 in X and S_0 in Σ such that, for bounded subsets $B \subset X$ and $S \subset \Sigma$ there exists $t_0 = t_0(B,S) \geqslant 0$ such that $\Pi(t)(B \times S) \subset B_0 \times S_0$ for all $t \geqslant t_0$. This implies that for any bounded subset $S \subset \Sigma$ there exists $t_0 = t_0(S) \geqslant 0$ such that $\Theta(t)S \subset S_0$ for all $t \geqslant t_0$, and also that for bounded subsets $B \subset X$ and $S \subset \Sigma$ there exists $t_0 = t_0(B,S) \geqslant 0$ such that $\mathcal{K}(t,\sigma)B \subset B_0$, for all $t \geqslant t_0$ and $\sigma \in S$.

The asymptotic compactness property of Π is equivalent to saying that given a sequence $\{t_n\}$ in $\mathbb{R}^+$ such that $t_n \to \infty$ and a bounded sequence $\{(x_n,\sigma_n)\}$ in $\mathbb{X}$, $\{\Pi(t_n)(x_n,\sigma_n)\}$ has a convergent subsequence in $\mathbb{X}$, which is equivalent to saying that given a sequence $\{t_n\}$ in $\mathbb{R}^+$ such that $t_n \to \infty$, and bounded sequences $\{x_n\}$ in X and $\{\sigma_n\}$ in Σ we have

- $\{\Theta(t_n)\sigma_n\}$ has a convergent subsequence in Σ and
- $\{\mathcal{K}(t_n,\sigma_n)x_n\}$ has a convergent subsequence in X.

This reasoning proves the following result on the characterization of the skew-product semigroups that possess a global attractor.

THEOREM 6.5. *Let (X, d_X), (Σ, d_Σ) be metric spaces and $\mathbb{X} = X \times \Sigma$ with the product metric. Consider Θ a semigroup in Σ and $\mathcal{K}$ a cocycle relative to Θ. If Π is the associated skew-product semigroup, then it has a global attractor $\mathbb{A}$ in $\mathbb{X}$ if and only if*

(i) *Θ has a global attractor $\mathscr{S}$ in Σ;*
(ii) *$\mathcal{K}$ is* **bounded dissipative**, *that is, there is a bounded subset B_0 of X such that, for bounded subsets $B \subset X$ and $S \subset \Sigma$, there exists $t_0 = t_0(B, S) \geqslant 0$ such that $\mathcal{K}(t, \sigma)B \subset B_0$, for all $t \geqslant t_0$ and $\sigma \in S$;*
(iii) *$\mathcal{K}$ is* **asymptotically compact**, *that is, given a sequence $\{t_n\}$ in $\mathbb{R}^+$ with $t_n \overset{n\to\infty}{\longrightarrow} \infty$, and bounded sequences $\{x_n\}$ in X and $\{\sigma_n\}$ in Σ, the sequence $\{\mathcal{K}(t_n, \sigma_n)x_n\}$ has a convergent subsequence in X.*

If we now return to our example (6.1), it is easy to see that any function in $\Sigma(f)$ must satisfy (6.2) and (6.3). Now if $\sigma \in \Sigma(f)$ and $u_0 \in \mathbb{R}^n$, the solution $\mathcal{K}(\cdot, \sigma)u_0$ of (6.4) must satisfy

$$\frac{d}{dt}\|\mathcal{K}(t,\sigma)u_0\|^2 = 2\langle \mathcal{K}(t,\sigma)u_0, \sigma(t, \mathcal{K}(t,\sigma)u_0)\rangle \leqslant -2\delta$$

and consequently, integrating both sides from 0 to t, we obtain $\|\mathcal{K}(t,\sigma)u_0\|^2 \leqslant \|u_0\|^2 - 2\delta t$, for all $t \geqslant 0$, such that $\|\mathcal{K}(t,\sigma)u_0\| > M$. Using this fact, we conclude that the solutions of (6.4) exist for all $t \geqslant 0$. Now if $B_M = \{u_0 \in \mathbb{R}^n : \|u_0\| \leqslant M\}$, then given a bounded subset B of $\mathbb{R}^n$ we have

$$\sup_{\sigma \in \Sigma(f)} \operatorname{dist}_H(\mathcal{K}(t,\sigma)B, B_M) \to 0, \qquad \text{as } t \to \infty,$$

and since we have already seen that Θ has a global attractor in $\Sigma = \Sigma(f)$, all the conditions of Theorem 6.5 are satisfied.

Another important family of operators that naturally appears in the study of non-autonomous problems are the *evolution processes*, which we define next.

DEFINITION 6.6. A family $\mathcal{S} = \{S(t,s) : t \geqslant s\} \subset C(X)$ is called an **evolution process** in X if it satisfies:

(i) $S(t,t)x = x$ for all $t \in \mathbb{R}$ and $x \in X$;
(ii) $S(t,s) = S(t,r)S(r,s)$ for all $t \geqslant r \geqslant s$;
(iii) the map $\mathcal{P} \times X \ni (t,s,x) \mapsto S(t,s)x \in X$ is continuous, where $\mathcal{P} = \{(t,s) \in \mathbb{R}^2 : t \geqslant s\}$.

In fact if $\eta \colon \mathbb{R} \to \Sigma$ is a global solution for Θ, the family $\mathcal{S}_\eta = \{S_\eta(t,s) : t \geqslant s\}$ defined by

$$S_\eta(t,s) = \mathcal{K}(t-s, \eta(s)) \qquad \text{for } t \geqslant s,$$

is an evolution process in X.

We say that an evolution process $\mathcal{S} = \{S(t,s) : t \geqslant s\}$ is **autonomous** if the evolution depends only on the elapsed time $t - s$ rather than on t and s explicitly, that is, if $S(t,s) = S(t-s, 0)$ for all $t \geqslant s$. If an evolution process is not autonomous it is said to be **non-autonomous** or simply an **evolution process**. When $t, s \in \mathbb{Z}$ instead of $\mathbb{R}$, we say that $\mathcal{S} = \{S(t,s) : t \geqslant s \in \mathbb{Z}\}$ is a **discrete evolution process**.

Let us now define the notion of *invariance* for an evolution process $\mathcal{S}$. Such a notion is fundamentally connected to the possibility of constructing global solutions. In the autonomous case, a set is invariant if it is fixed by the evolution process and that notion is sufficient to study the asymptotic behavior. However, in the non-autonomous case, it may happen that no set is fixed by the evolution process. Note that simple objects which are very helpful in the description of the asymptotic behavior, like equilibria, may not exist. The notion of invariance for non-autonomous evolution processes which seems most suitable is for a *family of sets*, which is the following.

DEFINITION 6.7. Let X be a metric space and let $\mathcal{S}$ be an evolution process in X. A family of sets $\{B(t)\}_{t\in\mathbb{R}}$ in X is said to be **$\mathcal{S}$-invariant** if $S(t,s)B(s) = B(t)$ for all $t \geqslant s$.

This notion does not reduce to the notion of invariance for an autonomous evolution process, but it has been shown to be suitable for the purpose of studying non-autonomous problems. A thorough discussion can be seen in [**78**].

For an autonomous evolution process, the asymptotic behavior can be studied making $t \to \infty$, called the *forward dynamics*, or fixing t and making $s \to -\infty$, called the *pullback dynamics*. In the non-autonomous case, these two dynamic regimes give rise to completely different scenarios. In what follows we will pursue the second scenario, defining *pullback attraction* for an evolution process.

DEFINITION 6.8. Let X be a metric space and let $\mathcal{S}$ be an evolution process in X. We say that a set B_0 **$\mathcal{S}$-pullback attracts** a set B, at time t if

$$\lim_{s\to-\infty} \operatorname{dist}_H(S(t,s)B, B_0) = 0.$$

With these notions of invariance and pullback attraction, we can define the notion of *pullback attractor* for an evolution process, but here we call the readers' attention to the fact that there are several distinct possible definitions of a pullback attractor in the literature. We choose to present the following.

DEFINITION 6.9. Let X be a metric space and let $\mathcal{S}$ be an evolution process in X. A family $\{A(t)\}_{t\in\mathbb{R}}$ is called a **pullback attractor** for $\mathcal{S}$ if $\bigcup_{t\in\mathbb{R}} A(t)$ is bounded, $A(t)$ is compact in X for each $t \in \mathbb{R}$, and $\{A(t)\}_{t\in\mathbb{R}}$ is $\mathcal{S}$-invariant and $\mathcal{S}$-pullback attracts bounded subsets of X at time t, for each $t \in \mathbb{R}$.

The pullback attractor, with this definition, is unique. In fact if $\{\tilde{A}(t)\}_{t\in\mathbb{R}}$ and $\{A(t)\}_{t\in\mathbb{R}}$ are pullback attractors for $\mathcal{S}$, since $B = \bigcup_{t\in\mathbb{R}} \tilde{A}(t)$ is bounded and $\{A(t)\}_{t\in\mathbb{R}}$ pullback attracts bounded subsets of X at time t, the $\mathcal{S}$-invariance property of pullback attractors implies that

$$\operatorname{dist}_H(\tilde{A}(t), A(t)) = \operatorname{dist}_H(S(t,s)\tilde{A}(s), A(t)) \leqslant \operatorname{dist}_H(S(t,s)B, A(t)) \overset{s\to-\infty}{\longrightarrow} 0.$$

This implies that $\tilde{A}(t) \subset A(t)$, for each $t \in \mathbb{R}$. The other inclusion follows in the same way, interchanging $\tilde{A}(t)$ and $A(t)$.

There is a vast literature on pullback attractors with results that characterize the evolution processes that possess pullback attractors. The interested reader can consult, for example, [**78**] and [**153**]. In the context of this book, the conditions for existence of pullback attractors will be drawn from the conditions for existence of a global attractor for the skew-product semigroup.

Now, as in the autonomous case, using *bounded global solutions*, one can characterize a pullback attractor, when it exists.

DEFINITION 6.10. Let X be a metric space and let $\mathcal{S}$ be an evolution process in X. We say that a function $\xi\colon \mathbb{R} \to X$ is a **global solution for** $\mathcal{S}$ if $S(t,s)\xi(s) = \xi(t)$ for all $t \geqslant s$, that is, if the family $\{\xi(t)\}_{t\in\mathbb{R}}$ is $\mathcal{S}$-invariant. If $\xi(\mathbb{R})$ is bounded in X, we say that ξ is a **bounded global solution**.

Within that context we have the following result on the characterization of pullback attractors.

THEOREM 6.11. *Let X be a metric space and let $\mathcal{S}$ be an evolution process in X. If $\mathcal{S}$ has a pullback attractor $\{A(t)\}_{t\in\mathbb{R}}$, then for all $t \in \mathbb{R}$ we have*

$$A(t) = \{\xi(t)\colon \xi \text{ is a bounded global solution of } \mathcal{S}\}.$$

PROOF. From the $\mathcal{S}$-invariance of $\{A(t)\}_{t\in\mathbb{R}}$, proceeding as in Theorem 1.9, for each $x \in A(t_0)$, we can construct a global solution ξ such that $\xi(t_0) = x$ and $\xi(t) \in A(t)$ for all $t \in \mathbb{R}$.

On the other hand, if ξ is a bounded global solution of $\mathcal{S}$, then, since $\{A(t)\}_{t\in\mathbb{R}}$ $\mathcal{S}$-pullback attracts $B = \overline{\xi(\mathbb{R})}$ at time t, for each $n \in \mathbb{N}^*$, there exists $t_n \leqslant t$ such that $\operatorname{dist}_H(S(t,s)B, A(t)) \leqslant \frac{1}{n}$ for all $s \leqslant t_n$. In particular

$$\operatorname{dist}_H(S(t,t_n)\xi(t_n), A(t)) = \operatorname{dist}_H(\xi(t), A(t)) \leqslant \frac{1}{n} \quad \text{for all } n \in \mathbb{N}^*.$$

It follows that $\xi(t) \in A(t)$ and completes the proof. □

DEFINITION 6.12. Let X be a metric space and $\mathcal{S}$ an evolution process in X. Given a subset $B \subset X$ and $t \in \mathbb{R}$, the **pullback ω-limit set** at time t of B is defined by

$$\omega(B,t) = \bigcap_{\sigma\leqslant t} \overline{\bigcup_{s\leqslant\sigma} S(t,s)B}.$$

Clearly $\omega(B,t)$ is closed for each $t \in \mathbb{R}$, and it is easy to see (cf. Proposition 1.13) that

$$\begin{aligned} \omega(B,t) = \{y \in X\colon\ &\text{there are sequences } \{s_k\} \text{ in } (-\infty,t],\ s_k \to -\infty, \\ &\text{and } \{x_k\} \text{ in } B, \text{ such that } y = \lim_{k\to\infty} S(t,s_k)x_k\}. \end{aligned} \tag{6.9}$$

THEOREM 6.13. *Let X be a metric space and $\mathcal{S}$ an evolution process in X. Assume that there exists a compact subset K of X such that K $\mathcal{S}$-pullback attracts bounded subsets of X at time t, for each $t \in \mathbb{R}$. Then, $\mathcal{S}$ has a bounded pullback attractor $\{A(t) = \omega(K,t)\}_{t\in\mathbb{R}}$ and $\overline{\bigcup_{t\in\mathbb{R}} A(t)}$ is a compact subset of X.*

PROOF. Let us first show that $\omega(K,t)$ is nonempty. Given sequences $\{s_k\}$ in $(-\infty,t]$, with $s_k \to -\infty$, and $\{x_k\}$ in K, since K is compact and $\mathcal{S}$-pullback attracts itself at time t, $\{S(t,s_k)x_k\}$ must have a convergent subsequence, and consequently $\omega(K,t)$ is nonempty.

Clearly $\omega(K,t) \subset K$ for all $t \in \mathbb{R}$, since K $\mathcal{S}$-pullback attracts itself at time t, and $\overline{\bigcup_{t\in\mathbb{R}} \omega(K,t)} \subset K$. Since $\omega(K,t)$ is closed, we conclude that $\omega(K,t)$ is compact for all $t \in \mathbb{R}$ and that $\overline{\bigcup_{t\in\mathbb{R}} \omega(K,t)}$ is compact.

We now prove the $\mathcal{S}$-invariance of $\{\omega(K,t)\}_{t\in\mathbb{R}}$. If $t \geqslant s$ and $x \in \omega(K,t)$, there are sequences $\{s_k\}$ in $(-\infty,t]$, with $s_k \to -\infty$, and $\{x_k\}$ in K such that $S(t,s_k)x_k \to x$. Since $s_k \to -\infty$, we may assume that $s_k \leqslant s$ for all $k \in \mathbb{N}$. Now since K $\mathcal{S}$-pullback attracts itself at time s, $\{S(s,s_k)x_k\}$ has a convergent subsequence (which we denote the same) with limit, say, y. Clearly, $y \in \omega(K,s)$ and from the continuity of $S(t,s)$ we have $S(t,s)y = x$. It follows that $S(t,s)\omega(K,s) \supset \omega(K,t)$. The proof of the other inclusion is simpler and it is left to the reader.

Next we prove that $\omega(K,t)$ $\mathcal{S}$-pullback attracts bounded subsets of X at time t. If there is a bounded subset B of X that is not $\mathcal{S}$-pullback attracted by $\omega(K,t)$ at time t, then there would exist $\epsilon > 0$ and sequences $\{s_k\}$ in $(-\infty, t]$, with $s_k \to -\infty$, and $\{x_k\}$ in B such that $\operatorname{dist}_H(S(t,s_k)x_k, \omega(K,t)) \geqslant \epsilon$ for all k. Since K is compact and $\mathcal{S}$-pullback attracts B, $\{S(t,s_k)x_k\}$ must have a convergent subsequence with limit in $\omega(K,t)$. That contradiction proves that $\omega(K,t)$ $\mathcal{S}$-pullback attracts bounded subsets of X, and the proof is complete. □

The next corollary follows directly from Theorems 6.5 and 6.13.

COROLLARY 6.14. *Let (X,d_X), (Σ, d_Σ) be metric spaces and consider $\mathbb{X} = X \times \Sigma$ with the product metric. Consider also a semigroup Θ in Σ and a cocycle $\mathcal{K}$ relative to Θ. If the associated skew-product semigroup Π has a global attractor $\mathbb{A}$ in $\mathbb{X}$, then for any bounded global solution $\eta\colon \mathbb{R} \to \Sigma$ of Θ, the evolution process $\mathcal{S}_\eta = \{S_\eta(t,s) = \mathcal{K}(t-s, \eta(s))\colon t \geqslant s\}$ has a pullback attractor $\{A_\eta(t)\}_{t\in\mathbb{R}}$ and $\overline{\bigcup_{t\in\mathbb{R}} A_\eta(t)} \subset \pi_X(\mathbb{A})$.*

We can now characterize the global attractor of the skew-product semigroup in terms of the pullback attractors of evolution processes associated to bounded global solutions of the driving semigroup.

THEOREM 6.15. *Let (X,d_X), (Σ, d_Σ) be metric spaces and consider $\mathbb{X} = X \times \Sigma$ with the product metric. Consider also a semigroup Θ in Σ and a cocycle $\mathcal{K}$ relative to Θ. If the associated skew-product semigroup Π has a global attractor $\mathbb{A}$ in $\mathbb{X}$, then the driving semigroup Θ has a global attractor $\mathscr{S}$ in Σ. If $\eta\colon \mathbb{R} \to \mathscr{S}$ is a bounded global solution for Θ, then the evolution process $\mathcal{S}_\eta = \{S_\eta(t,s)\colon t \geqslant s\}$, given by $S_\eta(t,s) = \mathcal{K}(t-s, \eta(s))$, possesses a pullback attractor $\{A_\eta(t)\}_{t\in\mathbb{R}}$ with the property that $A_\eta(t) = \{x \in X\colon (x, \eta(t)) \in \mathbb{A}\}$. Moreover,*

$$\mathbb{A} = \bigcup_{\eta} \bigcup_{t\in\mathbb{R}} A_\eta(t) \times \{\eta(t)\},$$

where the first union is taken over all bounded global solutions η of Θ.

PROOF. Note that a global bounded solution for Π is a continuous function $(x(\cdot), \eta(\cdot))\colon \mathbb{R} \to \mathbb{A}$ that satisfies

$$(x(t+s), \eta(t+s)) = \Pi(t)(x(s), \eta(s)) = (\mathcal{K}(t, \eta(s))x(s), \Theta(t)\eta(s)),$$

for all $t \geqslant 0$ and $s \in \mathbb{R}$. Thus $x(t+s) = \mathcal{K}(t,\eta(s))x(s) = S_\eta(t+s,s)x(s)$ and $\Theta(t)\eta(s) = \eta(t+s)$, for all $t \geqslant 0$ and $s \in \mathbb{R}$. It follows that $x\colon \mathbb{R} \to X$ is a bounded global solution for $\mathcal{S}_\eta$ and η is a bounded global solution for Θ. Hence, any point of $\mathbb{A}$ is of the form $(x(t), \eta(t))$, where $x(t) \in A_\eta(t)$ and $\eta\colon \mathbb{R} \to \Sigma$ is a bounded global solution for Θ, and therefore

$$\mathbb{A} \subset \bigcup_{\eta} \bigcup_{t\in\mathbb{R}} A_\eta(t) \times \{\eta(t)\},$$

where the first union is taken over all global bounded solutions η of Θ.

On the other hand, if $\eta\colon \mathbb{R} \to \mathscr{S}$ is a bounded global solution for Θ and $x_0 \in A_\eta(t_0)$, there exists a global solution $x\colon \mathbb{R} \to X$ of $\mathcal{S}_\eta$ such that $x(t_0) = x_0$ and $x(t) \in A_\eta(t)$ for all $t \in \mathbb{R}$. Since $\bigcup_{t\in\mathbb{R}} A_\eta(t)$ is bounded, we have $x\colon \mathbb{R} \to X$ a bounded global solution of $\mathcal{S}_\eta$, and consequently $(x(\cdot), \eta(\cdot))\colon \mathbb{R} \to \mathbb{X}$ is a bounded global solution of Π, showing that $(x(t), \eta(t)) \in \mathbb{A}$ for all $t \in \mathbb{R}$. Then, in particular, $(x_0, \eta(t_0)) \in \mathbb{A}$ and

$$\bigcup_{\eta} \bigcup_{t\in\mathbb{R}} A_\eta(t) \times \{\eta(t)\} \subset \mathbb{A},$$

where the first union is taken over all bounded global solutions η for Θ, which completes the proof. □

An important object that appears in Theorem 6.15 is $\pi_X(\mathbb{A})$, which has the property of $\mathcal{K}$-attraction of bounded subsets of X, uniformly in bounded subsets of Σ. This object turns out to be, in our context, what is known in the literature as *uniform attractor* for the cocycle $\mathcal{K}$, which we explain next.

DEFINITION 6.16. Let (X, d_X), (Σ, d_Σ) be metric spaces and $\mathbb{X} = X \times \Sigma$ with the product metric. Consider Θ a semigroup in Σ and $\mathcal{K}$ a cocycle relative to Θ. The **uniform attractor** for the cocycle $\mathcal{K}$ is the minimal closed subset $\mathcal{A}$ of X such that for each bounded subset B of X and each bounded subset Γ of Σ we have

$$\lim_{t\to\infty} \sup_{\sigma\in\Gamma} \operatorname{dist}_H(\mathcal{K}(t,\sigma)B, \mathcal{A}) = 0,$$

where the minimality condition means that if C is any closed subset of X such that

$$\lim_{t\to\infty} \sup_{\sigma\in\Gamma} \operatorname{dist}_H(\mathcal{K}(t,\sigma)B, C) = 0,$$

for each bounded subset B of X and each bounded subset Γ of Σ, then we have $\mathcal{A} \subset C$.

From Theorem 6.5 the following result holds.

THEOREM 6.17. *Let (X, d_X), (Σ, d_Σ) be metric spaces and $\mathbb{X} = X \times \Sigma$ with the product metric. Consider Θ a semigroup in Σ and $\mathcal{K}$ a cocycle relative to Θ. If Θ has a global attractor $\mathscr{S}$ in Σ and $\mathcal{K}$ is bounded dissipative and asymptotically compact, then the cocycle $\mathcal{K}$ has a compact uniform attractor $\mathcal{A}$, and if $\mathbb{A}$ is the global attractor for the associated skew-product semigroup Π, then $\mathcal{A} = \pi_X(\mathbb{A})$. Furthermore*

$$\mathcal{A} = \bigcup_{\eta} A_\eta(0),$$

where the union is taken over all bounded global solutions η for Θ.

This result is a straightforward application of Theorem 6.5. Just note that if η is a bounded global solution for Θ, then so is $\eta_r(\cdot) = \eta(\cdot + r)$, for each $r \in \mathbb{R}$, and we can conclude that $A_{\eta_r}(0) = A_\eta(r)$.

Notes. The theories of pullback, cocycle, skew-product, and uniform attractors have been developed in the literature as more or less different subjects and frameworks. This is why we can find many papers with a single one of these approaches, independently of the others. Related to the *existence of attractors* we can find [**41, 51, 78, 150**] for pullback attractors, [**89, 153**] for cocycle attractors and [**96–98, 206**] for uniform attractors. Note that all the theories could have been written without any reference to cocycle attractors, but only to pullback, uniform and attractors for skew-product semigroups. A nice set of results has also proved the finiteness of the box-counting dimension of pullback attractors; see, for instance, [**28, 78, 160**]. The relationship among all of these concepts is newer, and can be found in [**29, 30, 50, 50**]. We can also find nice papers on applications of the theory of pullback attractors to non-autonomous delay equations ([**151**, **169**]), differential equations without uniqueness of solutions, or even the more general case of multivalued dynamical systems ([**45**, **49**, **107**, **205**]). Applications to models of differential or partial differential equations systems (also considering delays or nonuniqueness of solutions) has also received a lot of attention in the last years, and, for instance, we can refer to [**32**, **46**, **51**, **99**, **154**, **154**, **196**, **199**, **207**] for applications to PDEs for the pullback attractor theory, or to [**100**, **117**, **128**, **165**, **198**] for the case of uniform attractors applied to different phenomena from Physics, Chemistry or Biology. Finally, some papers have focused on the relation and differences between pullback and forward behavior on invariant sets for non-autonomous evolution processes and cocycles ([**46**, **50**, **56**, **90**, **149**, **155**, **162**, **188**]), a crucial insight for a correct understanding of the rich dynamics of this kind of systems. More recently, interesting papers have gone to the definition of exponential pullback attractors ([**87**, **110**, **161**]), and the description of attractors for non-autonomous equations with random terms ([**53**, **109**, **203**]). In all of these papers it is clear that the concept of pullback attractor can be generalized for families of sets which are unbounded, but in this case they are not related to the attractor of the associated skew-product semigroup (that may not exist in this scenario).

CHAPTER 7

Upper and Lower Semicontinuity

As we have explained in Chapter 2, a fundamental question in the study of the asymptotic dynamics (attractors) of dynamical systems is the continuity with respect to perturbations. This question can be seen from several different points of view, related to what the *metric* is that determines proximity of attractors and to the kind of *small perturbations* that are considered.

In any case, an appropriate concept of attractor must not explode in *size* under perturbations, a property which is called *upper semicontinuity.* A less common property, with a more intricate proof which is also true in many situations, is that attractors do not implode in *size* under perturbations, called *lower semicontinuity.* In what follows we state and prove results that provide conditions under which upper and lower semicontinuity of pullback and uniform attractors hold.

Throughout this chapter (X, d) and (M, d_M) will be fixed metric spaces, and we consider the product space $\mathbb{X}_M = X \times M$ with the product metric. Assume that $\Theta \subset C(M)$ is a semigroup, $\mathcal{K}$ is a cocyle relative to Θ and let $(\mathcal{K}, \Theta)_{(X,M)}$ and $\Pi \subset C(\mathbb{X}_M)$ be, respectively, the associated non-autonomous dynamical system and the skew-product semigroup.

Let $\Lambda = [0, 1]$ and let $\{\Sigma_\lambda\}_{\lambda\in\Lambda}$ be a family of closed and positively Θ-invariant subsets of M, and assume that

$$M = \bigcup_{\lambda\in\Lambda} \Sigma_\lambda.$$

Let d_λ be the restriction of d_M to Σ_λ, let $\Theta_\lambda = \{\Theta_\lambda(t)\colon t \geqslant 0\} \subset C(\Sigma_\lambda)$ be the restriction of Θ to Σ_λ, and let $\{\mathcal{K}(t, \sigma)\colon t \geqslant 0,\ \sigma \in \Sigma_\lambda\} \subset C(X)$ be the cocycle relative to Θ_λ (which we will refer to as $\mathcal{K}_\lambda$).

For each $\lambda \in \Lambda$, let $\mathbb{X}_\lambda = X \times \Sigma_\lambda$ and let $(\mathcal{K}_\lambda, \Theta_\lambda)_{(X,\Sigma_\lambda)}$ and $\{\Pi_\lambda(t)\colon t \geqslant 0\} \subset C(\mathbb{X}_\lambda)$ be the associated non-autonomous dynamical system and skew-product semigroup, respectively. We will refer to $\{\Pi_\lambda(t)\colon t \geqslant 0\}$ simply as Π_λ. Clearly, Π_λ is also the restriction of Π to $\mathbb{X}_\lambda$.

We assume throughout this chapter that:

(C1) $\qquad\qquad$ Π has a global attractor $\mathbb{A}$ in $\mathbb{X}_M$.

In what follows, to avoid pollution with notation, we will use dist_H^E for the Hausdorff semidistance in the metric space E, where E can be either X, Σ, Σ_λ, $\mathbb{X}_\lambda$ or $\mathbb{X}_M$. Condition (C1) implies some properties regarding global attractors for Π_λ and Θ_λ for each $\lambda \in \Lambda$.

PROPOSITION 7.1. *If* (C1) *holds, then* Π_λ *has a global attractor* $\mathbb{A}_\lambda$ *for each* $\lambda \in \Lambda$ *and* $\bigcup_{\lambda\in\Lambda} \mathbb{A}_\lambda \subset \mathbb{A}$.

PROOF. Let $\mathbb{K}_\lambda = \mathbb{A} \cap \mathbb{X}_\lambda$, for each $\lambda \in \Lambda$. We claim that for each bounded subset $\mathbb{B}_\lambda$ of $\mathbb{X}_\lambda$ we have

$$\operatorname{dist}_H^{\mathbb{X}_\lambda}(\Pi_\lambda(t)\mathbb{B}_\lambda, \mathbb{K}_\lambda) \to 0 \quad \text{as } t \to \infty.$$

If that is not the case, there exist $\epsilon > 0$, sequences $t_n \to \infty$ and $\{(x_n, \sigma_n)\}$ in $\mathbb{B}_\lambda$ such that

$$\operatorname{dist}_H^{\mathbb{X}_\lambda}(\Pi_\lambda(t_n)(x_n, \sigma_n), \mathbb{K}_\lambda) \geqslant \epsilon \quad \text{for all } n \in \mathbb{N}. \tag{7.1}$$

Since $\{(x_n, \sigma_n)\}$ is also bounded in $\mathbb{X}_M$, $\Pi_\lambda(t)(x_n, \sigma_n) = \Pi(t)(x_n, \sigma_n)$ for all $t \geqslant 0$, and $\mathbb{A}$ is the global attractor of Π in $\mathbb{X}_M$, up to a subsequence $\Pi_\lambda(t_n)(x_n, \sigma_n) \to (x_0, \sigma_0) \in \mathbb{A}$. Since $\Pi_\lambda(t_n)(x_n, \sigma_n) \in \mathbb{X}_\lambda$ for all $n \in \mathbb{N}$ and $\mathbb{X}_\lambda$ is closed, $(x_0, \sigma_0) \in \mathbb{A} \cap \mathbb{X}_\lambda = \mathbb{K}_\lambda$, which contradicts (7.1) when making $n \to \infty$. Since $\mathbb{K}_\lambda$ is compact, this proves that Π_λ is both bounded dissipative and asymptotically compact in $\mathbb{X}_\lambda$. Hence Π_λ has a global attractor $\mathbb{A}_\lambda \subset \mathbb{K}_\lambda$ for each $\lambda \in \Lambda$. Taking the union in $\lambda \in \Lambda$ we conclude the last assertion, and complete the proof. □

PROPOSITION 7.2. *If* (C1) *holds, then* Θ_λ *has a global attractor* $\mathscr{S}_\lambda$ *in* Σ_λ, *for each* $\lambda \in \Lambda$.

PROOF. Define $\Gamma_\lambda = \pi_\Sigma(\mathbb{A}) \cap \Sigma_\lambda$, which is compact in Σ_λ. We claim that for each $J \subset \Sigma_\lambda$ bounded, we have

$$\operatorname{dist}_H^{\Sigma_\lambda}(\Theta_\lambda(t)J, \Gamma_\lambda) \to 0 \quad \text{as } t \to \infty.$$

If that is not the case, there exist $\epsilon > 0$ and sequences $t_n \to \infty$ and $\{\sigma_n\} \subset \Sigma_\lambda$ bounded such that

$$\operatorname{dist}_H^{\Sigma_\lambda}(\Theta_\lambda(t_n)\sigma_n, \Gamma_\lambda) \geqslant \epsilon \quad \text{for all } n \in \mathbb{N}. \tag{7.2}$$

Choose any $x \in X$ and consider $\mathbb{B} = \{x\} \times \{\sigma_n \colon n \in \mathbb{N}\}$, which is bounded in $\mathbb{X}_M$. From the fact that $\mathbb{A}$ is the global attractor of Π in $\mathbb{X}_M$, we have

$$\operatorname{dist}_H^{\mathbb{X}_M}(\Pi(t_n)(x, \sigma_n), \mathbb{A}) \leqslant \operatorname{dist}_H^{\mathbb{X}_M}(\Pi(t_n)\mathbb{B}, \mathbb{A}) \to 0 \text{ as } n \to \infty.$$

Hence, up to a subsequence, $\Pi(t_n)(x, \sigma_n) \to (z, \omega) \in \mathbb{A}$. Thus $\Theta_\lambda(t_n)\sigma_n = \Theta(t_n)\sigma_n \to \omega \in \pi_\Sigma(\mathbb{A})$. Since Σ_λ is closed and positively Θ-invariant and $\sigma_n \in \Sigma_\lambda$ for all $n \in \mathbb{N}$, we obtain $\omega \in \Sigma_\lambda$. Hence $\omega \in \Gamma_\lambda$, which contradicts (7.2) when making $n \to \infty$.

Since Γ_λ is compact, this proves that Θ_λ is both bounded dissipative and asymptotically compact in Σ_λ, which proves that Θ_λ has a global attractor $\mathscr{S}_\lambda \subset \Gamma_\lambda$ for each $\lambda \in \Lambda$. □

With this last proposition in mind, we obtain a family $\{\mathscr{S}_\lambda\}_{\lambda \in \Lambda}$ of global attractors of $\{\Theta_\lambda\}_{\lambda \in \Lambda}$, and we also assume throughout this chapter the following:

(C2) the family $\{\mathscr{S}_\lambda\}_{\lambda \in \Lambda}$ is continuous at each $\lambda \in \Lambda$.

PROPOSITION 7.3. *If* (C1) *and* (C2) *hold, then:*

(i) *given* $\lambda_0 \in \Lambda$, $\{\sigma_\lambda\}_{\lambda \in \Lambda}$ *with* $\sigma_\lambda \in \Sigma_\lambda$ *for each* $\lambda \in \Lambda$ *and* $\sigma_\lambda \to \sigma_{\lambda_0}$ *as* $\lambda \to \lambda_0$ *we have*

$$d_{\mathbb{X}_M}(\Pi_\lambda(t)(x, \sigma_\lambda), \Pi_{\lambda_0}(t)(x, \sigma_{\lambda_0})) \to 0 \quad \text{as } \lambda \to \lambda_0,$$

uniformly for (t, x) *in compact subsets of* $\mathbb{R}^+ \times X$;

(ii) *given sequences* $\{\lambda_n\} \subset \Lambda$ *with* $\lambda_n \to \lambda_0$, $t_n \to \infty$ *and* $\{(x_n, \sigma_n)\}$ *bounded in* $\mathbb{X}_M$ *with* $\sigma_n \in \mathscr{S}_{\lambda_n}$ *for each* $n \in \mathbb{N}$, *then* $\{\Pi_{\lambda_n}(t_n)(x_n, \sigma_n)\}$ *has a convergent subsequence to a point* (z, ω) *with* $\omega \in \mathscr{S}_{\lambda_0}$;

(iii) *there exists a bounded subset* $\mathbb{B}_0$ *of* $\mathbb{X}_M$ *such that the set* $\mathbb{B}_\lambda = \mathbb{B}_0 \cap \mathbb{X}_\lambda$ Π_λ*-attracts bounded subsets of* $\mathbb{X}_\lambda$ *for each* $\lambda \in [0,1]$.

PROOF. To prove (i), we first recall that Π_λ is simply the restriction of Π to $\mathbb{X}_\lambda$, hence

$$\begin{aligned} d_{\mathbb{X}_M}(\Pi_\lambda(t)(x,\sigma_\lambda),&\Pi_{\lambda_0}(t)(x,\sigma_{\lambda_0})) \\ &= d(\mathcal{K}(t,\sigma_\lambda)x, \mathcal{K}(t,\sigma_{\lambda_0})x) + d_\Sigma(\Theta(t)\sigma_\lambda, \Theta(t)\sigma_{\lambda_0}), \end{aligned}$$

which, since $\sigma_\lambda \to \sigma_{\lambda_0}$, tends to 0 as $\lambda \to \lambda_0$, uniformly for (t,x) in compact subsets of $\mathbb{R}^+ \times X$.

Now we prove (ii). The sequence $\{\Pi_{\lambda_n}(t)(x_n,\sigma_n)\}$ has a convergent subsequence to a point in $(z,\omega) \in \mathbb{X}_M$, since Π is asymptotically compact in $\mathbb{X}_M$. But $\Pi(t)(x_n,\sigma_n) = (\mathcal{K}(t,\sigma_n)x_n, \Theta(t)\sigma_n)$, which implies that, up to the convergent subsequence, $\Theta(t)\sigma_n \to \omega$. Since $\sigma_n \in \mathscr{S}_{\lambda_n}$ and $\mathscr{S}_{\lambda_n}$ is Θ-invariant, we obtain $\Theta(t)\sigma_n \in \mathscr{S}_{\lambda_n}$ for each $n \in \mathbb{N}$. Using (C2), we obtain $\omega \in \mathscr{S}_{\lambda_0}$. Thus $(z,\omega) \in \mathbb{X}_{\lambda_0}$, which proves (ii).

For (iii), just take $\mathbb{B}_0 = \mathbb{A}$, and the proof is complete. □

REMARK 7.4. Compare items (i), (ii) and (iii) of this last proposition with Definition 2.3.

1. Upper semicontinuity

1.1. Upper semicontinuity of global attractors for skew-product semigroups. Proceeding as in Chapter 2 (see Theorem 2.5), and using Proposition 7.3, we obtain the following result.

THEOREM 7.5. *Assume that conditions* (C1) *and* (C2) *hold, and let* $\mathbb{A}_\lambda$ *be the global attractor of* Π_λ *in* $\mathbb{X}_\lambda$ *for each* $\lambda \in \Lambda$, *with* $\bigcup_{\lambda\in\Lambda} \mathbb{A}_\lambda \subset \mathbb{A}$.

If $\lambda_0 \in \Lambda$, $\lambda_k \to \lambda_0$, $\{s_k\}$ *is a sequence in* $\mathbb{R}$ *and, for each* $k \in \mathbb{N}$, (v_k,η_k) *is a global solution of* Π_{λ_k} *in* $\mathbb{A}_{\lambda_k}$, *then there exists a subsequence* $\{(v_{k_j},\eta_{k_j})\}_{j\in\mathbb{N}}$ *of* $\{(v_k,\eta_k)\}_{k\in\mathbb{N}}$ *such that* $\{(v_{k_j}(\cdot+s_{k_j}), \eta_{k_j}(\cdot+s_{k_j}))\}$ *converges uniformly in bounded intervals of* $\mathbb{R}$ *to a global solution* (v,η) *of* Π_{λ_0} *in* $\mathbb{A}_{\lambda_0}$. *In particular, the family* $\{\mathbb{A}_\lambda\}_{\lambda\in\Lambda}$ *is upper semicontinuous at* λ_0.

PROOF. Define $\psi_k(t) = (v_k(t+s_k), \eta_k(t+s_k))$ for each $k \in \mathbb{N}$ and $t \in \mathbb{R}$, which is a bounded global solution of Π_{λ_k} in $\mathbb{X}_{\lambda_k}$ for each $k \in \mathbb{N}$. Since $\{\psi_k(0)\} \subset \mathbb{A}$, we can extract a subsequence $\{n_{0,k}\}$ of $\mathbb{N}$ such that $\psi_{n_{0,k}}(0) \to (x,\sigma) \in \mathbb{A}$. Since $\eta_k(t+s_k) \in \mathscr{S}_{\lambda_k}$ for all k, using (C2) we obtain $\sigma \in \mathscr{S}_{\lambda_0}$, and hence $(x,\sigma) \in \mathbb{A}_{\lambda_0}$.

Define $\psi(t) = \Pi_{\lambda_0}(t)(x,\sigma)$ for all $t \geqslant 0$. Then

$$\psi_{n_{0,k}}(t) = \Pi_{\lambda_{n_{0,k}}}(t)\psi_{n_{0,k}}(0) \to \Pi_{\lambda_0}(t)(x,\sigma) = \psi(t),$$

uniformly for t in bounded intervals of $\mathbb{R}^+$. Now proceeding inductively as in the proof of Lemma 2.4 we conclude the result. □

We now translate the above conditions to the cocycle and the driving semigroups so that they can be applied to our model of non-autonomous differential equations.

Since Π has a global attractor, it is bounded dissipative and asymptotically compact. Note that this implies that the driving semigroup Θ has a global attractor $\mathscr{S}$ in M. The bounded dissipativity of Π can be translated into the notion of *collective bounded dissipativity* for the cocycle family $\{\mathcal{K}_\lambda\}_{\lambda\in\Lambda}$ and for the driving semigroup family $\{\Theta_\lambda\}_{\lambda\in\Lambda}$, which are precisely defined next.

DEFINITION 7.6. The cocycle family $\{\mathcal{K}_\lambda\}_{\lambda\in\Lambda}$ and $\{\Theta_\lambda\}_{\lambda\in\Lambda}$ are said to be **collectively bounded dissipative** if they satisfy, respectively:

(a) there exists a bounded subset B_0 of X such that given a bounded subset $B \subset X$ and a $S_\lambda \subset \Sigma_\lambda$ for each $\lambda \in \Lambda$ with $\bigcup_{\lambda\in\Lambda} S_\lambda$ bounded in M, we have

$$\lim_{t\to\infty} \sup_{\lambda\in[0,1]} \sup_{\sigma_\lambda\in S_\lambda} \operatorname{dist}_H^X(\mathcal{K}(t,\sigma_\lambda)B, B_0) = 0;$$

(b) there exists a bounded subset Ξ of M such that, if $\Xi_\lambda = \Xi \cap \Sigma_\lambda$ and $S_\lambda \subset \Sigma_\lambda$ for each $\lambda \in \Lambda$ is such that $\bigcup_{\lambda\in\Lambda} S_\lambda$ is bounded in M, then

$$\lim_{t\to\infty} \sup_{\lambda\in[0,1]} \operatorname{dist}_H^{\Sigma_\lambda}(\Theta_\lambda(t)S_\lambda, \Xi_\lambda) = 0.$$

The asymptotic compactness of Π can be translated into the notion of *collective asymptotic compactness* for the cocycle family $\{\mathcal{K}_\lambda\}_{\lambda\in\Lambda}$ and for the driving semigroups $\{\Theta_\lambda\}_{\lambda\in\Lambda}$, which are defined next.

DEFINITION 7.7. The cocycle family $\{\mathcal{K}_\lambda\}_{\lambda\in\Lambda}$ and $\{\Theta_\lambda\}_{\lambda\in\Lambda}$ are **collectively asymptotically compact** if they satisfy, respectively:

(a) given sequences $t_n \to \infty$, $\{x_n\}$ bounded in X and $\{\sigma_n\}$ bounded in M, with $\sigma_n \in \Sigma_{\lambda_n}$, then $\{\mathcal{K}(t_n,\sigma_n)x_n\}$ has a convergent subsequence in X;
(b) given sequences $t_n \to \infty$ and $\{\sigma_n\}$ bounded in M, with $\sigma_n \in \Sigma_{\lambda_n}$, then $\{\Theta_{\lambda_n}(t_n)\sigma_n\}$ has a convergent subsequence.

REMARK 7.8. If both $\{\mathcal{K}_\lambda\}_{\lambda\in\Lambda}$ and $\{\Theta_\lambda\}_{\lambda\in\Lambda}$ are collectively bounded dissipative and collectively asymptotically compact, then condition (C1) holds.

These are the standing conditions for this chapter, unless stated otherwise. Theorem 7.5 can now be stated with the conditions on the cocycle family and on the driving semigroup family, instead of on the conditions of the skew-product semigroups.

1.2. Upper semicontinuity of uniform attractors. As we have seen in Chapter 6, the uniform attractor $\mathcal{A}_\lambda = \pi_X(\mathbb{A}_\lambda)$ is the projection in the first coordinate of the global attractor $\mathbb{A}_\lambda$ for the skew-product semigroup Π_λ in $\mathbb{X}_\lambda$. It follows directly that:

THEOREM 7.9. *If the cocycle family $\{\mathcal{K}_\lambda\}_{\lambda\in\Lambda}$ and the driving semigroup family $\{\Theta_\lambda\}_{\lambda\in\Lambda}$ are collectively bounded dissipative and collectively asymptotically compact, then, for each $\lambda \in \Lambda$, the cocycle $\mathcal{K}_\lambda$ has a uniform attractor $\mathcal{A}_\lambda$ in X. Furthermore $\bigcup_{\lambda\in\Lambda} \mathcal{A}_\lambda$ is relatively compact in X and $\{\mathcal{A}_\lambda\}_{\lambda\in\Lambda}$ is upper semicontinuous at $\lambda = \lambda_0$, for each $\lambda_0 \in \Lambda$.*

1.3. Upper semicontinuity of pullback attractors. As we have seen in Chapter 6, pullback attractors appear in a natural way in the study of global attractors for skew-product semigroups and of uniform attractors for cocycles. We will first prove general results on upper semicontinuity of pullback attractors for evolution processes and later we will state them specifically to the case of pullback attractors arising in the context of skew-product attractors.

For each $\lambda \in \Lambda$ we consider an evolution process $\mathcal{S}_\lambda = \{S_\lambda(t,s)\colon (t,s) \in \mathcal{P}\}$ in X with a pullback attractor $\hat{A}_\lambda = \{A_\lambda(t)\}_{t\in\mathbb{R}}$. Throughout this section, we assume that for each compact subset K of X we have

$$\sup_{\tau\in[0,T]} \sup_{x\in K} d(S_\lambda(t,t-\tau)x, S_{\lambda_0}(t,t-\tau)x) \overset{\lambda\to\lambda_0}{\longrightarrow} 0, \quad t \in \mathbb{R}, T > 0, \tag{7.3}$$

also that

$$\bigcup_{\lambda\in\Lambda} \bigcup_{s\leqslant t} A_\lambda(s) \text{ is bounded for each } t \in \mathbb{R}, \tag{7.4}$$

and lastly that given $\lambda_0 \in \Lambda$ we have

$$\overline{\bigcup_{n\in\mathbb{N}} A_{\lambda_n}(t)} \text{ is compact for each } t \in \mathbb{R} \text{ and } \{\lambda_n\}_{n\in\mathbb{N}} \text{ in } \Lambda \text{ with } \lambda_n \to \lambda_0. \tag{7.5}$$

Let us first state what we mean by *upper semicontinuity of pullback attractors.*

DEFINITION 7.10. If $I \subset \mathbb{R}$, we say that $\{A_\lambda(t)\colon t \in I\}_{\lambda\in\Lambda}$ is **upper semicontinuous** at λ_0 if

$$\lim_{\lambda\to\lambda_0} \sup_{t\in I} \operatorname{dist}_H^X(A_\lambda(t), A_{\lambda_0}(t)) = 0.$$

The following lemma plays a fundamental role in the proof of the results on upper semicontinuity of pullback attractors.

LEMMA 7.11. *Assume that* (7.3), (7.4) *and* (7.5) *are satisfied. If* $\lambda_n \to \lambda_0$ *and* $\xi_n\colon \mathbb{R} \to X$ *is a global solution of* $\mathcal{S}_{\lambda_n}$ *with* $\xi_n(t) \in A_{\lambda_n}(t)$ *for all* $t \in \mathbb{R}$*, then there exist a global solution* $\xi_0\colon \mathbb{R} \to X$ *of* $\mathcal{S}_{\lambda_0}$*, with* $\xi_0(t) \in A_{\lambda_0}(t)$ *for all* $t \in \mathbb{R}$*, and a subsequence* $\{\xi_{n_k}\}$ *of* $\{\xi_n\}$ *such that* ξ_{n_k} *converges to* ξ_0 *uniformly in bounded intervals of* $\mathbb{R}$*.*

PROOF. Note that $\{\xi_n(0)\}$ is a sequence in the compact set $\overline{\bigcup_{n\in\mathbb{N}} A_{\lambda_n}(0)}$. Thus there is an infinite subset $\mathbb{N}_0$ of $\mathbb{N}$ such that $\{\xi_n(0)\}_{n\in\mathbb{N}_0}$ is convergent with limit $x_0 \in X$. Let $\xi_0(t) = S_{\lambda_0}(t,0)x_0$ for $t \geqslant 0$. It is clear from (7.3) that ξ_n, for $n \in \mathbb{N}_0$, converges to ξ_0 uniformly in bounded intervals of $\mathbb{R}^+$.

Suppose by induction that we have constructed a solution $\xi_0\colon [-k,\infty) \to X$, $k \in \mathbb{N}$, of $\mathcal{S}_{\lambda_0}$ and infinite subsets $\mathbb{N}_k$ of $\mathbb{N}_{k-1}$ such that ξ_n, $n \in \mathbb{N}_k$, converges to ξ_0 uniformly in bounded intervals of $[-k,\infty)$. Since the sequence $\{\xi_n(-k-1)\}_{n\in\mathbb{N}_k}$ is in the compact set $\overline{\bigcup_{n\in\mathbb{N}} A_{\lambda_n}(-k-1)}$, there is an infinite subset $\mathbb{N}_{k+1}$ of $\mathbb{N}_k$ such that $\{\xi_n(-k-1)\}_{n\in\mathbb{N}_{k+1}}$ is convergent with limit x_{-k-1}. Let $\xi_0(t) = S_{\lambda_0}(t,-k-1)x_{-k-1}$ for $t \in [-k-1,-k)$. Clearly, $\xi_0\colon [-k-1,\infty) \to X$ is a solution of $\mathcal{S}_{\lambda_0}$ and ξ_n, $n \in \mathbb{N}_{k+1}$, converges to ξ_0 uniformly in bounded intervals of $[-k-1,\infty)$.

This procedure gives us a global solution $\xi_0\colon \mathbb{R} \to X$ of $\mathcal{S}_{\lambda_0}$ such that $\{\xi_0(s)\colon s \leqslant t\}$ is bounded for each $t \in \mathbb{R}$, and therefore $\xi_0(t) \in A_{\lambda_0}(t)$. Besides this, if n_k is the k-th element of $\mathbb{N}_k$, $k \in \mathbb{N}$, then $\{\xi_{n_k}\}$ converges to ξ_0, uniformly in bounded subsets of $\mathbb{R}$, which completes the proof. □

Under the above assumptions, the upper semicontinuity of pullback attractors is immediate as shown in the result that follows.

THEOREM 7.12. *Suppose that* (7.3), (7.4) *and* (7.5) *are satisfied and let* $\lambda_n \to \lambda_0$*. Then for each* $t \in \mathbb{R}$ *we have*

$$\operatorname{dist}_H^X(A_{\lambda_n}(t), A_{\lambda_0}(t)) \to 0.$$

PROOF. From Lemma 7.11, given a sequence $\{x_n\}$ in X with $x_n \in A_{\lambda_n}(t)$, there exist $y \in X$, a subsequence of $\{x_n\}$ (which we denote the same), a global solution $\xi_n\colon \mathbb{R} \to X$ of $\{S_{\lambda_n}(t,s)\colon t \geqslant s\}$ with $\xi_n(t) = x_n$ and $\xi_{\lambda_n}(s) \in A_{\lambda_n}(s)$ for all $s \in \mathbb{R}$ and a global solution $\xi_0\colon \mathbb{R} \to X$ of $\{S_{\lambda_0}(t,s)\colon t \geqslant s\}$ such that $\xi_{\lambda_n}(\cdot)$ converges to ξ_0 uniformly in compact subsets of $\mathbb{R}$. In particular $\xi_0(t) \in A_{\lambda_0}(t)$ and the result follows from Lemma 2.2. □

This last result proves the upper semicontinuity for the family $\{A_\lambda(t)\}_{\lambda\in\Lambda}$ for each $t \in \mathbb{R}$ fixed. The next result will allow us to ensure the upper semicontinuity of the family $\{A_\lambda(t)\colon t \in I\}_{\lambda\in\Lambda}$, with $I \subset \mathbb{R}$ a bounded interval.

THEOREM 7.13. *Suppose that* (7.3), (7.4) *and* (7.5) *are satisfied. Then, if* $I \subset \mathbb{R}$ *is a bounded interval, the family* $\{A_\lambda(t)\colon t \in I\}_{\lambda\in\Lambda}$ *is upper semicontinuous at* $\lambda = \lambda_0$, *for each* $\lambda_0 \in \Lambda$.

PROOF. Let us show firstly that if $\lambda_n \to \lambda_0$ and $\{t_n\}$ is a sequence in I with $t_n \to t^*$, then

$$\operatorname{dist}_H^X(A_{\lambda_n}(t^*), A_{\lambda_n}(t_n)) \to 0, \tag{7.6}$$

and

$$\operatorname{dist}_H^X(A_{\lambda_n}(t_n), A_{\lambda_n}(t^*)) \to 0. \tag{7.7}$$

We prove (7.6) by contradiction. If there exist $\epsilon > 0$ and a sequence $\{x_n\}$ with $x_n \in A_{\lambda_n}(t^*)$ with $\operatorname{dist}_H^X(x_n, A_{\lambda_n}(t_n)) \geqslant \epsilon$ for all $n \in \mathbb{N}$, we obtain, from Theorem 7.12, global solutions $\xi_n\colon \mathbb{R} \to X$ of $\mathcal{S}_{\lambda_n}$, for $n \in \mathbb{N}$, and $\xi_0\colon \mathbb{R} \to X$ of $\mathcal{S}_{\lambda_0}$ such that $\xi_n(t) \in A_{\lambda_n}(t)$, $\xi_0(t) \in A_{\lambda_0}(t)$ for all $t \in \mathbb{R}$, $\xi_n(t) \to \xi_0(t)$, uniformly in bounded subsets of $\mathbb{R}$ and $\xi_n(t^*) = x_n$. Hence

$$d(x_n, \xi_n(t_n)) \leqslant d(x_n, \xi_0(t^*)) + d(\xi_0(t^*), \xi_0(t_n)) + d(\xi_0(t_n), \xi_n(t_n)) \to 0,$$

and

$$d(x_n, \xi_n(t_n)) \geqslant \operatorname{dist}_H^X(x_n, A_{\lambda_n}(t_n)) \geqslant \epsilon \quad \text{for all } n \in \mathbb{N},$$

which is a contradiction. The proof of (7.7) is similar and it is left to the reader.

Secondly, note that if $\lambda_n \to \lambda_0$ and $\{t_n\}$ is a sequence in I with $t_n \to t^*$, then

$$\begin{aligned}\operatorname{dist}_H^X(A_{\lambda_n}(t_n), A_{\lambda_0}(t_n)) \leqslant \operatorname{dist}_H^X(A_{\lambda_n}(t_n), A_{\lambda_n}(t^*)) + \operatorname{dist}_H^X(A_{\lambda_n}(t^*), A_{\lambda_0}(t^*))\\ + \operatorname{dist}_H^X(A_{\lambda_0}(t^*), A_{\lambda_0}(t_n)),\end{aligned}$$

and from (7.6), (7.7) and the upper semicontinuity of $\{A_\lambda(t^*)\}_{\lambda\in\Lambda}$ at $\lambda = \lambda_0$, it follows that

$$\operatorname{dist}_H^X(A_{\lambda_n}(t_n), A_{\lambda_0}(t_n)) \to 0. \tag{7.8}$$

Now if $\{A_\lambda(t)\colon t \in I\}_{\lambda\in\Lambda}$ is not upper semicontinuous at $\lambda = \lambda_0$, there exist $\epsilon > 0$ and sequences $\lambda_n \to \lambda_0$ and $\{t_n\}$ in I such that

$$\operatorname{dist}_H^X(A_{\lambda_n}(t_n), A_{\lambda_0}(t_n)) \geqslant \epsilon \quad \text{for all } n \in \mathbb{N},$$

and since I is bounded, $\{t_n\}$ has a convergent subsequence to a point t^* in $\mathbb{R}$, which contradicts (7.8) and completes the proof. □

From these results, the following holds.

THEOREM 7.14. *Suppose the cocycle family* $\{\mathcal{K}_\lambda\}_{\lambda\in\Lambda}$ *and the driving semigroups* $\{\Theta_\lambda\}_{\lambda\in\Lambda}$ *are collectively bounded dissipative and collectively asymptotically compact. Assume also that* (C2) *holds. Then, for each* $\lambda \in \Lambda$ *and bounded global solution* $\eta_\lambda\colon \mathbb{R} \to \mathscr{S}_\lambda$ *of* Θ_λ *in* Σ_λ, *the evolution process* $\mathcal{S}_{\eta_\lambda}$, *given by*

$$S_{\eta_\lambda}(t,s) = \mathcal{K}(t-s, \eta_\lambda(s)) \quad \text{for } t \geqslant s,$$

has a pullback attractor $\hat{A}_{\eta_\lambda}$. *Also, if* $\lambda_n \to \lambda_0$, *there exist a subsequence of* $\{\eta_{\lambda_n}\}$ *(which we denote the same) and a bounded global solution* η_{λ_0} *of* Θ_{λ_0} *such that* $\eta_{\lambda_n}(t) \to \eta_{\lambda_0}(t)$, *uniformly for* t *in bounded subsets of* $\mathbb{R}$.

Moreover, if I *is a bounded interval of* $\mathbb{R}$, *we have*

$$\lim_{n\to\infty} \sup_{t\in I} \operatorname{dist}_H^X(A_{\eta_{\lambda_n}}(t), A_{\eta_{\lambda_0}}(t)) = 0.$$

PROOF. The existence of $\hat{A}_{\eta_\lambda}$ is guaranteed by Theorem 6.15, since Π_λ has a global attractor in $\mathbb{X}_\lambda$. The convergence (up to a subsequence) of $\{\eta_{\lambda_n}\}$ to a bounded global solution $\{\eta_{\lambda_0}\}$ of $\mathcal{S}_{\eta_{\lambda_0}}$, uniformly in bounded subsets of $\mathbb{R}$, follows as in the proof of Theorem 2.5.

Now if $\lambda_n \to \lambda_0$, η_{λ_n} is a global bounded solution of Θ_{λ_n}, for each $n \in \mathbb{N}$, that converges to a global bounded solution η_{λ_0} of Θ_{λ_0}, uniformly in bounded intervals of $\mathbb{R}$, it follows that, for $\mathcal{S}_{\eta_{\lambda_n}}$ and $\mathcal{S}_{\eta_{\lambda_0}}$, our hypotheses ensure that (7.3), (7.4) and (7.5) are satisfied, and the result follows from Theorem 7.13. □

2. Lower semicontinuity

2.1. Lower semicontinuity of global attractors of skew-product semigroups. Let us restate the results of Chapter 2 - Section 3, specifically to the case of a family of skew-product semigroups. To that end, as before, assume that conditions (C1) and (C2) hold. We already know that the family $\{\mathbb{A}_\lambda\}_{\lambda\in\Lambda}$ of global attractors of Π_λ in $\mathbb{X}_\lambda$, for each $\lambda \in \Lambda$, is upper semicontinuous at λ_0 for any $\lambda_0 \in \Lambda$.

In order to prove the lower semicontinuity, we will need to reproduce inside $\mathbb{A}_\lambda$ the same *structures* that are present in $\mathbb{A}_{\lambda_0}$. To that end, the reader should recall the notions of *isolated invariant sets* and *disjoint collections of isolated invariant sets* for a semigroup (see Definitions 2.15 and 2.16).

From the form of the skew-product semigroup Π_λ, the existence of a Π_λ-invariant set $\mathbb{L}$ in $\mathbb{X}_\lambda$ implies the existence of the Θ_λ-invariant set $\pi_{\Sigma_\lambda}(\mathbb{L})$, as we can see in the next result.

PROPOSITION 7.15. *If* $\mathbb{L}$ *is a* Π_λ*-invariant set in* $\mathbb{X}_\lambda$, *then* $\pi_{\Sigma_\lambda}(\mathbb{L})$ *is a* Θ_λ*-invariant set in* Σ_λ.

PROOF. Define $\mathscr{L} = \pi_{\Sigma_\lambda}(\mathbb{L})$. If $\sigma \in \mathscr{L}$ and $t \geqslant 0$, then choosing $x \in X$ such that $(x,\sigma) \in \mathbb{L}$, we have $\Pi_\lambda(t)(x,\sigma) = (\mathcal{K}(t,\sigma)x, \Theta_\lambda(t)\sigma) \in \mathbb{L}$, and hence $\Theta(t)\sigma \in \mathscr{L}$. The other inclusion is analogous, and we leave it to the reader. Thus we conclude that $\mathscr{L}$ is Θ_λ-invariant. □

In the situations presented in this book, given an isolated Π_λ-invariant set $\mathbb{L}$ in $\mathbb{X}_\lambda$, the set $\pi_{\Sigma_\lambda}(\mathbb{L})$ will always coincide with $\mathscr{S}_\lambda$, since our goal is to apply this theory to small suitable perturbations of a given semigroup. However, in what follows, we leave open the possibility that $\pi_{\Sigma_\lambda}(\mathbb{L})$ is a strict subset of $\mathscr{S}_\lambda$, so we can also deal with the inner structure of the global attractor $\mathscr{S}_\lambda$, which is not very common in the literature.

Now we can rewrite the *unstable sets* for isolated Π_λ-invariant sets, just as in Chapter 2 - Section 3. That is, given an isolated invariant set $\mathbb{L}$, its *unstable set* is given by

$$W^u(\mathbb{L}) = \{(x, \sigma_\lambda) \in \mathbb{X}_\lambda \colon \text{ there is a global solution } \psi \text{ of } \Pi_\lambda \\ \text{through } (x, \sigma_\lambda) \text{ such that } \lim_{t \to -\infty} \operatorname{dist}_H^{\mathbb{X}_\lambda}(\psi(t), \mathbb{L}) = 0\},$$

and the (δ)-*local unstable set* of $\mathbb{L}$ is given by

$$W^u_\delta(\mathbb{L}) = \{(x, \sigma_\lambda) \in W^u(\mathbb{L}) \colon \text{ there is a global solution } \psi \text{ of } \Pi_\lambda \\ \text{through } (x, \sigma_\lambda) \text{ with } \operatorname{dist}_H^{\mathbb{X}_\lambda}(\psi(t), \mathbb{L}) \leqslant \delta \text{ for all } t \leqslant 0 \\ \text{and } \lim_{t \to -\infty} \operatorname{dist}_H^{\mathbb{X}_\lambda}((\xi(t), \eta(t)), \mathbb{L}) = 0\}.$$

Now we can restate the result on lower semicontinuity of global attractors for skew-product semigroups.

THEOREM 7.16. *Assume that conditions* (C1) *and* (C2) *hold, and suppose that:*

(a) *there exists a disjoint collection of isolated* Π_λ*-invariant sets* $\{\mathbb{L}_{1,\lambda}, \cdots, \mathbb{L}_{n,\lambda}\}$ *for each* $\lambda \in [0,1]$*;*
(b) $\operatorname{dist}_H^{\mathbb{X}_\lambda}(W^u_\delta(\mathbb{L}_{j,\lambda_0}), W^u_\delta(\mathbb{L}_{j,\lambda})) \overset{\lambda \to \lambda_0}{\longrightarrow} 0$*;*
(c) $\mathbb{A}_{\lambda_0} = \bigcup_{j=1}^n W^u(\mathbb{L}_{j,\lambda_0})$.

Then $\operatorname{dist}_H^{\mathbb{X}_\lambda}(\mathbb{A}_\lambda, \mathbb{A}_{\lambda_0}) + \operatorname{dist}_H^{\mathbb{X}_\lambda}(\mathbb{A}_{\lambda_0}, \mathbb{A}_\lambda) \overset{\lambda \to \lambda_0}{\longrightarrow} 0$.

The proof of this result follows, step by step, the proof of Theorem 2.14 and will be left to the reader; just recall that $\bigcup_{\lambda \in \Lambda} \mathbb{A}_\lambda \subset \mathbb{A}$, and hence is bounded (in fact, it is relatively compact) in $\mathbb{X}_M$.

2.2. Lower semicontinuity of pullback attractors. To state a result on lower semicontinuity of pullback attractors, we first define a few concepts.

DEFINITION 7.17. Let ξ^* be a global solution for an evolution process $\mathcal{S}$. The **unstable set** of ξ^* is the set defined as the set

$$W^u(\xi^*) = \{(\tau, \zeta) \in \mathbb{R} \times X \colon \text{ there exists a global solution } \xi \text{ of } \mathcal{S} \\ \text{with } \xi(\tau) = \zeta \text{ and } \lim_{t \to -\infty} d(\xi(t), \xi^*(t)) = 0\}.$$

The **section of the unstable set of** ξ^* **at time** τ is denoted by

$$W^u(\xi^*)(\tau) = \{\zeta \colon (\tau, \zeta) \in W^u(\xi^*)\}.$$

A **local unstable set of** ξ^* **at time** τ is defined by

$$W^u_\delta(\xi^*)(\tau) = \{\zeta \in W^u(\xi^*)(\tau) \colon \text{ there exists a global solution } \xi \text{ of } \mathcal{S} \\ \text{with } \xi(\tau) = \zeta \text{ and } d(\xi(t), \xi^*(t)) < \delta \text{ for all } t \leqslant \tau \\ \text{such that } \lim_{t \to -\infty} d(\xi(t), \xi^*(t)) = 0\}.$$

REMARK 7.18. We note that if ξ^* is a global solution of $\mathcal{S}$, then the family $\{W^u(\xi^*)(t)\}_{t \in \mathbb{R}}$ is $\mathcal{S}$-invariant. Moreover if ξ^* is *backwards bounded*, that is, $\bigcup_{s \leqslant t} \xi(s)$ is bounded in X for each $t \in \mathbb{R}$, and $\mathcal{S}$ has a pullback attractor $\hat{A}$, then for each $t \in \mathbb{R}$ we obtain $W^u(\xi^*)(t) \subset A(t)$. We also note that if $\mathcal{S}$ is an autonomous evolution process and ξ^* is an equilibrium, then $W^u(\xi^*)(\tau)$ is independent of τ.

DEFINITION 7.19. We say that two global solutions $\xi_1^*,\ \xi_2^*\colon \mathbb{R} \to X$ of $\mathcal{S}$ are **backwards separated** if $\limsup_{t\to-\infty} d(\xi_1^*(t), \xi_2^*(t)) > 0$.

REMARK 7.20. We observe that if two global solutions ξ_1^* and ξ_2^* are not backwards separated, then $d(\xi_1^*(t),\ \xi_2^*(t)) \to 0$ as $t \to -\infty$, and the unstable sets of these solutions coincide, that is, $W^u(\xi_1^*) = W^u(\xi_2^*)$.

With this in mind we will be interested in *maximal sets of backwards separated global solutions.* Clearly a maximal set of backwards separated global solutions is not unique, since any solution ξ in a maximal set of backwards separated global solutions may be replaced by any other in the unstable set of ξ and maintaining maximality.

If an evolution process $\mathcal{S}$ has a pullback attractor $\hat{A}$ and $\mathfrak{S}$ denotes a maximal set of backwards separated and backwards bounded global solutions, then

$$A(t) = \bigcup_{\xi^*\in\mathfrak{S}} W^u(\xi^*)(t) \qquad \text{for each } t \in \mathbb{R}.$$

Since many of the unstable sets of global solutions are equal, there is hope that the family of backwards separated global solutions is *small.*

THEOREM 7.21. *Let $\mathcal{S}$ be an evolution process in X with a pullback attractor $\hat{A}_\lambda$, for each $\lambda \in \Lambda$, and suppose* (7.3), (7.4) *and* (7.5) *are satisfied. If for some $t_0 \in \mathbb{R}$, the family $\{A_\lambda(t_0)\}_{\lambda\in\Lambda}$ is lower semicontinuous at λ_0 and I is a bounded interval in $[t_0, \infty)$, then the family $\{A_\lambda(t)\colon t \in I\}_{\lambda\in\Lambda}$ is lower semicontinuous at λ_0.*

PROOF. We may assume that $I = [t_0, t_0 + T]$ for some $T > 0$. We show firstly that if $\{A_\lambda(t_0)\}_{\lambda\in\Lambda}$ is lower semicontinuous at λ_0 and $a > t_0$, then $\{A_\lambda(a)\}_{\lambda\in\Lambda}$ is also lower semicontinuous.

For each $x_0 \in A_{\lambda_0}(a)$, from the $\mathcal{S}_{\lambda_0}$-invariance of the pullback attractor $\hat{A}_{\lambda_0}$, there exists a bounded global solution $\xi_0\colon \mathbb{R} \to X$ of $\mathcal{S}_{\lambda_0}$ with $\xi_0(a) = x_0$. If $\lambda_n \to \lambda_0$, take $y_n \in A_{\lambda_n}(t_0)$ with $y_n \to \xi_0(t_0)$. Then, there exist a subsequence of $\{y_n\}$ (which we denote the same) and a sequence $\xi_n\colon \mathbb{R} \to X$ of global solutions of $\mathcal{S}_{\lambda_n}$ such that $\xi_n(t_0) = y_n$ and $\xi_n(t) \in A_{\lambda_n}(t)$ for all $t \in \mathbb{R}$, and global solutions $\tilde{\xi_0}\colon \mathbb{R} \to X$ of $\mathcal{S}_{\lambda_n}$ such that $\xi_n(t) \to \tilde{\xi_0}(t)$ uniformly for t in bounded subsets of $\mathbb{R}$ and $\xi_0(t) = \tilde{\xi_0}(t)$ for all $t \geqslant t_0$. It follows that $\xi_n(a) \to x_0$ proving that $\{A_\lambda(a)\}_{\lambda\in\Lambda}$ is lower semicontinuous at λ_0.

The uniformity follows as in Theorem 7.13 using (7.6) and (7.7). □

THEOREM 7.22. *Let $\lambda_n \to \lambda_0$ and consider a family $\{\mathcal{S}_{\lambda_n}\}_{n\in\mathbb{N}}$ of evolution processes. Suppose that for each $n \in \mathbb{N}$, $\mathcal{S}_{\lambda_n}$ has a pullback attractor $\hat{A}_{\lambda_n}$, and that* (7.3), (7.4) *and* (7.5) *are satisfied. Suppose also that:*

(a) *there is a sequence of backwards separated backwards bounded global solutions $\{\xi_j^*\}$ in $\hat{A}_{\lambda_0}$ such that*

$$A_{\lambda_0}(t) = \overline{\bigcup_{j=1}^{\infty} W^u(\xi_j^*(\cdot))(t)} \qquad \textit{for each } t \in \mathbb{R};$$

(b) *for each $j \in \mathbb{N}$ there is a sequence $\{\xi_{j,n}^*\}_{n\in\mathbb{N}}$ of backwards bounded global solutions of $\mathcal{S}_{\lambda_n}$ such that the local unstable set of each $\xi_{j,n}^*$ behaves* *continuously as $n \to \infty$, that is, for each j there are $\delta_j > 0$ and $t_j \in \mathbb{R}$ such that*

$$\mathrm{dist}_H^X(W^u_{\delta_j}(\xi_j^*)(t), W^u_{\delta_j}(\xi_{j,n}^*)(t)) \stackrel{n\to\infty}{\longrightarrow} 0, \quad \textit{for all } t \leqslant t_j.$$

Then the family $\{\hat{A}_{\lambda_n}\}_{n\in\mathbb{N}}$ is upper and lower semicontinuous at $\lambda = \lambda_0$, that is, for each bounded interval I of $\mathbb{R}$ we obtain

$$\lim_{n\to\infty} \sup_{t\in I} \left[\mathrm{dist}_H^X(A_{\lambda_n}(t), A_{\lambda_0}(t)) + \mathrm{dist}_H^X(A_{\lambda_0}(t), A_{\lambda_n}(t))\right] = 0.$$

PROOF. First note that the upper semicontinuity is a direct consequence of Theorem 7.14. To prove that $\mathrm{dist}_H^X(A_{\lambda_0}(t), A_{\lambda_n}(t)) \to 0$ as $n \to \infty$, it is sufficient to show that for $t \in \mathbb{R}$ and $x \in A_{\lambda_0}(t)$ there exists a sequence $x_n \in A_{\lambda_n}(t)$ such that $x_n \to x$.

Given $x \in A_{\lambda_0}(t)$ and $\epsilon > 0$ there is $x_\epsilon \in \bigcup_{j=1}^\infty W^u(\xi_j^*)(t)$ such that $d(x, x_\epsilon) < \epsilon/2$. Let $j \in \mathbb{N}$ be such that $x_\epsilon \in W^u(\xi_j^*)(t)$ and $\xi_j \colon \mathbb{R} \to X$ is a global solution of $\mathcal{S}_{\lambda_0}$ in $\hat{A}_{\lambda_0}$ with $\xi_j(t) = x_\epsilon$ and $d(\xi_j(s), \xi_j^*(s)) \to 0$ as $s \to -\infty$.

Choose $\tau > 0$ such that $z_j = \xi_j(t-\tau) \in W^u_{\delta_j}(\xi_j^*)(t-\tau)$, and using (b) there exists a sequence $\{z_j^n\}$ with $z_j^n \in W^u_{\delta_j}(\xi_{j,n}^*)(t-\tau)$ such that $z_j^n \to z_j$ as $n \to \infty$ and $n_0 \in \mathbb{N}$ such that

$$d(S_{\lambda_n}(t, t-\tau)z_j^n, S_{\lambda_0}(t, t-\tau)z_j) < \frac{\epsilon}{2}, \quad \text{for all } n \geqslant n_0.$$

Since $S_{\lambda_n}(t, t-\tau)z_j^n \in A_{\lambda_n}(t)$, $x_\epsilon = S_{\lambda_0}(t, t-\tau)z_j$ and $\epsilon > 0$ is arbitrary, the lower semicontinuity of $\{A_{\lambda_n}(t)\}_{n\in\mathbb{N}}$ follows. The result now follows easily from Theorem 7.21. □

2.3. Lower semicontinuity of uniform attractors. Since the uniform attractor $\mathcal{A}_\lambda$ for the non-autonomous dynamical system $(\mathcal{K}_\lambda, \Theta)_{X,\Sigma_\lambda}$ is equal to $\pi_X(\mathbb{A}_\lambda)$, that is, the projection on the first coordinate of the global attractor for the skew-product semigroup Π_λ in $\mathbb{X}_\lambda$, the lower semicontinuity of uniform attractors follows from Theorem 7.16, but we will work to restate it in this subsection.

DEFINITION 7.23. Let $\lambda \in \Lambda$. If $(\mathcal{K}_\lambda, \Theta_\lambda)_{(X,\Sigma_\lambda)}$ is a non-autonomous dynamical system, which we will denote simply by $\mathcal{K}_\lambda$, $\mathcal{L} \subset X$ is said to be a **$\mathcal{K}_\lambda$-lifted invariant set** if for each $x \in \mathcal{L}$ there exists a global solution $\eta \colon \mathbb{R} \to \Sigma_\lambda$ of Θ_λ and a global solution of $\xi \colon \mathbb{R} \to X$ of the evolution process $\{\mathcal{K}(t-s, \eta(s)) \colon (t,s) \in \mathcal{P}\}$ such that $\xi(0) = x$ and $\xi(\mathbb{R}) \subset \mathcal{L}$.

DEFINITION 7.24. Fix $\lambda \in \Lambda$, let $(\mathcal{K}_\lambda, \Theta_\lambda)_{(X,\Sigma_\lambda)}$ be a non-autonomous dynamical system and let $\mathcal{L}$ be a $\mathcal{K}_\lambda$-lifted invariant set. Given an open neighborhood U of $\mathcal{L}$ in X we say that $\mathcal{L}$ is the **maximal $\mathcal{K}_\lambda$-lifted invariant set in** U if $V \subset U$ is a $\mathcal{K}_\lambda$-lifted invariant, and then $V \subset \mathcal{L}$.

Also, we say that $\mathcal{L}$ is an **isolated $\mathcal{K}_\lambda$-lifted invariant set** when there exists $\delta > 0$ such that $\mathcal{L}$ is the maximal $\mathcal{K}_\lambda$-lifted invariant set in $\mathcal{O}_\delta(\mathcal{L})$.

We remark that the uniform attractor $\mathcal{A}_\lambda$ is the maximal bounded $\mathcal{K}_\lambda$-lifted invariant set, that is, if $\mathcal{L}$ is a bounded $\mathcal{K}_\lambda$-lifted invariant set, then $\mathcal{L} \subset \mathcal{A}_\lambda$. This can be seen, for instance, using Theorem 6.15.

DEFINITION 7.25. We say that $\{\mathcal{L}_1, \cdots, \mathcal{L}_n\}$ is a **disjoint collection of isolated $\mathcal{K}_\lambda$-lifted invariant sets** if each $\mathcal{L}_i$ is an isolated $\mathcal{K}_\lambda$-lifted invariant set and there exists $r > 0$ such that $\mathcal{O}_r(\mathcal{L}_i) \cap \mathcal{O}_r(\mathcal{L}_j) = \varnothing$ for $1 \leqslant i < j \leqslant n$.

Now we can define *unstable sets* for isolated $\mathcal{K}_\lambda$-lifted invariant sets, just as in Chapter 2 - Section 3.

DEFINITION 7.26. Given an isolated $\mathcal{K}_\lambda$-lifted invariant set $\mathcal{L}$ we define the **unstable set** of $\mathcal{L}$ by

$$\begin{aligned} W^u(\mathcal{L}) = \{x \in X \colon & \text{ there are global solutions } \eta \text{ of } \Theta_\lambda \text{ and} \\ & \xi \text{ of } \{\mathcal{K}(t-s,\eta(s)) \colon t \geqslant s\} \text{ such that } \xi(0) = x \\ & \text{and } \lim_{t\to-\infty} \operatorname{dist}_H^X(\xi(t),\mathcal{L}) = 0\}, \end{aligned}$$

and a **local unstable set** of $\mathcal{L}$ by

$$\begin{aligned} W^u_\delta(\mathcal{L}) = \{x \in W^u(\mathcal{L}) \colon & \text{ there are global solutions } \eta \text{ of } \Theta_\lambda \text{ and} \\ & \xi \text{ of } \{\mathcal{K}(t-s,\eta(s)) \colon t \geqslant s\} \text{ such that } \xi(0) = x, \\ & \operatorname{dist}_H^X(\xi(t),\mathcal{L}) < \delta \text{ for all } t \leqslant 0 \text{ and } \lim_{t\to-\infty} \operatorname{dist}_H^X(\xi(t),\mathcal{L}) = 0\}. \end{aligned}$$

The main difference with respect to the result on lower semicontinuity of global attractors for skew-product semigroups is the interpretation of the meaning of the *lower semicontinuity of local unstable manifolds.* That is expressed in the following definition.

DEFINITION 7.27. Let $\{\mathcal{L}_\lambda\}_{\lambda\in\Lambda}$ be a family of $\mathcal{K}_\lambda$-lifted invariant sets. Then the family of local unstable sets $\{W^u_\delta(\mathcal{L}_\lambda)\}_{\lambda\in\Lambda}$ is **lower semicontinuous** at $\lambda = \lambda_0$ if and only if for each convergent sequence $\lambda_n \to \lambda_0$ in Λ and $x \in W^u_\delta(\mathcal{L}_{\lambda_0})$, there are bounded global solutions η_n of Θ_{λ_n}, converging to a bounded global solution η_0 of Θ_{λ_0}, uniformly in compact subsets of $\mathbb{R}$, and global solutions ξ_n of $\{\mathcal{K}(t-s,\eta_n(s)) : (t,s) \in \mathcal{P}\}$ converging to a global solution ξ_0 of $\{\mathcal{K}(t-s,\eta_0(s)) : (t,s) \in \mathcal{P}\}$, uniformly in compact subsets of $\mathbb{R}$, with $\xi_0(0) = x$, $\lim_{t\to-\infty} \operatorname{dist}_H^X(\xi_n(t),\mathcal{L}_{\lambda_n}) = 0$ and $\operatorname{dist}_H^X(\xi_n(t),\mathcal{L}_{\lambda_n}) < \delta$, for all $t \leqslant 0$ and $n = 0$ or $n \in \mathbb{N}^*$.

With this we can state the result on lower semicontinuity of uniform attractors.

THEOREM 7.28. *If the cocycle family $\{\mathcal{K}_\lambda\}_{\lambda\in\Lambda}$ and the family of driving semigroups $\{\Theta_\lambda\}_{\lambda\in\Lambda}$ are collectively bounded dissipative and collectively asymptotic compact, and* (C2) *holds, then $\mathcal{K}_\lambda$ has a uniform attractor $\mathcal{A}_\lambda$ for each $\lambda \in \Lambda$ with $\bigcup_{\lambda\in\Lambda} \mathcal{A}_\lambda$ relatively compact in X. Suppose also that:*

(a) *there exists a disjoint collection $\{\mathcal{L}_{1,\lambda}, \cdots, \mathcal{L}_{n,\lambda}\}$ of isolated $\mathcal{K}_\lambda$-lifted invariant sets for each $\lambda \in \Lambda$;*
(b) *the family of local unstable sets $\{W^u_\delta(\mathcal{L}_{j,\lambda})\}_{\lambda\in\Lambda}$ is lower semicontinuous at $\lambda = \lambda_0$ for each $j = 1, \cdots, n$;*
(c) $\mathcal{A}_{\lambda_0} = \bigcup_{j=1}^n W^u(\mathcal{L}_{j,\lambda_0})$.

Then

$$\operatorname{dist}_H^X(\mathcal{A}_\lambda,\mathcal{A}_{\lambda_0}) + \operatorname{dist}_H^X(\mathcal{A}_{\lambda_0},\mathcal{A}_\lambda) \overset{\lambda\to\lambda_0}{\longrightarrow} 0.$$

The proof of this results is analogous to the proof of Theorem 2.14, and is left to the reader.

3. Non-autonomous perturbations of an autonomous equation

In this example we illustrate the upper semicontinuity of pullback, uniform and global attractors. Let $(X, \|\cdot\|)$ be a Banach space and consider, for each $\lambda \in [0,1]$, the non-autonomous evolution equation

$$\begin{cases} \dot{x} = f_\lambda(t,x), \ t \in \mathbb{R}^+, \\ x(0) = x_0 \in X, \end{cases} \tag{7.9}$$

where $f_\lambda \in \mathcal{C} = C_b(\mathbb{R}^+ \times X, X)$, the set of continuous functions from $\mathbb{R}^+ \times X$ to X that are bounded in subsets of the form $\mathbb{R} \times B$, for $B \subset X$ bounded.

For positive integers j and n, define the seminorm $\|\cdot\|_{n,j}$ in $\mathcal{C}$ by

$$\|g\|_{j,n} = \sup_{t\in[0,j]} \sup_{x\in B^n} \|g(t,x)\|,$$

where $B^n = \{x \in X \colon \|x\| \leqslant n\}$. Now define a metric $d_\mathcal{C}$ in $\mathcal{C}$ by

$$d_\mathcal{C}(g,h) = \sum_{j,n=1}^{\infty} 2^{-(j+n)} \frac{\|g-h\|_{j,n}}{1+\|g-h\|_{j,n}}. \tag{7.10}$$

In this way $(\mathcal{C}, d_\mathcal{C})$ is a Frechét space. Consider Σ_λ the positive hull of f_λ for each $\lambda \in [0,1]$, that is, Σ_λ is the closure of the set $\{\Theta_t f_\lambda\}_{t\in\mathbb{R}^+}$ in the metric $d_\mathcal{C}$. By continuity, Σ_λ is invariant for Θ for each λ.

We assume first that the limiting problem at $\lambda = 0$ is *autonomous*, that is,

$$f_0(t,x) = f_0(x) \text{ for each } (t,x) \in \mathbb{R}^+ \times X.$$

And also we assume that $d_\mathcal{C}(f_\lambda, f_0) \to 0$ as $\lambda \to 0^+$ and

$$\bigcup_{\lambda\in[0,1]} \Sigma_\lambda \text{ is precompact in } \mathcal{C}. \tag{7.11}$$

Note that for $\lambda = 0$ we have $\Sigma_0 = \{f_0\}$ and hence $\Xi_0 = \Sigma_0 = \{f_0\}$. Clearly if $t \geqslant 0$ we have $d_\mathcal{C}(\Theta_t f_\lambda, f_0) = d_\mathcal{C}(f_\lambda, f_0)$ and moreover $d_\mathcal{C}(\sigma_\lambda, f_0) \to 0$ as $\lambda \to 0^+$, if $\sigma_\lambda \in \Sigma_\lambda$ for each $\lambda > 0$. With these considerations we have the following result.

LEMMA 7.29. *The family of phase spaces $\{\Sigma_\lambda\}_{\lambda\in[0,1]}$ is continuous at $\lambda = 0$.*

PROOF. Since for any sequence $\lambda_n \to 0^+$ and any choice of $\sigma_{\lambda_n} \in \Sigma_{\lambda_n}$ we have $d_\mathcal{C}(\sigma_{\lambda_n}, f_0) \to 0$ as $n \to \infty$, the result is straightforward. □

We have for each $\lambda \in [0,1]$ a non-autonomous dynamical system $(\mathcal{K}_\lambda, \Theta_\lambda)_{(X,\Sigma_\lambda)}$, where $\mathcal{K}_\lambda(t,\sigma_\lambda)x$ is the solution at time $t \geqslant 0$ of the problem

$$\begin{cases} \dot{x} = \sigma_\lambda(t,x), \quad t > 0, \\ x(0) = x \in X \end{cases} \tag{7.12}$$

for each $\sigma_\lambda \in \Sigma_\lambda$.

If we assume that each non-autonomous dynamical system $\mathcal{K}_\lambda$ satisfies the conditions of Theorem 6.17, then there exist a uniform attractor $\mathcal{A}_\lambda$ for $\mathcal{K}_\lambda$ and global attractor $\mathbb{A}_\lambda$ for the associated skew-product semigroup Π_λ. Moreover $\mathcal{A}_\lambda = \pi_X(\mathbb{A}_\lambda)$ for each $\lambda \in [0,1]$.

Also, the existence of $\mathbb{A}_\lambda$ implies, for a given global solution η_λ on Σ_λ of Θ, the existence of a pullback attractor $\hat{A}_{\eta_\lambda}$ for each $\lambda \in [0,1]$, with $A_{\eta_\lambda}(t) = \{x \in X \colon (x, \eta_\lambda(t)) \in \mathbb{A}_\lambda\}$ and

$$\mathcal{A}_\lambda = \bigcup_{\eta_\lambda} A_{\eta_\lambda}(0) \quad \text{for each } \lambda \in [0,1].$$

We finally assume that

$$\bigcup_{\lambda\in[0,1]} \mathcal{A}_\lambda \text{ is precompact in } X. \tag{7.13}$$

Clearly for $\lambda = 0$, since the equation is autonomous, we have $\mathbb{A}_0 = \mathcal{A}_0 \times \{f_0\} = A_0(f_0) \times \{f_0\}$, where $\mathcal{A}_0 = A_0(f_0)$ is the global attractor of (7.9) with $\lambda = 0$.

THEOREM 7.30. *Under all the previous hypotheses and notations, the families* $\{\mathcal{A}_\lambda\}_{\lambda\in[0,1]}$, $\{\hat{A}_{\eta,\lambda}\}_{\lambda\in[0,1]}$, *and* $\{\mathbb{A}_\lambda\}_{\lambda\in[0,1]}$ *are upper semicontinuous at* $\lambda = 0$, *that is, if* $\lambda \to 0^+$ *we have, for all* $I \subset \mathbb{R}$ *bounded interval,*

$$\operatorname{dist}_H^X(\mathcal{A}_\lambda,\mathcal{A}_0) \to 0, \qquad \sup_{t\in I} \operatorname{dist}_H^X(A_{\eta,\lambda}(t), A_0(f_0)) \to 0,$$
$$\textit{and} \quad \operatorname{dist}_H^{X\times\mathcal{C}}(\mathbb{A}_\lambda, \mathbb{A}_0) \to 0.$$

PROOF. Note firstly that our hypotheses imply conditions (7.4) and (7.5). For the continuity of the map $[0,1] \ni \lambda \to f_\lambda \in \mathcal{C}$ at $\lambda = 0$, for each $T > 0$ and K a compact subset of X we have

$$\sup_{\sigma_\lambda\in\Sigma_\lambda}\ \sup_{(t,x)\in[0,T]\times K} d_X(\mathcal{K}_\lambda(t,\sigma_\lambda)x, \mathcal{K}_0(t,f_0)x) \to 0 \quad \text{as } \lambda \to 0^+,$$

which is (7.3). Hence the result follows analogously as in the proofs of Theorem 7.9 for $\{\mathcal{A}_\lambda\}_{\lambda\in[0,1]}$ and Theorem 7.14 for $\{\hat{A}_{\eta,\lambda}\}_{\lambda\in[0,1]}$.

We now prove the assertion for the family $\{\mathbb{A}_\lambda\}_{\lambda\in[0,1]}$. Let $\lambda_n \to 0^+$ and $(x,\sigma_n) \in X \times \Sigma_{\lambda_n}$ for each $n \geqslant 1$. From Lemma 7.29 we have $d_{\mathcal{C}}(\sigma_n, f_0) \to 0$ as $n \to \infty$. Hence

$$\begin{aligned} d_{X\times\mathcal{C}}(\Pi_{\lambda_n}(x,\sigma_n), \Pi_0(x,f_0)) &= d(\mathcal{K}_{\lambda_n}(t,\sigma_n)x, \mathcal{K}_0(t,f_0)x) + d_{\mathcal{C}}(\Theta_t\sigma_n, f_0) \\ &= d(\mathcal{K}_{\lambda_n}(t,\sigma_n)x, \mathcal{K}_0(t,f_0)x) + d_{\mathcal{C}}(\sigma_n, f_0) \to 0 \text{ as } n \to \infty, \end{aligned}$$

uniformly for (t,x) in compact subsets of $\mathbb{R}^+ \times X$ and uniformly for $\sigma_n \in \Sigma_{\lambda_n}$. From (7.11) and (7.13) we obtain $\bigcup_{\lambda\in[0,1]} \mathbb{A}_\lambda$ precompact in $X \times \mathcal{C}$, and hence we can use Remark 7.8 to prove the desired result. ☐

4. Hyperbolic perturbation of parabolic equations

In this example, taken from [**122**], we study the continuity of pullback, uniform, and cocycle attractors for the non-autonomous dynamical system given by the hyperbolic equation:

$$\epsilon u_{tt} + u_t - \Delta u = f_\epsilon(t,u). \tag{7.14}$$

The unknown is the function $u\colon (t_0,\infty) \times \Omega \to \mathbb{R}$, where $\Omega \subset \mathbb{R}^3$ is an open and bounded domain with smooth boundary. For each $\epsilon > 0$, we impose conditions on this equation so that it has uniform and pullback attractors in $H_0^1(\Omega) \times L^2(\Omega)$ and, for $\epsilon = 0$, the limit autonomous parabolic equation

$$u_t - \Delta u = f_0(u) \tag{7.15}$$

has a global attractor A_0 in $H_0^1(\Omega)$, which can be naturally embedded into a compact set $\mathcal{A}_0$ in $H_0^1(\Omega) \times L^2(\Omega)$. We show that these attractors converge, both upper and lower semicontinuously, to $\mathcal{A}_0$. The study of the detailed structure of the attractors under perturbation plays the crucial role in the argument. For the autonomous case, the positive answer, given in terms of Hausdorff convergence of global attractors for both problems, was given by Hale and Raugel in their two papers [**134**], [**136**].

Let $\|\cdot\|_0$ be the usual norm in $L^2(\Omega)$ and $\|u\|_1 = \|\nabla u\|_0$. We denote $H = H_0^1(\Omega) \times L^2(\Omega)$ with the product norm

$$\|(u,v)\|_H^2 = \|v\|_0^2 + \|u\|_1^2.$$

We now provide assumptions on the non-autonomous term f_ϵ. To follow [**122**], we assume that f is defined for all $t \in \mathbb{R}$ and that it holds the backwards uniqueness in σ_ϵ, so that in this case we could even work with the *cocycle attractor*, as in Remark 6.3. Assume that $f_\epsilon \colon \mathbb{R}^2 \to \mathbb{R}$ is continuously differentiable with respect to the first variable and twice differentiable in the second variable, and satisfies the following conditions:

$$\frac{\partial f_\epsilon}{\partial u} \colon \mathbb{R}^2 \to \mathbb{R} \quad \text{is a continuous function,}$$

$$\left|\frac{\partial^2 f_\epsilon}{\partial u^2}(t,u)\right| \leqslant c(1+|u|) \quad \text{for all } u \in \mathbb{R} \text{ and } t \in \mathbb{R}, \tag{7.16}$$

with a constant $c > 0$ independent of $\epsilon \in [0,1]$ (for f_0 we assume that $|f_0''(u)| \leqslant c(1+|u|)$). For the upper semicontinuous convergence of attractors we need the assumption

$$\sup_{t\in\mathbb{R}} |f_\epsilon(t,u) - f_0(u)| \xrightarrow{\epsilon\to 0} 0 \quad \text{for all } u \in \mathbb{R}. \tag{7.17}$$

To obtain the lower semicontinuity we need to assume also that

$$\sup_{t\in\mathbb{R}} \left(|f_\epsilon(t,u) - f_0(u)| + \left|\frac{\partial f_\epsilon}{\partial u}(t,u) - f_0'(u)\right| \right) \xrightarrow{\epsilon\to 0} 0 \quad \text{for every} \quad u \in \mathbb{R}. \tag{7.18}$$

We observe that the growth condition (7.16) implies that

$$|f_\epsilon(t,u+v) - f_\epsilon(t,u)| \leqslant C(1+|u|^2+|v|^2)|v|, \tag{7.19}$$

and also that

$$\left|\frac{\partial f_\epsilon}{\partial u}(t,u+v) - \frac{\partial f_\epsilon}{\partial u}(t,u)\right| \leqslant C(1+|u|+|v|)|v|, \tag{7.20}$$

for all $u, v, t \in \mathbb{R}$ and $\epsilon \in [0,1]$.

We will treat f_ϵ as a function which maps a time to a real valued function of the variable $u \in \mathbb{R}$. Then we have $f_\epsilon \in \mathcal{C} = C_{loc}(\mathbb{R}; C_{loc}(\mathbb{R}))$, the latter being a metric space with the metric of the uniform convergence on compact sets given by

$$\rho(f,g) = \sum_{T\in\mathbb{N}} \frac{1}{2^T} \frac{\max_{t\in[-T,T]} d(f(t),g(t))}{1+\max_{t\in[-T,T]} d(f(t),g(t))},$$

where $d(\phi,\psi)$ is a metric on $C_{loc}(\mathbb{R})$ given by

$$d(\phi,\psi) = \sum_{R\in\mathbb{N}} \frac{1}{2^R} \frac{\max_{u\in[-R,R]} |\phi(u)-\psi(u)|}{1+\max_{u\in[-R,R]} |\phi(u)-\psi(u)|}.$$

For $\epsilon \in (0,1]$ we define $\Sigma_\epsilon = \overline{\{\Theta_t f_\epsilon : t \in \mathbb{R}\}}^{\rho}$ where $\Theta_t \colon \Sigma_\epsilon \to \Sigma_\epsilon$ is the translation operator defined by $\Theta_t f(\cdot) = f(t+\cdot)$. Note that if $\sigma_\epsilon \in \Sigma_\epsilon$, then $\frac{\partial\sigma_\epsilon}{\partial u}(t,u)$ exists. Indeed, for $s \neq 0$, we have

$$\frac{\sigma_\epsilon(t,u+s) - \sigma_\epsilon(t,u)}{s} = \lim_{n\to\infty} \frac{f_\epsilon(t+t_n,u+s) - f_\epsilon(t+t_n,u)}{s},$$

where $\{t_n\} \subset \mathbb{R}$, and

$$\frac{\partial f_\epsilon}{\partial u}(t+t_n,u) = \lim_{s\to 0} \frac{f_\epsilon(t+t_n,u+s) - f_\epsilon(t+t_n,u)}{s}.$$

By (7.20) the last limit is uniform with respect to n. Hence

$$\frac{\partial \sigma_\epsilon}{\partial u}(t,u) \quad \text{exists and} \quad \frac{\partial \sigma_\epsilon}{\partial u}(t,u) = \lim_{n\to\infty} \frac{\partial f_\epsilon}{\partial u}(t+t_n,u).$$

We also deduce that

$$\frac{\partial^2 \sigma_\epsilon}{\partial u^2}(t,u) \quad \text{exists, and} \quad \frac{\partial^2 \sigma_\epsilon}{\partial u^2}(t,u) = \lim_{n\to\infty} \frac{\partial^2 f_\epsilon}{\partial u^2}(t+t_n,u).$$

It follows that (7.19) and (7.20) hold for σ_ϵ in place of f_ϵ, and that

$$\left|\frac{\partial^2 \sigma_\epsilon}{\partial u^2}(t,u)\right| \leqslant c(1+|u|).$$

Observe that also (7.17) and (7.18) stay valid if we replace f_ϵ by any $\sigma_\epsilon \in \Sigma_\epsilon$.

By [**98**, Proposition V.2.5] it follows that f_ϵ is translation compact in $\mathcal{C}$. We conclude that Σ_ϵ is compact in this metric space and invariant with respect to the translation group.

Under the above assumptions, for every $R > 0$ and for every $\sigma_\epsilon \in \Sigma_\epsilon$, we have

$$\lim_{\epsilon\to 0} \sup_{|s|\leqslant R} \sup_{t\in\mathbb{R}} \left(|\sigma_\epsilon(t,s) - f_0(s)| + \left|\frac{\partial \sigma_\epsilon}{\partial s}(t,s) - f_0'(s)\right|\right) = 0.$$

First we observe that the problem (7.14) defines a non-autonomous Dynamical System $(\mathcal{K}_\epsilon, \Theta)_{(H,\Sigma_\epsilon)}$, where Θ_t are the shift operators on Σ_ϵ and $\mathcal{K}_\epsilon(t,\sigma_\epsilon)$ is the map that assigns to the initial data $(u_0,u_1) \in H$ taken at time $t_0 = 0$ the $(u(t),u_t(t))$ of solutions to (7.14) where f_ϵ is replaced by σ_ϵ. For any $\epsilon \in (0,1]$ the NDS defined by the solutions of (7.14) is uniformly asymptotically compact (see [**122**]).

From the results in [**9**], equation (7.14) has existence and uniqueness of solutions in H, so that we can define a nonlinear process $\{S_\epsilon(t,s) \colon t \geqslant s\}$ on H, where

$$S_\epsilon(t,s)(u_0,u_1) = (u(t),u_t(t)) \quad \text{for } t \geqslant s,$$

with $(u(t),u_t(t))$ being the solution of problem (7.14) such that $u(s) = u_0$ and $u_t(s) = u_1$. From [**9, 122**] we can deduce that $\{S_\epsilon(t,s) \colon (t,s) \in \mathcal{P}\}$ has a pullback attractor $\hat{A}_\epsilon$ in H. Moreover, for each $\sigma_\epsilon \in \Sigma_\epsilon$ there exists the pullback attractor associated to $\{S_{\sigma_\epsilon}(t,s) \colon (t,s) \in \mathcal{P}\}$, implying the existence of the uniform attractor $\mathcal{A}_\epsilon$ for the associated non-autonomous dynamical system and the global attractor $\mathbb{A}_\epsilon$ for the associated skew-product semigroup.

If $\{S_0(t) \colon t \geqslant 0\}$ is the semigroup on $H_0^1(\Omega)$, defined by $S_0(t)u_0 = u(t)$, where $u(t)$ is a solution of (7.15), we denote by A_0 its global attractor. We can use regularity properties of A_0, namely that $A_0 \subset H_0^1(\Omega) \cap H^2(\Omega)$, to define the natural embedding of attractor $A_0 \subset H_0^1(\Omega)$ into H, that is,

$$\mathcal{A}_0 = \{(\phi,\psi) \in H \colon \psi = \Delta\phi + f_0(\phi),\ \phi \in A_0\}.$$

The set $\mathcal{A}_0$ is bounded in $H^2(\Omega) \cap H_0^1(\Omega) \times H_0^1(\Omega)$ (see [**134**]) and hence it is compact in H.

It is known (see [**132**, Theorem 3.8.5] for instance) that the fact that all equilibria of (7.15) are hyperbolic implies that the number of these equilibria is finite.

We can denote them by $E = \{e_1, \cdots, e_n\}$, and the global attractor has the representation

$$\mathcal{A}_0 = \bigcup_{i=1}^{n} W^u(e_i),$$

where $W^u(e_i)$ is the unstable set of the equilibrium e_i. We are interested in the long-time behavior of solutions of (7.15) after hyperbolic and non-autonomous perturbation and, in particular, the way the structure of the global attractor for (7.15) is transferred to the pullback and uniform attractors after perturbation. We have the following theorem as the main result of this section (see [**122**]).

THEOREM 7.31. *Uniform and pullback attractors for the non-autonomous dynamical system, and the global attractor for the associated skew-product semigroup are continuous at $\epsilon = 0$. More specifically, we have*

$$\lim_{\epsilon\to 0}\left[\mathrm{dist}_H^H(\mathcal{A}_\epsilon, \mathcal{A}_0) + \mathrm{dist}_H^H(\mathcal{A}_0, \mathcal{A}_\epsilon)\right] = 0,$$

$$\lim_{\epsilon\to 0}\sup\nolimits_{t\in I}\left[\mathrm{dist}_H^H(\mathcal{A}_0, A_\epsilon(t)) + \mathrm{dist}_H^H(A_\epsilon(t), \mathcal{A}_0)\right] = 0,$$

and

$$\lim_{\epsilon\to 0}\left[\mathrm{dist}_H^{H\times\mathcal{C}}(\mathbb{A}_\epsilon, \mathcal{A}_0\times\{f_0\}) + \mathrm{dist}_H^{H\times\mathcal{C}}(\mathcal{A}_0\times\{f_0\}, \mathbb{A}_\epsilon)\right] = 0.$$

The lower semicontinuous convergence result essentially follows the methodology in [**136**], where the key ingredient relies on showing that the local unstable set of the equilibrium is the graph of a Lipschitz function. Since the lower semicontinuous convergence must hold for both u and u_t, the variables (u, u_t) are changed into $(u, \epsilon u_t)$, thus enforcing that the velocities tend to zero and use the upper semicontinuity to recover the convergence of velocities u_t from the convergence of u. Although the main idea of the proof on the existence of the local unstable set as a graph for the problem with changed variables follows the same Banach fixed point scheme as in [**73**], the difficulty that arises is that the constant in the linear exponential dichotomy blows up to infinity as $\epsilon \to 0$ (see [**122**]).

Notes. Upper and lower semicontinuity of attractors for non-autonomous dynamical systems show the robustness of the concepts of pullback and uniform attractors, and global attractors for the skew-product semigroup. Note that upper semicontinuity is typically easier to obtain than lower semicontinuity, and the key ingredients of the proof are the a priori estimates on the attractor and no knowledge on the attractor structure is needed. Clearly, the uniform attractor contains all fibers of pullback (and cocycle) attractors, and hence it is sufficient to obtain the upper semicontinuity result for uniform attractors and the corresponding result for pullback (and cocycle) attractors will follow trivially.

Results on the lower semicontinuity of attractors need a careful description of the structure of the attractor for the limiting equation, which is then transferred to the attractors under perturbation. The description of the uniform attractor as the maximal compact lifted invariant set of a non-autonomous dynamical system is the key ingredient to prove the result on the lower semicontinuity of uniform attractors for PDEs (see [**30**]; see also [**79**]). The characterization of attractors as dynamically gradient attractors has been proved in a series of papers, first focused on pullback attractors ([**73, 74**]), and then generalized to cocycle attractors ([**29, 44**]). Both properties are the crucial facts to prove the lower semicontinuity of the uniform attractors. Upper semicontinuity of uniform attractor also appears in [**3, 204**]. Continuity properties for pullback and cocycle attractors can be seen in

[**10**, **47**, **49**, **75**, **78**, **80**]. A nice set of applications to systems of ODEs and PDEs can be found in these papers (see also [**42**, **122**]). Equi-attraction for pullback, cocycle and uniform attractors and its equivalence to the lower semicontinuity of attractors can be learnt in [**39**, **163**].

The theory of global attractors for the wave equation with damping dependent on u_t has been developed in [**21**], [**127**], [**131**], and [**138**, **139**]. An overview of the theory can be found in, among others, the monographs [**22**], [**140**] and [**104**]. We also mention the work [**9**] where the authors proved the existence of the global attractor for the cubic case (with exponent three in the growth condition for the semilinear term), [**67**] for the supercubic (but subquintic) case, and the generalization to the quintic case (with exponent five) in [**148**]. Original work of Hale and Raugel on the hyperbolic perturbation of the semilinear heat equation has been the subject of vivid study since the appearance of their papers [**134**, **136**]. G. Raugel has observed that the introduced type of perturbation is, in fact, regular [**186**, **187**].

CHAPTER 8

Topological Structural Stability

In this chapter we seek to reinterpret the concept of *dynamically gradient semigroups*, presented in Chapter 2, for non-autonomous dynamical systems, and study the behavior of this structure under suitable perturbations. More precisely, we will see that such perturbations are also dynamically gradient non-autonomous dynamical systems. We will also investigate the behavior of these perturbations for uniform attractors and pullback attractors.

1. Dynamically gradient skew-product semigroups

We start our study by describing the structure of global attractors of skew-product semigroups. To that end let (X, d) and (M, d_M) be metric spaces, let $\mathbb{X} = X \times M$ with the product metric, and let Θ be a semigroup in M. Assume that there is a cocycle $\mathcal{K}$ in X relative to Θ and let $(\mathcal{K}, \Theta)_{(X,M)}$ and Π be, respectively, the associated non-autonomous dynamical system in X and the skew-product semigroup in $\mathbb{X}$. Assume condition (C1) holds.

Some additional assumptions and new concepts will be required, and we start by assuming that there exists a disjoint collection of isolated Π-invariant sets, namely $\mathfrak{L} = \{\mathbb{L}_1, \cdots, \mathbb{L}_n\}$, in $\mathbb{X}$.

We recall that a **homoclinic structure** in $\mathbb{A}$ (see Definition 3.16) is a subcollection $\{\mathbb{L}_{i_1}, \cdots, \mathbb{L}_{i_k}\}$ of $\mathfrak{L}$, with $k \leqslant n$, together with a set of bounded global solutions $\{(\xi_p, \eta_p)\}_{p=1}^{k}$ of Π such that

$$\mathbb{L}_{i_p} \overset{t\to-\infty}{\longleftarrow} (\xi_p(t), \eta_p(t)) \overset{t\to\infty}{\longrightarrow} \mathbb{L}_{i_{p+1}} \qquad \text{for } 1 \leqslant p \leqslant k,$$

with $\mathbb{L}_{i_{k+1}} = \mathbb{L}_{i_1}$. In the case $k = 1$, we additionally require that the solution (ξ_1, η_1) is not entirely contained in L_{i_1}.

REMARK 8.1. Note that a global solution (ξ, η) of Π consists of a global solution $\eta\colon \mathbb{R} \to M$ of the driving semigroup Θ and a global solution $\xi\colon \mathbb{R} \to X$ of the evolution process $\{\mathcal{K}(t-s, \eta(s))\colon (t,s) \in \mathcal{P}\}$. Hence, in a homoclinic structure of a skew-product semigroup, the first coordinates of the global solutions may be associated to different evolution processes.

As in Chapter 3, the skew-product semigroup Π, satisfying condition (C1), is said to be **dynamically gradient** relative to the family $\mathfrak{L} = \{\mathbb{L}_1, \cdots, \mathbb{L}_n\}$ of isolated Π-invariant sets if:

(1) given a bounded global solution (ξ, η) of Π, there exist $\mathbb{L}_i, \mathbb{L}_j \in \mathfrak{L}$ such that
$$\mathbb{L}_i \overset{t\to-\infty}{\longleftarrow} (\xi(t), \eta(t)) \overset{t\to\infty}{\longrightarrow} \mathbb{L}_j \quad \text{and}$$
(2) there are no homoclinic structures in $\mathbb{A}$.

REMARK 8.2. In the applications that arise from non-autonomous differential equations, for which the vector fields are only defined for $t \geqslant 0$, the only possible gradient structure for the attractor of the driving semigroup Θ is the trivial one, that is, when $n = 1$ and Ξ_1 is equal to the global attractor $\mathscr{S}$ of Θ in M. In that case it might be interesting to look at connections between isolated invariant sets of $\mathscr{S}$ beyond the gradient structure. Also, if the vector fields are defined for all $t \in \mathbb{R}$, or in some other more general situations, the decomposition of $\mathscr{S}$ may contain another interesting gradient structure, as in the case where the system is asymptotically autonomous, with different end points when $t \to -\infty$ and $t \to \infty$.

As an immediate consequence of Theorem 3.35 we have the following result.

THEOREM 8.3. *Let Π be a dynamically gradient skew-product semigroup relative to the disjoint collection of isolated invariant sets $\mathfrak{L} = \{\mathbb{L}_1, \cdots, \mathbb{L}_n\}$. Then, there is a reordering of $\mathfrak{L}$ which is a Morse decomposition.*

With that we have:

THEOREM 8.4. *Let Π be a dynamically gradient skew-product semigroup relative to the disjoint collection of isolated invariant sets $\mathfrak{L} = \{\mathbb{L}_1, \cdots, \mathbb{L}_n\}$, reordered in such a way that it is a Morse decomposition. Then Π is a gradient semigroup relative to $\mathfrak{L}$. That is, there exists a continuous function $V\colon \mathbb{X} \to \mathbb{R}^+$ which is nonincreasing along solutions and such that $V(\mathcal{K}(t,\sigma)x, \Theta(t)\sigma) = V(x,\sigma)$ for all $t \geqslant 0$ if and only if $(x,\sigma) \in \mathbb{L}_i$ for some $1 \leqslant i \leqslant n$. Such a function is called a* **Lyapunov function** *for Π with respect to $\mathfrak{L}$.*

For what follows, for simplicity of notation, given a disjoint collection of isolated Π-invariant sets $\mathfrak{L} = \{\mathbb{L}_1, \cdots, \mathbb{L}_n\}$, we write

$$\mathcal{L}_j = \pi_X(\mathbb{L}_j) \quad \text{and} \quad \Xi_j = \pi_M(\mathbb{L}_j) \quad \text{for } 1 \leqslant j \leqslant n.$$

REMARK 8.5. We note that if Π is a dynamically gradient skew-product semigroup relative to the disjoint collection of isolated invariant sets $\mathfrak{L} = \{\mathbb{L}_1, \cdots, \mathbb{L}_n\}$ ordered in such a way that it is a Morse decomposition, and $\eta\colon \mathbb{R} \to \mathscr{S}$ is a fixed bounded global solution of Θ, then for any bounded global solution $\xi\colon \mathbb{R} \to X$ of $\{\mathcal{K}(t-s,\eta(s))\colon (t,s) \in \mathcal{P}\}$ there exist $1 \leqslant i < j \leqslant n$ such that

$$\mathcal{L}_j \overset{t\to-\infty}{\longleftarrow} \xi(t) \overset{t\to\infty}{\longrightarrow} \mathcal{L}_i \quad \text{and} \quad \Xi_j \overset{t\to-\infty}{\longleftarrow} \upsilon(t) \overset{t\to\infty}{\longrightarrow} \Xi_i.$$

If $V(\mathcal{K}(t,\sigma)x, \Theta(t)\sigma) = V(x,\sigma)$ for all $t \geqslant 0$ we must have that $(x,\sigma) \in \mathbb{L}_i$ for some $1 \leqslant i \leqslant n$. In particular $\sigma \in \Xi_i$ and there exist a bounded global solution η of Θ with $\eta(0) = \sigma$ and a bounded global solution ξ of $\{\mathcal{K}(t-s,\eta(s))\colon (t,s) \in \mathcal{P}\}$, with $\xi(0) = x$, such that $(\xi(t), \eta(t)) \in \mathbb{L}_i$ for all $t \in \mathbb{R}$, and hence $\xi(t) \in \mathcal{L}_i$ for all $t \in \mathbb{R}$.

The fact that the collection of isolated Π-invariant sets $\mathfrak{L} = \{\mathbb{L}_1, \cdots, \mathbb{L}_n\}$ is disjoint in $\mathbb{X}$ may come in many different ways. It may be that the Ξ_i's are disjoint in M, and hence it does not matter what happens to the $\mathcal{L}_i$'s. It may also be that the Ξ_i's are all equal to $\mathscr{S}$ and that $\{\mathcal{L}_1, \cdots, \mathcal{L}_n\}$ are all disjoint. Each of these possibilities, as well as many others, are interesting, but we will only pursue the second one. That is, we will assume for the rest of this chapter the following condition:

(8.1) the Ξ_i's are all equal to $\mathscr{S}$ and the $\mathcal{L}_i$'s are all disjoint.

This reflects the situations we are interested in in this book, which corresponds to a small non-autonomous perturbation of an autonomous problem.

In particular, this says that the uniform attractors also enjoy some gradient-like structure, that can be described as follows: if the skew-product semigroup Π associated to the non-autonomous dynamical system $(\mathcal{K},\Theta)_{(X,M)}$ satisfies (C1) and is dynamically gradient relative to a disjoint collection of isolated Π-invariant sets $\{\mathbb{L}_1,\cdots,\mathbb{L}_n\}$ ordered in such a way that it becomes a Morse decomposition, and condition (8.1) holds, then the uniform attractor $\mathcal{A}$ of the cocycle $\mathcal{K}$ is dynamically gradient relative to the disjoint collection of isolated $\mathcal{K}$-lifted invariant sets $\{\mathcal{L}_1,\cdots,\mathcal{L}_n\}$. That is, for each point $x\in\mathcal{A}$ there is a bounded global solution η of Θ and a bounded global solution ξ of the evolution process $\{\mathcal{K}(t-s,\eta(s))\colon (t,s)\in\mathcal{P}\}$ such that

$$\mathcal{L}_j \overset{t\to-\infty}{\longleftarrow} \xi(t) \overset{t\to\infty}{\longrightarrow} \mathcal{L}_i$$

for some $1\leqslant i<j\leqslant n$. Furthermore there are no *homoclinic structures* in the uniform attractor $\mathcal{A}$, where by **homoclinic structure** in the uniform attractor $\mathcal{A}$ we mean a subset $\{\mathcal{L}_{i_1},\cdots,\mathcal{L}_{i_k}\}$ of $\{\mathcal{L}_1,\cdots,\mathcal{L}_n\}$, with $k\leqslant n$, together with bounded global solutions η_j of Θ and bounded global solutions ξ_j of the evolution processes $\{\mathcal{K}(t-s,\eta_j(s))\colon (t,s)\in\mathcal{P}\}$ with $1\leqslant j\leqslant k$, such that

$$\mathcal{L}_{i_j} \overset{t\to-\infty}{\longleftarrow} \xi_j(t) \overset{t\to\infty}{\longrightarrow} \mathcal{L}_{i_{j+1}} \quad \text{for } 1\leqslant j\leqslant k,$$

with $\mathcal{L}_{i_{k+1}}=\mathcal{L}_{i_1}$.

2. Dynamically gradient skew-product semigroups under perturbations

We now restate the results of Chapter 3 to perturbations of dynamically gradient skew-product semigroups, and to do that we start with the general setup of Chapter 7, that is, throughout this chapter (X,d) and (M,d_M) will be fixed metric spaces, $\mathbb{X}_M=X\times M$ with the product metric, and Θ will be a semigroup in M. Assume that there is a cocycle $\mathcal{K}$ in X relative to Θ and let $(\mathcal{K},\Theta)_{(X,M)}$ be the associated non-autonomous dynamical system and Π the associated skew-product semigroup in $\mathbb{X}_M$.

Let $\Lambda=[0,1]$ and let $\{\Sigma_\lambda\}_{\lambda\in\Lambda}$ be the family of closed and Θ-positively invariant subsets of M. We will assume that $M=\bigcup_{\lambda\in\Lambda}\Sigma_\lambda$. Let d_λ be the restriction of d_M to $\Sigma_\lambda\times\Sigma_\lambda$, let Θ_λ be the restriction of Θ to Σ_λ and let $\{\mathcal{K}(t,\sigma_\lambda)\colon t\geqslant 0,\sigma_\lambda\in\Sigma_\lambda\}\subset C(X)$ be the cocycle relative to Θ_λ, which we will refer to as $\mathcal{K}_\lambda$. For each $\lambda\in\Lambda$, let $\mathbb{X}_\lambda=X\times\Sigma_\lambda$ with the product metric, let $(\mathcal{K}_\lambda,\Theta_\lambda)_{(X,\Sigma_\lambda)}$ be the associated non-autonomous system and let Π_λ be the skew-product semigroup in $\mathbb{X}_\lambda$ (which is also the restriction of Π to $\mathbb{X}_\lambda$).

Recalling that condition (C1) implies that each semigroup Π_λ possesses a global attractor $\mathbb{A}_\lambda$ in $\mathbb{X}_\lambda$, with $\bigcup_{\lambda\in\Lambda}\mathbb{A}_\lambda$ bounded (in fact, relatively compact) in $\mathbb{X}_M$, we can restate Theorem 3.42 specifically to this case.

THEOREM 8.6. *Assume that condition* (C1) *holds, and also that* $\{\Pi_\lambda\}_{\lambda\in\Lambda}$ *is continuous and collectively asymptotically compact at* $\lambda=\lambda_0$, *and collectively bounded dissipative. Suppose also that:*

(a) *for each* $\lambda\in\Lambda$, $\mathbb{A}_\lambda$ *has a disjoint collection of isolated invariant sets* $\mathfrak{L}_\lambda=\{\mathbb{L}_{1,\lambda},\cdots,\mathbb{L}_{n,\lambda}\}$ *such that*

$$\lim_{\lambda\to\lambda_0}\operatorname{dist}_H^{\mathbb{X}_M}(\mathbb{L}_{i,\lambda},\mathbb{L}_{i,\lambda_0})=0 \quad \textit{for } 1\leqslant i\leqslant n;$$

(b) *there exist* $\delta > 0$ *and* $\epsilon > 0$ *such that* $\mathbb{L}_{i,\lambda}$ *is the maximal isolated invariant set in* $\mathcal{O}_\delta(\mathbb{L}_{i,\lambda})$ *for* $1 \leqslant i \leqslant n$ *and for all* $\lambda \in \Lambda$ *such that* $|\lambda - \lambda_0| < \epsilon$;

(c) Π_{λ_0} *is dynamically gradient relative to the disjoint collection of isolated* Π_{λ_0}*-invariant sets* $\mathfrak{L}_{\lambda_0}$.

Then there exists $\epsilon_1 > 0$ *such that for all* $\lambda \in \Lambda$ *with* $|\lambda - \lambda_0| < \epsilon_1$, Π_λ *is dynamically gradient relative to the disjoint collection of isolated* Π_λ*-invariant sets* $\mathfrak{L}_\lambda$.

Conditions (a) and (b), together with the assumption that Π_{λ_0} is dynamically gradient relative to the disjoint collection of isolated invariant sets $\mathfrak{L}_{\lambda_0}$, are the most difficult conditions to check in order to apply Theorem 8.6. These assumptions are usually checked for particular gradient systems, for example autonomous gradient systems with hyperbolic equilibria. Conditions (a) and (b) are usually a consequence of hyperbolicity (robustness of exponential dichotomy), which will be presented in the next chapter. In any case, the verification of these conditions requires that we step down from the skew-product structure and look at each individual global solution $\eta_\lambda\colon \mathbb{R} \to \mathscr{S}_\lambda$ of the driving semigroup Θ_λ and the corresponding evolution process $\{\mathcal{K}(t-s, \eta_\lambda(s))\colon (t,s) \in \mathcal{P}\}$. In the next section we present the *dynamically gradient structure for evolution processes.*

The robustness of the gradient structure of uniform attractors can also be attained as an immediate consequence of Theorem 8.6, when condition (8.1) is satisfied.

3. Gradient structure of pullback attractors

In this section we define the concept of *dynamically gradient evolution processes*, and we start by extending the notion of isolated invariant sets.

DEFINITION 8.7. Consider an evolution process $\mathcal{S}$ in X with a pullback attractor $\{A(t)\}_{t\in\mathbb{R}}$. A family $\{E(t)\}_{t\in\mathbb{R}}$, with $E(t) \subset A(t)$ for each $t \in \mathbb{R}$, is called an **isolated $\mathcal{S}$-invariant family** if $\{E(t)\}_{t\in\mathbb{R}}$ is $\mathcal{S}$-invariant and if there exists $r > 0$ such that if $\xi\colon \mathbb{R} \to X$ is a global solution of $\mathcal{S}$ with $\xi(t) \in \mathcal{O}_r(E(t))$ for all $t \in \mathbb{R}$, then $\xi(t) \in E(t)$ for all $t \in \mathbb{R}$.

From now on, we write $\hat{E}$ to represent a family $\{E(t)\}_{t\in\mathbb{R}}$.

DEFINITION 8.8. Consider an evolution process $\mathcal{S}$ in X with a pullback attractor $\{A(t)\}_{t\in\mathbb{R}}$. A collection $\{\hat{E}_1, \cdots, \hat{E}_n\}$ is called a **disjoint collection of isolated $\mathcal{S}$-invariant families** if $\hat{E}_i$ is an isolated $\mathcal{S}$-invariant family for $1 \leqslant i \leqslant n$, and there exists $\delta > 0$ such that $\mathcal{O}_\delta(E_j(t)) \cap \mathcal{O}_\delta(E_k(t)) = \varnothing$ for each $1 \leqslant j < k \leqslant n$ and all $t \in \mathbb{R}$.

Exactly as in the autonomous case we define *homoclinic structures* in the following way.

DEFINITION 8.9. Let $\mathcal{S}$ be an evolution process with a pullback attractor $\hat{A}$ and a disjoint collection of isolated $\mathcal{S}$-invariant families $\mathfrak{E} = \{\hat{E}_1, \cdots, \hat{E}_n\}$. A **homoclinic structure** in $\hat{A}$ is a subcollection $\{\hat{E}_{i_1}, \cdots, \hat{E}_{i_k}\}$ of $\mathfrak{E}$ together with global solutions $\{\xi_1, \cdots, \xi_k\}$ of $\mathcal{S}$ such that

$$\lim_{t\to-\infty} \operatorname{dist}_H^X(\xi_j(t), E_{i_j}(t)) = 0 \quad \text{and} \quad \lim_{t\to\infty} \operatorname{dist}_H^X(\xi_j(t), E_{i_{j+1}}(t)) = 0,$$

for $1 \leqslant j \leqslant k$, with $\hat{E}_{i_{k+1}} = \hat{E}_{i_1}$.

With this we can define the *dynamically gradient evolution processes* as follows.

DEFINITION 8.10. Let $\mathcal{S}$ be an evolution process with a pullback attractor $\hat{A}$ and a disjoint collection of isolated $\mathcal{S}$-invariant families $\mathfrak{E} = \{\hat{E}_1, \cdots, \hat{E}_n\}$. We say that $\mathcal{S}$ is a **dynamically gradient evolution process** relative to $\mathfrak{E}$ if:

(DGEP1) given a global bounded solution $\xi\colon \mathbb{R} \to X$ of $\mathcal{S}$ there are $1 \leqslant i, j \leqslant n$ such that

$$\lim_{t\to-\infty} \operatorname{dist}_H^X(\xi(t), E_i(t)) = 0 \quad \text{and} \quad \lim_{t\to\infty} \operatorname{dist}_H^X(\xi(t), E_j(t)) = 0;$$

(DGEP2) there are no homoclinic structures in $\hat{A}$.

3.1. Non-autonomous perturbations of gradient semigroups. Now we consider small non-autonomous perturbations of dynamically gradient semigroups. To this end we consider a family $\{\mathcal{S}_\lambda\}_{\lambda\in\Lambda}$ of evolution processes in X.

DEFINITION 8.11. The family $\{\mathcal{S}_\lambda\}_{\lambda\in\Lambda}$ of evolution processes is said to be:

(a) **collectively asymptotically compact** at $\lambda = \lambda_0$ if given sequences $\lambda_n \to \lambda_0$, $\{x_n\}$ bounded in X, $t_n \to \infty$ and $\{\tau_n\}$ in $\mathbb{R}$, then the sequence $\{S_{\lambda_n}(t_n + \tau_n, \tau_n)x_n\}$ is precompact;

(b) **continuous** at $\lambda = \lambda_0$ if for each compact subset K of X and $T > 0$ we have

$$\sup_{x\in K}\sup_{\tau\in[0,T]} d(S_\lambda(t, t-\tau)x, S_{\lambda_0}(t, t-\tau)x) \overset{\lambda\to\lambda_0}{\longrightarrow} 0 \quad \text{for all } t \in \mathbb{R}.$$

In the results that follow, we will need to construct global solutions for the *limiting* evolution process $\mathcal{S}_{\lambda_0}$ using *partial* solutions of the family $\{\mathcal{S}_\lambda\}_{\lambda\in\Lambda}$ and the following lemma is a crucial result in this direction.

LEMMA 8.12. *Let $\{\mathcal{S}_\lambda\}_{\lambda\in\Lambda}$ be a family of evolution processes, which is continuous at $\lambda = \lambda_0$. Let $\lambda_n \to \lambda_0$, a_n, $b_n \to \infty$, and set $\mathbb{J}_n = [-a_n, b_n]$. Let $\xi_n\colon \mathbb{J}_n \to X$ be a solution of $\mathcal{S}_{\lambda_n}$ for each $n \in \mathbb{N}$, and assume that:*

(i) *for some $\tau_{\lambda_0} \in \mathbb{R}$, there exists $n_0 = n_0(t) \in \mathbb{N}$ such that $\bigcup_{n\geqslant n_0(t)} \xi_n(t)$ is precompact for each $t \in (-\infty, \tau_{\lambda_0}]$;*

(ii) *$\bigcup_n \xi_n(\mathbb{J}_n)$ is bounded.*

Then there exists a subsequence $\{n_k\}$ of $\mathbb{N}$ and a bounded global solution $\xi_0\colon \mathbb{R} \to X$ of $\mathcal{S}_{\lambda_0}$ such that $\xi_{n_k} \to \xi_0$ uniformly on compact subintervals of $\mathbb{R}$.

PROOF. Without loss of generality we may assume $\tau_{\lambda_0} = 0$. As a first step, we find a subsequence $n_{0,k}$ of $\mathbb{N}$ such that $\xi_{n_{0,k}}(0) \overset{k\to\infty}{\longrightarrow} x_0$ for some $x_0 \in X$. Define $\xi_0\colon [0,\infty) \to X$ by $\xi_0(t) = S_{\lambda_0}(t, 0)x_0$ for $t \geqslant 0$. Then

$$\xi_0(t) = S_{\lambda_0}(t, 0)x_0 = \lim_{k\to\infty} S_{\lambda_{n_{0,k}}}(t, 0)\xi_{n_{0,k}}(0) = \lim_{k\to\infty} \xi_{n_{0,k}}(t),$$

with the convergence being uniform in compact subsets of $\mathbb{R}$ and, in particular, $\xi_0([0,\infty)) \subset \overline{\bigcup_n \xi_n(\mathbb{J}_n)}$.

Suppose that we have already constructed a subsequence $\{n_{j,k}\}$ of $\{n_{j-1,k}\}$ for $0 \leqslant j \leqslant m-1$ such that $\xi_{n_{m-1,k}}(t) \overset{k\to\infty}{\longrightarrow} \xi_0(t)$, uniformly for t on compact subintervals of $[-m+1, \infty)$, where $\xi_0\colon [-m+1,\infty) \to X$ is a solution of $\mathcal{S}_{\lambda_0}$ that lies in $\overline{\bigcup_n \xi_n(\mathbb{J}_n)}$.

Now find a subsequence $n_{m,k}$ such that $\xi_{n_{m,k}}(-m) \overset{k\to\infty}{\longrightarrow} x_m$ for some $x_m \in X$, and extend the definition of ξ_0 to $[-m,\infty)$ by setting $\xi_0(s) = S_{\lambda_0}(s,-m)x_m$ for $s \in [-m, -(m-1))$. Then

$$\begin{aligned} S_{\lambda_0}(-m+1,-m)x_m &= \lim_{k\to\infty} S_{\lambda_{n_{m,k}}}(-m+1,-m)\xi_{n_{m,k}}(-m) \\ &= \lim_{k\to\infty} \xi_{n_{m,k}}(-m+1) = \xi_0(-m+1), \end{aligned}$$

and so $\xi_0\colon [-m,\infty)$ is a solution of $\mathcal{S}_{\lambda_0}$. Clearly $\xi_{n_{m,k}}(t) \to \xi_0(t)$ uniformly for t in compact subintervals of $[-m,\infty)$, and $\xi_0([-m,\infty)) \subset \overline{\bigcup_n \xi_n(\mathbb{J}_n)}$.

Finally, if we let $n_k = n_{k,k}$, the sequence ξ_{n_k} has all the properties stated in the theorem. □

COROLLARY 8.13. *Let $\{\mathcal{S}_\lambda\}_{\lambda\in\Lambda}$ be a family of evolution processes, which is continuous at $\lambda = \lambda_0$. Assume that $\mathcal{S}_\lambda$ has a pullback attractor $\hat{A}$, $S_{\lambda_0}(t,s) = T(t-s)$, $t \geqslant s$, where $\mathcal{T} = \{T(t)\colon t \geqslant 0\}$ is a semigroup with a global attractor $\mathcal{A}$. Also assume that*

$$\overline{\bigcup_{\lambda\in[0,1]} \bigcup_{t\in\mathbb{R}} A_\lambda(t)} \quad \textit{is compact.}$$

If $\lambda_k \to \lambda_0$ and, for each $k \in \mathbb{N}$, ξ_k is a global solution of $\mathcal{S}_{\lambda_k}$ in $\hat{A}_{\lambda_k}$, then there exists a subsequence $\{\xi_{k_j}\}_{j\in\mathbb{N}}$ of $\{\xi_k\}_{k\in\mathbb{N}}$ that converges uniformly in bounded intervals of $\mathbb{R}$ to a global solution ξ of $\mathcal{T}$ with $\xi(\mathbb{R}) \subset \mathcal{A}$.

THEOREM 8.14. *Let $\{\mathcal{S}_\lambda\}_{\lambda\in\Lambda}$ be a family of evolution processes, which is continuous and collectively asymptotically compact at $\lambda = \lambda_0$. Assume that:*

(a) *$\mathcal{S}_\lambda$ has a pullback attractor $\hat{A}_\lambda$, for each $\lambda \in \Lambda$, and $\overline{\bigcup_{\lambda\in\Lambda}\bigcup_{t\in\mathbb{R}} A_\lambda(t)}$ is compact;*

(b) *$S_{\lambda_0}(t,s) = T(t-s)$, $t \geqslant s$, where $\mathcal{T} = \{T(t)\colon t \geqslant 0\}$ is a dynamically gradient semigroup relative to the disjoint collection of* unitary *isolated invariant sets (equilibria) $\mathbf{E}_{\lambda_0} = \{e_1^*, \cdots, e_n^*\}$ and with global attractor $\mathcal{A}$;*

(c) *$\mathbf{E}_\lambda = \{\xi_{1,\lambda}^*, \cdots, \xi_{n,\lambda}^*\}$ is a disjoint collection of* unitary *isolated invariant families (isolated global solutions) for $\mathcal{S}_\lambda$, for each $\lambda \in \Lambda$, and*

$$\lim_{\lambda\to\lambda_0} \max_{1\leqslant i\leqslant n} \operatorname{dist}_H^X(\xi_{i,\lambda}^*(\mathbb{R}), e_i^*) = 0;$$

(d) *there exist $\delta > 0$ and $r > 0$ such that, if $\lambda \in \Lambda$ and $|\lambda - \lambda_0| < r$, ξ_λ is a global solution of $\mathcal{S}_\lambda$ in $\hat{A}_\lambda$, $\tau_{\lambda_0} \in \mathbb{R}$ and $\operatorname{dist}_H^X(\xi_\lambda(t), \xi_{i,\lambda}^*(\mathbb{R})) \leqslant \delta$ for all $t \leqslant \tau_{\lambda_0}$ $(t \geqslant \tau_{\lambda_0})$, then $d(\xi_\lambda(t), \xi_{i,\lambda}^*(t)) \to 0$ as $t \to -\infty$ $(t \to \infty)$.*

Then there exists an $\epsilon > 0$ such that $\mathcal{S}_\lambda$ is a dynamically gradient evolution process, for all $\lambda \in \Lambda$ such that $|\lambda - \lambda_0| < \epsilon$.

PROOF. Note that $\operatorname{dist}_H^X(\xi_{i,\lambda}^*(\mathbb{R}), e_i^*) \overset{\lambda\to\lambda_0}{\longrightarrow} 0$ and that there is a $\delta > 0$ for which, for all λ suitably close to λ_0, if a solution ξ_λ satisfies $d(\xi_\lambda(t), e_i^*) \leqslant \delta$ for all $t \geqslant \tau_{\lambda_0}$ and for some $\tau_{\lambda_0} \in \mathbb{R}$, then $\operatorname{dist}_H^X(\xi_\lambda(t), \xi_{i,\lambda}^*(t)) \overset{t\to\infty}{\longrightarrow} 0$.

We argue by contradiction to prove that for all λ suitably close to λ_0, $\mathcal{S}_\lambda$ satisfies (DGEP1). Assume that there exists a sequence $\lambda_k \to \lambda_0$ and corresponding global solutions ξ_k of $\mathcal{S}_{\lambda_k}$ in $\hat{A}_{\lambda_k}$ such that

$$\limsup_{t\to\infty} \operatorname{dist}_H^X(\xi_k(t), \xi_{i,\lambda_k}^*(\mathbb{R})) > \delta \qquad \text{for } 1 \leqslant i \leqslant n.$$

Consequently, for large values of k, we have

$$\limsup_{t\to\infty} d(\xi_k(t), e_i^*) > \delta \quad \text{for } 1 \leqslant i \leqslant n. \tag{8.2}$$

Taking subsequences, if necessary, we may assume that $\xi_k(t) \to \xi_0(t)$ uniformly for t in compact subsets of $\mathbb{R}$. Since $d(\xi_0(t), e_i^*) \overset{t\to\infty}{\longrightarrow} 0$ for some $1 \leqslant i \leqslant n$, given $\ell > \frac{1}{\delta}$ there are a $t_\ell > 0$ and $k_\ell \in \mathbb{N}$ such that $d(\xi_k(t_\ell), e_i^*) < \frac{1}{\ell}$, for each $k \geqslant k_\ell$. From (8.2), there exists $t'_\ell > t_\ell$ such that $d(\xi_{k_\ell}(t), e_i^*) < \delta$ for all $t \in [t_\ell, t'_\ell)$ and $d(\xi_{k_\ell}(t'_\ell), e_i^*) = \delta$. Taking subsequences, if necessary, let $\xi_1(t) = \lim_{\ell\to\infty} \xi_{k_\ell}(t + t'_\ell)$, which is a global solution of $\mathcal{T}$ in $\mathcal{A}$. Then, since $t'_\ell - t_\ell \to \infty$ as $\ell \to \infty$, $d(\xi_1(t), e_i^*) \leqslant \delta$ for all $t \leqslant 0$ and consequently $d(\xi_1(t), e_i^*) \overset{t\to-\infty}{\longrightarrow} 0$.

Furthermore, $d(\xi_1(t), e_j^*) \overset{t\to\infty}{\longrightarrow} 0$ with $j \neq i$. From the fact that $\xi_{k_\ell}(t + t_\ell) \to \xi_1(t)$ uniformly in compact subsets of $\mathbb{R}$ we have, for each $m \in \mathbb{N}$, a time $t_m > 0$ and $k_m \in \mathbb{N}$ such that $d(\xi_k(t_m), e_j^*) < \frac{1}{m}$ for all $k_\ell \geqslant k_m$. Again, from (8.2), there exists $t'_m > t_m$ such that $d(\xi_{k_m}(t), e_j^*) < \delta$ for all $t \in [t_m, t'_m)$ and $d(\xi_{k_m}(t'_m), e_j^*) = \delta$. Proceeding exactly as before we obtain a global solution ξ_2 of $\mathcal{T}$ in $\mathcal{A}$ such that $d(\xi_2(t), e_j^*) \overset{t\to-\infty}{\longrightarrow} 0$ and $d(\xi_2(t), e_r^*) \overset{t\to\infty}{\longrightarrow} 0$ with $r \notin \{i, j\}$. In a finite number of steps we arrive at a contradiction with the fact that $\mathcal{T}$ is dynamically gradient relative to $\mathbf{E}_{\lambda_0}$. This proves that there is an $\epsilon > 0$ such that, for all global solutions ξ_λ of $\mathcal{S}_\lambda$ in $\hat{A}_\lambda$ with $|\lambda - \lambda_0| < \epsilon$, we have

$$\lim_{t\to\infty} \operatorname{dist}_H^X(\xi_\lambda(t), \xi_{i,\lambda}^*(\mathbb{R})) = 0.$$

The proof of the convergence when $t \to -\infty$ is completely analogous and that proves (DGEP1). The proof of (DGEP2) follows the same reasoning, and it is left to the reader. □

4. Application to asymptotically autonomous evolution processes

In this section we will consider *asymptotically autonomous evolution processes.* Loosely speaking, an evolution process is *asymptotically autonomous* if it is very close to an autonomous evolution process when the initial times are taken further and further to the future (or to the past). This idea leads to the following definition (for a similar definition see [**172**]).

DEFINITION 8.15. Let $\mathcal{S} = \{S(t,s) \colon t \geqslant s\}$ be an evolution process and $\mathcal{T} = \{T(t) \colon t \geqslant 0\}$ a semigroup in a metric space (X, d). We say that $\{S(t,s) \colon t \geqslant s\}$ is an **asymptotically autonomous evolution process at** ∞ $(-\infty)$ if given a bounded interval I of $[0, \infty)$ and a compact subset K of X we have

$$\sup_{(t,x)\in I\times K} d(S(t+s,s)x, T(t)x) \to 0 \text{ as } s \to \infty \ (s \to -\infty).$$

We say that $\mathcal{T}$ is the semigroup **associated** with the asymptotically autonomous evolution process $\mathcal{S}$.

In order to obtain information about an asymptotically autonomous evolution process using the results of the previous sections, it is convenient to introduce a new evolution process, which is close, for *all initial times*, to a semigroup.

DEFINITION 8.16. Let $\mathcal{S}$ be an evolution process and let $\mathcal{T}$ be a semigroup in a metric space (X, d). Given $\tau \in \mathbb{R}$ we define the **forward truncation** of $\mathcal{S}$ at time

τ, denoted by $\mathcal{S}^\tau$, by

$$S^\tau(t,s) = \begin{cases} S(t,s) & \text{if } s \leqslant t \leqslant \tau, \\ T(t-\tau)S(\tau,s) & \text{if } s \leqslant \tau \leqslant t, \\ T(t-s) & \text{if } \tau \leqslant s \leqslant t. \end{cases} \tag{8.3}$$

Similarly we define the **backwards truncation** of $\mathcal{S}$ at time τ, denoted by $\mathcal{S}_\tau$, by

$$S_\tau(t,s) = \begin{cases} T(t-s) & \text{if } s \leqslant t \leqslant \tau, \\ S(t,\tau)T(\tau-s) & \text{if } s \leqslant \tau \leqslant t \\ S(t,s) & \text{if } \tau \leqslant s \leqslant t. \end{cases}$$

We have the following property for the truncations.

THEOREM 8.17. *Let $\mathcal{S}$ be an asymptotically autonomous evolution process at $-\infty$ with associated semigroup $\mathcal{T}$, in a metric space (X,d). Assume that for each bounded set $I \subset [0,\infty)$ and compact set $K \subset X$, the set*

$$\{S(t,s)K : t-s \in I\} \text{ is compact.} \tag{8.4}$$

Then the forward truncation $\mathcal{S}^\tau$ of $\mathcal{S}$ at time τ satisfies

$$\sup_{s\in\mathbb{R}} \sup_{(t,u)\in I\times K} d(S^\tau(t+s,s)u, T(t)u) \to 0 \text{ as } \tau \to -\infty.$$

If $\mathcal{S}$ is an asymptotically autonomous evolution process at ∞ with associated semigroup $\mathcal{T}$ in a metric space (X,d), then the backwards truncation $\mathcal{S}_\tau$ of $\mathcal{S}$ at time τ satisfies

$$\sup_{s\in\mathbb{R}} \sup_{(t,u)\in I\times K} d(S_\tau(t+s,s)u, T(t)u) \to 0 \text{ as } \tau \to \infty.$$

Furthermore, in each case, if $\mathcal{S}$ has a pullback attractor $\hat{A} = \{A(t)\}_{t\in\mathbb{R}}$ and $\mathcal{T}$ has a global attractor $\mathcal{A}$, then the forward (or backwards) truncation $\mathcal{S}^\tau$ (or $\mathcal{S}_\tau$) of $\mathcal{S}$ at time τ has a pullback attractor $\hat{A}^\tau$ (or $\hat{A}_\tau$).

PROOF. We will prove the result for the forward truncation, and the backwards case is proved analogously. Fix $\tau \in \mathbb{R}$ and consider the forward truncation $\mathcal{S}^\tau$ of $\mathcal{S}$, which is an asymptotically autonomous evolution process in $-\infty$.

Given $\epsilon > 0$, a compact interval $I \subset \mathbb{R}^+$ and a compact set $K_1 \subset X$ there exists $0 < \delta < \epsilon$ such that $d(T(t)u, T(t)v) < \epsilon$ if $u,v \in K_1$ with $d(u,v) < \delta$ and $t \in I$. Define $K = \bigcup_{r\in I} T(r)K_1 \cup \bigcup_{r-p\in I} S(r,p)K_1$, which is a compact subset of X and it contains K_1. Since $\mathcal{S}$ is asymptotically autonomous at $-\infty$, there exists $s_0 < 0$ such that

$$d(S(t+s,s)u, T(t)u) < \delta \text{ for all } (t,u) \in I \times K \text{ and } s \leqslant s_0.$$

Hence we have

$$d(S^\tau(t+s,s)u, T(t)u) = \begin{cases} d(S(t+s,s)u, T(t)u) & \text{if } t+s \leqslant \tau, \\ d(T(t+s-\tau)S(\tau,s)u, T(t+s-\tau)T(\tau-s)u) & \text{if } s \leqslant \tau \leqslant t+s, \\ 0 & \text{if } \tau \leqslant s, \end{cases}$$

which implies that $\sup_{s\in\mathbb{R}} d(S_\tau(t+s,s)u, T(t)u) < \epsilon$ for all $(t,u) \in I \times K_1$ and for $\tau \leqslant s_0$.

To prove the existence of the pullback attractor for $\mathcal{S}^\tau$, we define

$$A^\tau(t) = \begin{cases} A(t) & \text{if } t \leqslant \tau, \\ T(t-\tau)A(\tau) & \text{if } t > \tau. \end{cases}$$

Note that each $A^\tau(t)$ is compact, $\hat{A}^\tau$ is $\mathcal{S}^\tau$-invariant and if D is a bounded subset of X, for $s < \tau$, we have

$$\operatorname{dist}_H^X(S_\tau(t,s)D, B_\tau(t)) = \begin{cases} \operatorname{dist}_H^X(S(t,s)D, A(t)) & \text{if } t \leqslant \tau, \\ \operatorname{dist}_H^X(T(t-\tau)S(\tau,s)D, T(t-\tau)A(\tau)) & \text{if } t > \tau, \end{cases}$$

which implies that $\hat{A}^\tau$ $\mathcal{S}^\tau$-pullback attracts bounded subsets of X, using the compactness of $A(\tau)$ and the continuity of the map $T(t-\tau)$ for each fixed $t > \tau$. Since $\mathcal{T}$ has a global attractor, we obtain $\bigcup_{t>\tau} T(t-\tau)A(\tau)$ bounded in X, and hence $\bigcup_{t\in\mathbb{R}} A^\tau(t)$ bounded in X. Therefore $\hat{A}^\tau$ is the pullback attractor of $\mathcal{S}^\tau$. □

Directly from the results of Section 3.1, we obtain the next theorem for asymptotically autonomous evolution processes at $-\infty$.

THEOREM 8.18. *Let $\mathcal{S}$ be an asymptotically autonomous evolution process at $-\infty$ satisfying (8.4) and with associated semigroup $\mathcal{T}$ in a metric space (X,d). Suppose also that $\mathcal{S}$ has a pullback attractor $\hat{A}$, that $\mathcal{T}$ has a global attractor $\mathcal{A}$ and also that*

(a) *$\overline{\bigcup_{t\in\mathbb{R}} A(t)}$ is compact in X;*
(b) *$\mathcal{T}$ is a dynamically gradient semigroup with respect to a disjoint family of isolated invariants $\mathbf{E} = \{E_1, \cdots, E_n\}$;*
(c) *the forward truncation $\mathcal{S}^\tau$ of $\mathcal{S}$ has a disjoint collection of isolated invariant families $\mathscr{D}_\tau = \{\hat{D}_{1,\tau}, \cdots, \hat{D}_{n,\tau}\}$ with*

$$\max_{1\leqslant i\leqslant n} \sup_{s\in\mathbb{R}} \left[\operatorname{dist}_H^X(D_{i,\tau}(s), E_i) + \operatorname{dist}_H^X(E_i, D_{i,\tau}(s))\right] \to 0 \text{ as } \tau \to -\infty;$$

(d) *for all $-\tau$ suitably large, there exists $r > 0$ such that, if ξ_τ is a bounded global solution of $\mathcal{S}^\tau$ with $\operatorname{dist}_H^X(\xi_\tau(t), E_i) < r$ for all $t \leqslant t_0$, for some $t_0 \in \mathbb{R}$ and $E_i \in \mathbf{E}$, then $\lim_{t\to-\infty} \operatorname{dist}_H^X(\xi_\tau(t), D_{i,\tau}(t)) = 0$.*

Then there exists $\tau_0 < 0$ such that for all $\tau \leqslant \tau_0$ the forward truncation $\mathcal{S}^\tau$ of $\mathcal{S}$ is a dynamically gradient evolution process.

With this result we can look more closely at the behavior of solutions of the evolution process $\mathcal{S}$ at $-\infty$.

COROLLARY 8.19. *Assume that all the conditions of Theorem 8.18 hold. If ξ is a bounded global solution for $\mathcal{S}$, then there exists $1 \leqslant i \leqslant n$ such that $\xi(s) \to E_i$ as $s \to -\infty$.*

PROOF. Let ξ be a bounded global solution for the evolution process $\mathcal{S}$. Then for each $\tau \in \mathbb{R}$, the function $\xi_\tau \colon \mathbb{R} \to X$ defined by

$$\xi_\tau(t) = \begin{cases} \xi(t), & \text{if } t \leqslant \tau, \\ T(t-\tau)\xi(\tau), & \text{if } t > \tau, \end{cases}$$

is a bounded global solution for the forward truncated evolution process $\mathcal{S}^\tau$. From Theorem 8.18 there exist $\tau_0 < 0$ and $i = 1, \cdots, n$ such that

$$\operatorname{dist}_H^X(\xi_{\tau_0}(s), D_{i,\tau_0}(s)) \to 0 \quad \text{as } s \to -\infty.$$

Since $\sup_{s\in\mathbb{R}} \operatorname{dist}_H^X(D_{i,\tau}(s), E_i) \overset{\tau\to-\infty}{\longrightarrow} 0$ and $\xi_\tau(s) = \xi(s)$ for $s \leqslant \tau$, the result follows. □

Completely analogous results can be proved to treat the case of an asymptotically autonomous evolution process at ∞ and its backwards truncation. We leave their statements and proofs to the reader.

5. Example: a non-autonomous Chafee-Infante problem

This example is intended to show that, in a very specific case, it is still possible to give a good characterization of the attractor of the skew-product semigroup, exhibiting its gradient structure. Consider the semilinear non-autonomous parabolic problem

$$\begin{cases} u_t = u_{xx} + \lambda u - \beta(t)u^3, & t > s \geqslant 0,\ x \in (0,\pi), \\ u(0,t) = u(\pi,t) = 0, & t > s \geqslant 0, \\ u(x,s) = u_0(x), & u_0 \in H_0^1(0,\pi), \end{cases} \tag{8.5}$$

where $\lambda > 0$ and $\beta\colon \mathbb{R}^+ \to [\beta_1, \beta_2] \subset (0,\infty)$ is a globally Lipschitz continuous function.

Before we proceed with the characterization of the attractor for the skew-product semigroup associated to (8.5), we consider its autonomous counterpart to indicate the structures we will be looking for in the non-autonomous case.

5.1. The autonomous inspiration. We consider the autonomous Chafee-Infante problem, given by

$$\begin{cases} u_t = u_{xx} + \lambda u - \beta u^3, & t > 0,\ x \in (0,\pi), \\ u(0,t) = u(\pi,t) = 0, & t \geqslant 0, \\ u(x,0) = u_0(x), & u_0 \in H_0^1(0,\pi), \end{cases} \tag{8.6}$$

with $\lambda \in (0,\infty)$ and $\beta > 0$. First note that the solution operator $\mathcal{T}_\beta = \{T_\beta(t)\colon t \geqslant 0\}$ associated with (8.6) has a global attractor

$$\mathcal{A}_\beta = \{\xi(0) \mid \xi\colon \mathbb{R} \to H_0^1(0,\pi) \text{ is a bounded global solution of } \mathcal{T}_\beta\},$$

and also it is gradient. In fact the function $V\colon H_0^1(0,\pi) \to \mathbb{R}$ defined by

$$V(\phi) = \frac{1}{2}\int_0^\pi \phi'(x)^2 dx - \frac{\lambda}{2}\int_0^\pi \phi(x)^2 dx + \frac{\beta}{4}\int_0^\pi \phi(x)^4 dx$$

is continuous and if $u_0 \in H_0^1(0,\pi)$ and $u(t,x) = (T_\beta(t)u_0)(x)$, we have

$$\frac{d}{dt}V(T_\beta(t)u_0) = -\int_0^\pi u_t(t,x)^2 dx.$$

Hence $[0,\infty) \ni t \mapsto V(T_\beta(t)\phi) \in \mathbb{R}$ is nonincreasing, and if $V(T_\beta(t)\phi) = V(\phi)$ for all $t \geqslant 0$, then ϕ is an equilibrium solution for (8.6), that is, an element of the set $\mathcal{E}$ of solutions of the boundary value problem

$$\begin{cases} \phi'' + \lambda\phi - \beta\phi^3 = 0, \ x \in (0,\pi), \\ \phi(0) = \phi(\pi) = 0. \end{cases} \tag{8.7}$$

Thus V is a Lyapunov function for (8.6).

The equilibria, the most elementary bounded global solutions of $\mathcal{T}_\beta$, are clearly in the global attractor, that is, $\mathcal{E} \subset \mathcal{A}_\beta$. Given $\phi \in \mathcal{E}$, the unstable set of ϕ is given by

$$W^u(\phi) = \{u \in H_0^1(0,\pi)\colon \text{ there exists a global solution } \xi \text{ of } \mathcal{T}_\beta \text{ such that } \xi(0) = u \text{ and } \lim_{t\to-\infty} \xi(t) = \phi\}.$$

One can easily see that $W^u(\phi) \subset \mathcal{A}_\beta$ for all $\phi \in \mathcal{E}$.

It is well known that a gradient semigroup $\mathcal{T}_\beta$ with a global attractor $\mathcal{A}_\beta$ and a finite set $\mathcal{E}$ of equilibria has the following property: given a bounded global solution ξ of $\mathcal{T}_\beta$, there exist $\phi^-, \phi^+ \in \mathcal{E}$ such that

$$\phi^- \overset{t\to-\infty}{\longleftarrow} \xi(t) \overset{t\to+\infty}{\longrightarrow} \phi^+$$

and $V(\phi^-) > V(\phi^+)$. In particular, we obtain $\mathcal{A}_\beta = \bigcup_{\phi\in\mathcal{E}} W^u(\phi)$.

If $\lambda \in (0,1]$, the only solution of (8.7) is $\phi_{0,\beta} \equiv 0$, if $\lambda \in (1,4]$, there will be exactly three elements $\phi_{0,\beta} \equiv 0$, $\phi_{1,\beta}^+$ (positive in $(0,\pi)$) and $\phi_{1,\beta}^-$ (negative in $(0,\pi)$) in $\mathcal{E}$. For $\lambda \in (4,9]$ we will have two additional solutions $\phi_{2,\beta}^\pm$ which change sign at $x = \pi/2$. If $\lambda \in (N^2, (N+1)^2]$, the set $\mathcal{E}$ of solutions of (8.7) has exactly $2N+1$ elements (see [**88**]) given by

$$\phi_{0,\beta} \equiv 0 \quad \text{ and } \quad \phi_{j,\beta}^\pm \quad \text{ for } 1 \leqslant j \leqslant N.$$

The solutions $\phi_{j,\beta}^\pm$ of (8.7) bifurcate from $\phi_{0,\beta} \equiv 0$ at $\lambda = j^2$. $\phi_{j,\beta}^\pm$ has $j+1$ zeros in $[0,\pi]$ ($x_i = \frac{i\pi}{j}$, $0 \leqslant i \leqslant j$).

So, the set $\mathcal{E}$ will always be finite and the attractor $\mathcal{A}_\beta$ will always be $\mathcal{A}_\beta = \bigcup_{\phi\in\mathcal{E}} W^u(\phi)$. Furthermore, for any $u_0 \in H_0^1(0,\pi)$, $T_\beta(t)u_0 \overset{t\to\infty}{\longrightarrow} \phi$ for some $\phi \in \mathcal{E}$.

REMARK 8.20. In the non-autonomous case this analysis is no longer available and we must seek different ways to find the solutions that would play the role of equilibria. These solutions indeed exist but they are obtained in a completely different manner (see [**77**]).

When $\lambda \notin \{j^2\colon j \in \mathbb{N}\}$, all equilibria in $\mathcal{E}$ are hyperbolic (see [**88**]). The solutions $u\colon [0,\pi] \times [0,\infty) \to \mathbb{R}$ of (8.5), or a linear version of it, have yet another striking property called *lap-number* (see [**2, 170**]). This property is roughly described as follows: if $t_1 > t_2 > 0$, the number of times $u(\cdot,t_1)$ vanishes in the interval $[0,\pi]$ is at most the number of times $u(\cdot,t_2)$ vanishes in the interval $[0,\pi]$. As a consequence of this property we have the following:

(a) If $\xi\colon \mathbb{R} \to H_0^1(0,\pi)$ is a global solution of $\{S_\beta(t)\colon t \geqslant 0\}$ and $\phi^\pm \in \mathcal{E}$ are such that $\xi(t) \overset{t\to\pm\infty}{\longrightarrow} \phi^\pm$, then $W^u(\phi^-) \pitchfork W^s(\phi^+)$ and $\mathcal{T}_\beta$ is a Morse-Smale semigroup ([**143**],[**1**]).

(b) The connections between equilibria are given by the diagram in Figure 8.1 (see [**118, 143**]).

At this point it is natural to ask what happens in the non-autonomous case of (8.5). As we will see in Chapter 10, when $\beta(t)$ is a small non-autonomous perturbation of a constant we show that this structure remains the same. In [**33**] the authors show that much of this structure is still preserved even when $\beta(t)$ is not a small perturbation of a constant.

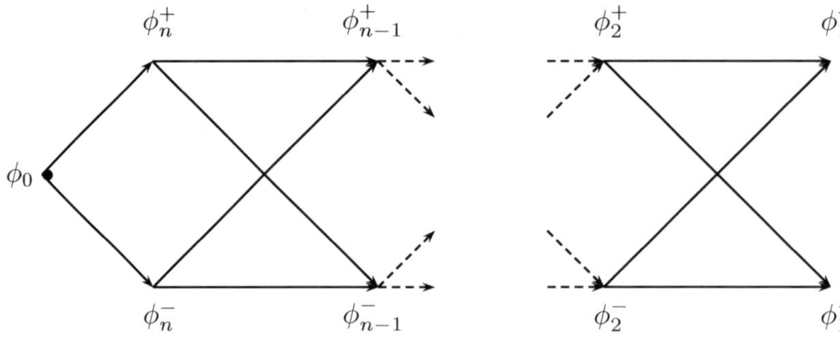

FIGURE 8.1. Diagram of connections.

5.2. The non-autonomous case. In $X = C([0,\infty),[\beta_1,\beta_2])$ consider the metric ρ of the uniform convergence in compact subsets of $[0,\infty)$ and let

$$\mathcal{H} = \text{closure (with respect to } \rho) \text{ of } \{\beta(s+\cdot) \in X : s \in \mathbb{R}^+\}.$$

For each $t \geqslant 0$, define the shift operator $\Theta_t : \mathcal{H} \to \mathcal{H}$ by

$$\Theta_t\gamma = \gamma(t+\cdot) \quad \text{for each } \gamma \in \mathcal{H}.$$

REMARK 8.21. Of course if β is independent of time, then $\mathcal{H} = \{\beta\}$ and $\Theta_t\beta = \beta$ for all $t \geqslant 0$. In this case (8.5) is autonomous.

For each $\gamma \in \mathcal{H}$ and $u_0 \in H_0^1(0,\pi)$, consider the initial value problem

$$\left\{\begin{array}{ll} u_t = u_{xx} + \lambda u - \gamma(t)u^3, & t > 0,\ x \in (0,\pi), \\ u(0,t) = u(\pi,t) = 0, & t > 0, \\ u(x,0) = u_0(x), & u_0 \in H_0^1(0,\pi), \end{array}\right. \tag{8.8}$$

and let $\mathbb{R}^+ \ni t \mapsto \mathcal{K}(t,\gamma)u_0 \in H_0^1(0,\pi)$ be its only solution. This defines the cocycle $\mathcal{K}$ associated to (8.5). If (8.5) is autonomous, then $\{\mathcal{K}(t,\beta) : t \geqslant 0\}$ is the semigroup $\mathcal{T}_\beta$.

The skew-product semigroup Π in $\mathbb{X} = H_0^1(0,\pi) \times \mathcal{H}$ (with the product metric) is given by

$$\Pi(t)(u,\gamma) = (\mathcal{K}(t,\gamma)u, \Theta_t\gamma) \quad \text{for all } t \geqslant 0,\ \gamma \in \mathcal{H} \text{ and } u \in H_0^1(0,\pi),$$

and again, if (8.5) is autonomous, we have $\mathbb{X} = H_0^1(0,\pi) \times \{\beta\}$ and $\Pi(t)(u,\beta) = (\mathcal{K}(t,\beta)u,\beta) = (\mathcal{T}_\beta(t)u,\beta)$ for all $t \geqslant 0$.

Note that if $\eta : \mathbb{R} \to \mathcal{H}$ is a global solution of Θ through γ, we define the evolution process associated to

$$\left\{\begin{array}{ll} u_t = u_{xx} + \lambda u - \eta(s)(t-s)u^3, & t > s,\ 0 < x < \pi, \\ u(0,t) = u(\pi,t) = 0, & t > s, \\ u(x,s) = u_0(x), & u_0 \in H_0^1(0,\pi), \end{array}\right. \tag{8.9}$$

by $S_\eta(t,s) = \mathcal{K}(t-s,\eta(s))$ for $t \geqslant s$. It can be proved (see [**33**]) that:

THEOREM 8.22. *The skew-product semigroup* Π *associated to* (8.5) *has a global attractor* $\mathbb{A}$ *(hence* Θ *has a global attractor* $\mathscr{S}$ *in* $\mathcal{H}$*). If* $\eta : \mathbb{R} \to \mathscr{S}$ *is a global solution for* Θ*, the evolution process*

$$S_\eta(t,s)u = \mathcal{K}(t-s,\eta(s))u \quad \textit{for } u \in H_0^1(0,\pi) \textit{ and } t \geqslant s$$

has a pullback attractor $\hat{A}_\eta$ *and*

$$\mathbb{A} = \bigcup_{\eta} \bigcup_{t \in \mathbb{R}} A_\eta(t) \times \{\eta(t)\},$$

where the first union is taken over all global solutions $\eta \colon \mathbb{R} \to \mathscr{S}$ *of* Θ.

The authors in [**33**] show that the gradient structure present in the skew-product attractor associated to (8.5) is analogous to the structure present in the autonomous attractor. In fact, it is shown that the skew-product semigroup associated to (8.5) is dynamically gradient.

Next we identify the isolated invariant sets associated to this gradient structure. For each global solution $\eta \colon \mathbb{R} \to \mathscr{S}$ of Θ for $N^2 < \lambda < (N+1)^2$, it is proved in [**77**], using the ideas of [**191**], that there are $2N$ non-autonomous equilibria $\xi^{\pm}_{j,\eta}$, $1 \leqslant j \leqslant N$, for (8.9), where the index η indicates the dependence on η.

It is also proved that for $u_0 \in H^1_0(0,\pi)$ and $s \in \mathbb{R}$, the solution of (8.5) starting at u_0 at time s converges, as $t \to \infty$, either to zero or to $\{\xi^{\pm}_{j,\eta}(0) \mid \eta \colon \mathbb{R} \to \mathscr{S}$ solution of $\Theta\}$, for some $1 \leqslant j \leqslant N$. It is proved in [**33**], using the results of [**95**], that any bounded global solution of (8.9) converges to zero or to $\{\xi^{\pm}_{j,\eta}(0) \mid \eta \colon \mathbb{R} \to \mathscr{S}$ solution of $\Theta\}$ as $t \to -\infty$, for some $1 \leqslant j \leqslant N$. Furthermore there are solutions connecting the zero equilibrium solution to all non-autonomous equilibria.

With this, the isolated invariant sets of the skew-product semigroup associated to (8.5) are given by $\mathbb{E}_0 = \{(0,\eta(0)) \colon \eta$ a global solution of Θ in $\mathscr{S}\}$ and

$$\mathbb{E}^{\pm}_j = \{(\xi^{\pm}_{j,\eta}(0), \eta(0)) \colon \eta \text{ a global solution of } \Theta \text{ in } \mathscr{S}\} \quad \text{for } 1 \leqslant j \leqslant N.$$

For each $1 \leqslant j \leqslant N$, $\xi^{\pm}_{j,\eta}(t)$ vanishes exactly at $i\pi/j$ for $0 \leqslant i \leqslant j$, $\xi^{+}_{j,\eta}(t)(x) = \xi^{+}_{j,\eta}(t)(\frac{\pi}{j} - x) > 0$ and $\xi^{-}_{j,\eta}(t)(x) = \xi^{-}_{j,\eta}(t)(\frac{\pi}{j} - x) < 0$ in $(0, \frac{\pi}{j})$. If $j \geqslant 2$, $\xi^{\pm}_{j,\eta}(t)(x) = -\xi^{\pm}_{j,\eta}(t)(x - \frac{\pi}{j})$ for $x \in [\frac{\pi}{j}, 1]$.

Also there is no homoclinic solution at zero and no homoclinic structures, as a consequence of the lap-number property of the solutions (see [**2, 170**]). In fact it is proved in [**77**] that the graph of $\xi^{\pm}_{j,\gamma}(t)$ lies between the graphs of the equilibrium solutions $\phi^{\pm}_{j,\beta_i}$ of $\mathcal{T}_{\beta_i}$, for $i = 1, 2$.

Therefore, we conclude that Π is dynamically gradient relative to the isolated invariant sets $\{\mathbb{E}_0, \mathbb{E}^{\pm}_j,\ 1 \leqslant j \leqslant N\}$.

6. Applications to ordinary differential equations

Before presenting some applications to ordinary differential equations we will look more closely at the positive hull of a function in $\mathbb{R}^+ \times \mathbb{R}^n$, to see how difficult it can be to treat such an object. Even for simple functions, the positive hull (or the hull) can be absolutely nontrivial and very complicated to describe.

6.1. Hull of functions and Morse decomposition. Given two continuous functions $r, g \colon \mathbb{R}^n \to \mathbb{R}^n$, we define $f \colon \mathbb{R}^+ \times \mathbb{R}^n \to \mathbb{R}^n$ by

$$f(t,x) = h(t)r(x) + (1 - h(t))g(x),$$

where $h \colon \mathbb{R}^+ \to \mathbb{R}^+$ is defined as follows: let $\{a_n\}$ be a sequence defined by $a_1 = 0$ and $a_{n+1} = a_n + 4n$ for each $n \geqslant 1$ and define $b_n = a_n + n$, $c_n = a_n + 2n$ and $d_n = a_n + 3n$ for all $n \geqslant 1$. Hence

- $a_n < b_n < c_n < d_n < a_{n+1}$ for all $n \in \mathbb{N}$;
- $b_n - a_n = c_n - b_n = d_n - c_n = a_{n+1} - d_n = n$.

Now, we define the function h in such a way that it is smooth, $0 \leqslant h \leqslant 1$, $h(t) = 1$ if $t \in [a_n, b_n]$ for some $n \in \mathbb{N}$ and $h(t) = 0$ if $t \in [c_n, d_n]$ for some $n \in \mathbb{N}$. Define $\Theta(t)f(\cdot,\cdot) = f(t+\cdot,\cdot)$ for each $t \geqslant 0$, and let Σ^+ be the positive hull of f in $\mathcal{C} = C_b(\mathbb{R}^+ \times \mathbb{R}^n, \mathbb{R}^n)$, with the metric $d_{\mathcal{C}}$ defined in (7.10). Then Θ is a semigroup in Σ^+.

First, we define the sequence $t_n^* = a_n + n/2$. Thus

$$\Theta(t_n^*)f(s,x) = f(s+t_n^*, x) = r(x),$$

if $s \in [0, n/2]$. Hence $\Theta(t_n^*)f \to r$ as $n \to \infty$ in the metric $d_{\mathcal{C}}$, which shows that $r \in \omega(f)$, the omega-limit set of f. Choosing $\tau_n^* = c_n + n/2$ we can see that $\Theta(\tau_n^*)f \to g$ as $n \to \infty$, which shows that $g \in \omega(f)$ in a similar way. Choosing $\gamma_n^* = b_n + n/2$ we see that

$$\Theta(\gamma_n^*)f(t,x) = f(t+\gamma_n^*, x).$$

Using the Arzelà-Ascoli Theorem, together with a diagonalization process, we can see that there exists a function $\xi^* \colon \mathbb{R} \times \mathbb{R}^n \to \mathbb{R}^n$ and a subsequence $\{\gamma_{n_k}^*\}$ of $\{\gamma_n^*\}$ such that $\Theta(\gamma_{n_k}^*)f \to \xi^*$ in the metric $d_{\mathcal{C}}$. Also, since $\Theta(\gamma_n^*)f(t,x) = f(t+\gamma_n^*, x)$, we can see that if $t \in [-n/2, n/2]$, then $\Theta(\gamma_n^*)f(t,x) = r(x)$ for all $x \in \mathbb{R}^n$. Thus $\xi(t) \overset{t\to\infty}{\longrightarrow} r$ where $\xi(t) = \xi^*(t,\cdot)$ and, analogously, $\xi(t) \overset{t\to-\infty}{\longrightarrow} g$. It is clear that ξ is a global solution of Θ, and therefore ξ is a connection between r and g.

With the same reasoning, we can construct a connection ψ between g and r. In this way we can have some description of what happens in Σ^+. We can see that if we want a Morse decomposition for the global attractor of Θ, then both r and g necessarily have to be in the same isolated invariant.

Consider now the following non-autonomous ordinary differential equation:

$$\begin{cases} \dot{x} = f(t,x), & t > 0, \\ x(0) = x_0 \in \mathbb{R}^n. \end{cases} \tag{8.10}$$

We have already shown that $g, r \in \Sigma^+$ and also that there is a connection between r and g, and a connection between g and r. Thus we cannot have a Morse decomposition in Σ^+ which separates r and g, since each one is connected to the other.

Nevertheless, we could choose one of these global solutions that connects g to r, for example ξ, and restrict ourselves to $\Xi = \{\text{closure of } \xi(\mathbb{R}) \text{ in the metric } d_{\mathcal{C}}\} \subset \Sigma^+$, which has a nontrivial Morse decomposition $D_1 = \{r\}$ and $D_2 = \{g\}$, and thus generates a Morse decomposition for the associated skew product semiflow and non-autonomous dynamical system related to (8.10), when the semigroup Θ is restricted to the invariant set Ξ.

6.2. Partially coupled ordinary differential systems. Consider the partially coupled system of autonomous differential equations

$$\begin{cases} \dot{v} = f(u,v), & t > 0, \\ \dot{u} = g(u), & t > 0, \\ u(0) = u_0 \in \mathbb{R}^n, \\ v(0) = v_0 \in \mathbb{R}^n, \end{cases} \tag{8.11}$$

where the u component is decoupled, and assume (8.11) generates a semigroup. Here, the u-component can represent an independent system that drives the v-component of the system in the sense that

$$\dot{v} = f(u(t), v)$$

for any given solution $u(t)$ of $\dot{u} = g(u)$. Assume that the system $\dot{u} = g(u)$ generates a semigroup $\Theta = \{\Theta(t) \colon t \geqslant 0\}$ in $\mathbb{R}^n$, that is, $\Theta(t)u_0 = u(t, u_0)$, where $u(\cdot, u_0)$ is the unique solution for $t > 0$ of the problem

$$\begin{cases} \dot{u} = g(u), & t > 0, \\ u(0) = u_0. \end{cases}$$

Assume also that Θ has a global attractor Ξ with a Morse decomposition $\{\Gamma_1, \cdots, \Gamma_n\}$, that the property of backwards uniqueness of Θ in Ξ holds (that is, $\Theta(t)$ is injective in Ξ for each $t \in \mathbb{R}$, and hence Θ is a group when restricted to Ξ) and also that the associated skew-product semigroup Π has a global attractor $\mathbb{A}$. Then for every pair of points $(u_0, v_0) \in \mathbb{R}^n \times \mathbb{R}^n$, the solution $v(t, v_0, u_0)$ of the problem

$$\begin{cases} \dot{v}(t) = f(\Theta(t)u_0, v(t)), & t > 0, \\ v(0) = v_0, \end{cases}$$

satisfies $v(t, v_0, u_0) \overset{t\to\infty}{\longrightarrow} \mathcal{A}_{\Gamma_i}$, for some $i = 1, \cdots, n$, where i is such that $\Theta(t)u_0 \overset{t\to\infty}{\longrightarrow} \Gamma_i$, and $\mathcal{A}_{\Gamma_i} = \pi_X(\mathbb{A} \cap \mathbb{R}^n \times \Gamma_i)$.

6.3. Asymptotically autonomous differential equations. Finally, we will consider a function $f \in \mathcal{C} = C_b(\mathbb{R} \times \mathbb{R}^n, \mathbb{R}^n)$, defined for all $t \in \mathbb{R}$, and the non-autonomous ordinary differential equation

$$\begin{cases} \dot{x} = f(t, x), & t \in \mathbb{R}, \\ x(\tau) = x_0 \in \mathbb{R}^n, \end{cases} \tag{8.12}$$

with $\tau \in \mathbb{R}$, and assume that there are functions $f_1, f_2 \colon \mathbb{R}^n \to \mathbb{R}^n$ such that

$$\sup_{x\in\mathbb{R}^n} \|f(t,x) - f_1(x)\| \overset{t\to\infty}{\longrightarrow} 0 \quad \text{and} \quad \sup_{x\in\mathbb{R}^n} \|f(t,x) - f_2(x)\| \overset{t\to-\infty}{\longrightarrow} 0. \tag{8.13}$$

Now considering $\mathcal{C} = C_b(\mathbb{R} \times \mathbb{R}^n, \mathbb{R}^n)$, with the metric $d_{\mathcal{C}}$ as in (7.10), we can see that

$$d_{\mathcal{C}}(\Theta(t)f, f_2) \overset{t\to-\infty}{\longrightarrow} 0 \quad \text{and} \quad d_{\mathcal{C}}(\Theta(t)f, f_1) \overset{t\to\infty}{\longrightarrow} 0.$$

Thus the hull Σ of f is given by $\Sigma = \{\Theta(t)f\}_{t\geqslant 0} \cup \{f_1, f_2\}$ and this set has a Morse decomposition $\{D_1, D_2\}$ given by $D_1 = \{f_1\}$ and $D_2 = \{f_2\}$.

If the skew-product semigroup Π associated to (8.12) has a global attractor $\mathbb{A}$, then $\{\mathbb{A}_{D_1}, \mathbb{A}_{D_2}\}$ is a Morse decomposition for $\mathbb{A}$, where in this case if $A_1 = \{x \in \mathbb{R}^n \colon (x, f_1) \in \mathbb{A}\}$ and $A_2 = \{x \in \mathbb{R}^n \colon (x, f_2) \in \mathbb{A}\}$, then $\mathbb{A}_{D_1} = A_1 \times \{f_1\}$ and $\mathbb{A}_{D_2} = A_2 \times \{f_2\}$. Therefore the solutions of (8.12) converge to A_1 as $t \to \infty$ and to A_2 as $t \to -\infty$. Moreover, if ξ is a bounded global solution of (8.12) we know that A_k is the global attractor of the problem $\dot{x} = f_k(x)$, $k = 1, 2$, and if A_k has a Morse decomposition $\{E_{k,1}, \cdots, E_{k,p_k}\}$, Corollary 8.19 guarantees that there exist $i = 1, \cdots, p_1$ and $j = 1, \cdots, p_2$ such that $\xi(t)$ converges to $E_{i,1}$ as $t \to \infty$ and $E_{j,2}$ as $t \to -\infty$.

For instance, let $f \colon \mathbb{R} \times \mathbb{R}^2 \to \mathbb{R}^2$ and consider the planar ordinary differential equation:

$$(\dot{x}, \dot{y}) = f(t, x, y), \quad t \in \mathbb{R}, \tag{8.14}$$

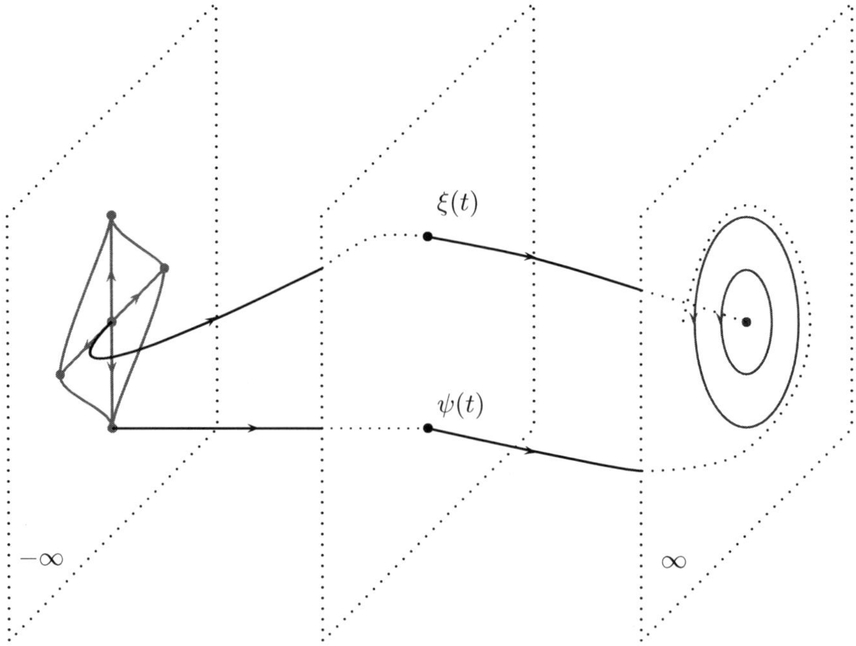

FIGURE 8.2. Asymptotic behavior of solutions.

and assume that (8.13) is satisfied with $f_1, f_2 \colon \mathbb{R}^2 \to \mathbb{R}^2$ given by

(i) $f_1(x,y) = (h(x), g(x,y))$, where $h(x) = x - x^3$ and $g(x,y) = (1-x^2)y - y^3$;
(ii) f_2 is given in polar coordinates by

$$f_2(r\cos\theta, r\sin\theta) = k(r,\theta) = (-r(r-1)(r-2), 1).$$

We have a Morse decomposition $\{\mathbb{A}_{\Gamma_0}, \mathbb{A}_{\Gamma_1}, \mathbb{A}_{\Gamma_2}\}$ for the global attractor $\mathbb{A}$ of the planar system

$$\begin{cases} \dot{x} = h(x), \\ \dot{y} = g(x,y), \end{cases}$$

given a Morse decomposition $\{\Gamma_0, \Gamma_1, \Gamma_2\}$ for the global attractor $\Xi = [-1,1] \subset \mathbb{R}$ of the equation $\dot{x} = h(x)$, where $\Gamma_0 = \{-1\}$, $\Gamma_1 = \{1\}$, $\Gamma_2 = \{0\}$. Indeed, it is not hard to see that the Morse decomposition in this case is given by $\mathbb{A}_{\Gamma_0} = \{(-1,0)\}$, $\mathbb{A}_{\Gamma_1} = \{(1,0)\}$ and $\mathbb{A}_{\Gamma_2} = \{(0,y) \colon y \in [-1,1]\}$. Also we know that the planar ODE $(\dot{x}, \dot{y}) = f_2(x,y)$ generates a dynamically gradient semigroup, with disjoint family of isolated invariants given by $\Upsilon_0 = \{0\}$, $\Upsilon_1 = \{(1,\theta) \colon \theta \in [0,2\pi]\}$ and $\Upsilon_2 = \{(2,\theta) \colon \theta \in [0,2\pi]\}$.

Thus we know that each bounded global solution $\xi \colon \mathbb{R} \to \mathbb{R}^2$ of (8.14) satisfies $\xi(t) \to \mathbb{A}_{\Gamma_i}$ for some $i = 1, \cdots, n$ as $t \to -\infty$ and $\xi(t) \to \Upsilon_j$ for some $j = 0, 1, 2$ as $t \to \infty$, which is described in Figure 8.2.

Notes. In this chapter we have described in detail the internal structure of pullback attractors of evolution processes, uniform attractors for non-autonomous dynamical systems and global attractors for skew-product semigroups. The possibility of this characterization came from a detailed study of the pullback attractors in a series of papers ([**28, 31, 40, 42, 73, 74, 76–78**]). The analysis on the relationship among the different concepts of attractors (see for instance [**79**]) allowed a study of the internal dynamics of pullback and uniform attractors, and global attractors for the skew-product semigroups ([**29, 30, 153, 185**]). This incursion into the inner description of this set of attractors led in the previous chapter, to prove upper and lower semicontinuity and, in this chapter, their topological structural stability, that is, the robustness of the dynamically gradient characterization of the attractors when we perform a non-autonomous perturbation to a dynamically gradient semigroup.

CHAPTER 9

Neighborhood of a Global Hyperbolic Solution

This chapter is devoted to the extension of the analysis done in Chapter 4 to the case when the critical element is a *bounded hyperbolic global solution*. The analysis is done in analogy to that of [**78**], and complemented with the study of the smoothness of the unstable and stable manifolds as well as their continuous dependence in the C^1 topology. Our approach is an extension of the results in [**141**] to the non-autonomous framework.

1. Introduction

We work in the *discrete non-autonomous framework*, in which we prove the results of this chapter. We recall the definition of a *discrete evolution process*, and from this point forward, unless stated otherwise, n and m will denote *integer numbers* and $\mathbb{Z}_m = \{n \in \mathbb{Z} \colon n \geqslant m\}$. As before, X denotes a Banach space over $\mathbb{K}$, where $\mathbb{K}$ is either $\mathbb{R}$ or $\mathbb{C}$, $C(X)$ is the set of continuous functions from X to X, $C^1(X)$ is the set of differentiable functions from X to X, and $\mathcal{L}(X)$ is the set of bounded linear operators from X to X.

DEFINITION 9.1. A family $\mathcal{S} = \{S_{n,m} \colon n \geqslant m\} \subset C(X)$ is called a **discrete evolution process** if

(i) $S_{n,n}x = x$ for all $n \in \mathbb{Z}$ and $x \in X$;
(ii) $S_{n,k}S_{k,m} = S_{n,m}$ for all $n \geqslant k \geqslant m \in \mathbb{Z}$.

If $\mathcal{S} \subset C^1(X)$ we say that it is a **C^1-discrete evolution process**. If $\mathcal{S} \subset \mathcal{L}(X)$ we say that it is a **linear discrete evolution process**.

A discrete evolution process determines a sequence $\{S_n\}_{n\in\mathbb{Z}}$ in $C(X)$, defined by $S_n = S_{n+1,n}$. Conversely, given a sequence of operators $\{S_n\}_{n\in\mathbb{Z}}$ in $C(X)$, we can define a discrete evolution process $\mathcal{S} = \{S_{n,m} \colon n \geqslant m\}$ in the following way:

(a) $S_{m,m}x = x$ for all $m \in \mathbb{Z}$ and $x \in X$;
(b) $S_{n,m} = S_{n-1} \cdots S_m$, for all $n > m \in \mathbb{Z}$.

Clearly, if $S_n = T$ for all $n \in \mathbb{Z}$, the discrete evolution process associated to the sequence $\{S_n\}_{n\in\mathbb{Z}}$ is $\{T^{n-m} \colon n \geqslant m\}$ which coincides with the *discrete semigroup* $\mathcal{T} = \{T^n \colon n \in \mathbb{N}\}$. In this sense, the class of the discrete evolution process is wider than the class of discrete semigroups.

For a given discrete evolution process $\mathcal{S}$ and a point $x \in X$, we define, for each $m \in \mathbb{Z}$, the function $\mathbb{Z}_m \ni n \mapsto S_{n,m}x \in X$, which is the **solution** through x of the discrete initial value problem:

$$
\begin{cases} x_{n+1} = S_n x_n, & n \geqslant m, \\ x_m = x, \end{cases} \tag{9.1}
$$

where $S_n = S_{n+1,n}$, $n \in \mathbb{Z}$. Recall that a **global solution** of $\{S_{n,m} \colon n \geqslant m\}$ is a sequence $\{x_n\}_{n\in\mathbb{Z}}$ such that $x_n = S_{n,m}x_m$ whenever $n \geqslant m$.

In the study of asymptotic properties of solutions, it is very useful to have an associated *variations of constants formula*. Next we derive such a formula for a discrete problem.

LEMMA 9.2 (Discrete variations of constants). *Suppose that $\{S_n\}_{n\in\mathbb{Z}}$ is a sequence in $\mathcal{L}(X)$ and that $\{y_n\}_{n\in\mathbb{Z}_m}$ is a sequence in X. If $x_{n+1} = S_n x_n + y_n$ for $n \geqslant m \in \mathbb{Z}$ we have*

$$x_n = S_{n,m}x_m + \sum_{k=m}^{n-1} S_{n,k+1}y_k \qquad \textit{for } n > m.$$

PROOF. Note that the result holds for $m = n-1$. If the result holds for $m = r < n$, that is,

$$x_n = S_{n,r}x_r + \sum_{k=r}^{n-1} S_{n,k+1}y_k,$$

let us show that it also holds for $m = r-1$. In fact, since $x_r = S_{r-1}x_{r-1} + y_{r-1}$, we have

$$x_n = S_{n,r}S_{r-1}x_{r-1} + S_{n,r}y_{r-1} + \sum_{k=r}^{n-1} S_{n,k+1}y_k = S_{n,r-1}x_{r-1} + \sum_{k=r-1}^{n-1} S_{n,k+1}y_k.$$

□

2. Discrete exponential dichotomy

The notion of hyperbolicity of an equilibrium for an autonomous problem is extended to the non-autonomous framework by the notion of *exponential dichotomy*, which is defined next.

DEFINITION 9.3. Consider a linear discrete evolution process $\mathscr{L} = \{L_{n,m} \colon n \geqslant m\} \subset \mathcal{L}(X)$ associated to a sequence $\{L_n\}_{n\in\mathbb{Z}} \subset \mathcal{L}(X)$. We say that $\mathcal{L}$ has **discrete exponential dichotomy**, with constant $M \geqslant 1$ and exponent $\omega > 0$, if there exists a sequence $\{Q_n\}_{n\in\mathbb{Z}} \subset \mathcal{L}(X)$ of projections, that is, $Q_n^2 = Q_n$ for each $n \in \mathbb{Z}$, such that

(i) $L_n Q_n = Q_{n+1}L_n$, for all $n \in \mathbb{Z}$;
(ii) $L_{n,m} \colon R(Q_m) \to R(Q_n)$ is an isomorphism with inverse $L_{m,n} \colon R(Q_n) \to R(Q_m)$, for all $n > m \in \mathbb{Z}$, where $R(Q)$ is the range of a projection $Q \in \mathcal{L}(X)$;
(iii) the following estimates hold:

$$\begin{aligned} \|L_{n,m}(I - Q_m)\|_{\mathcal{L}(X)} &\leqslant Me^{-\omega(n-m)}, \quad n \geqslant m, \\ \|L_{n,m}Q_m\|_{\mathcal{L}(X)} &\leqslant Me^{\omega(n-m)}, \quad n < m, \end{aligned} \tag{9.2}$$

where I denotes the identity operator in X.

We usually say that the sequence $\{L_n\}_{n\in\mathbb{Z}}$ has discrete exponential dichotomy instead of saying that the linear discrete evolution process $\mathcal{L}$ associated to $\{L_n\}_{n\in\mathbb{Z}}$ has discrete exponential dichotomy. In the autonomous framework, all the operators L_n are equal to a fixed operator L and the projections Q_n are all equal to a fixed projection Q which is given by the resolvent of L and the spectral set of L outside the unit circle. In the non-autonomous framework, the lack of tools to identify Q_n is a big handicap for the theory. Identifying many more examples for which one can prove that a bounded global solution is hyperbolic (the linearization around it has exponential dichotomy) is a challenge that must be tackled.

THEOREM 9.4. *If $\{L_n\}_{n\in\mathbb{Z}}$ has discrete exponential dichotomy, with constant $M \geqslant 1$ and exponent $\omega > 0$, and if $\{y_n\}_{n\in\mathbb{Z}}$ is a bounded sequence in X, there exists a unique bounded global solution $\{x_n\}_{n\in\mathbb{Z}}$ of*

$$x_{n+1} = L_n x_n + y_n, \tag{9.3}$$

and this solution is given by

$$x_n = \sum_{k=-\infty}^{\infty} G_{n,k+1} y_k, \quad n \in \mathbb{Z}, \tag{9.4}$$

where

$$G_{n,m} = \begin{cases} L_{n,m}(I - Q_m), & n \geqslant m, \\ -L_{n,m}Q_m, & n < m. \end{cases}$$

PROOF. To see that a global bounded solution of (9.3) is given by (9.4), note that, from Lemma 9.2, we have

$$x_n = L_{n,m}x_m + \sum_{k=m}^{n-1} L_{n,k+1}y_k \quad \text{for } n > m.$$

From this it follows that

$$(I - Q_n)x_n = L_{n,m}(I - Q_m)x_m + \sum_{k=m}^{n-1} L_{n,k+1}(I - Q_{k+1})y_k \quad \text{for } n > m.$$

From item (iii) of Definition 9.3, we obtain

$$\|L_{n,m}(I - Q_m)x_m\|_X \leqslant Me^{-\omega(n-m)}\|x_m\|_X \quad \text{for } n \geqslant m,$$

and, since $\{x_n\}_{n\in\mathbb{Z}}$ is bounded, $\|L_{n,m}(I - Q_m)x_m\|_X \to 0$ as $m \to -\infty$ and the series

$$\sum_{k=-\infty}^{n-1} L_{n,k+1}(I - Q_{k+1})y_k$$

converges. Thus, we have

$$(I - Q_n)x_n = \sum_{k=-\infty}^{n-1} L_{n,k+1}(I - Q_{k+1})y_k.$$

On the other hand, $Q_r x_r = L_{r,n}Q_n x_n + \sum_{k=n}^{r-1} L_{r,k+1}Q_{k+1}y_k$ for $r > n$, and therefore

$$L_{n,r}Q_r x_r = Q_n x_n + \sum_{k=n}^{r-1} L_{n,k+1}Q_{k+1}y_k \quad \text{for } r > n.$$

From item (iii) of Definition 9.3 we obtain

$$\|L_{n,r}Q_r x_r\|_X \leqslant Me^{-\omega(r-n)}\|x_r\|_X,$$

and from the fact that $\{x_n\}_{n\in\mathbb{Z}}$ is bounded, we have $\|L_{n,r}Q_r x_r\|_X \to 0$ as $r \to \infty$ and the series

$$\sum_{k=n}^{\infty} L_{n,k+1}Q_{k+1}y_k$$

converges. It follows that

$$Q_n x_n = -\sum_{k=n}^{\infty} L_{n,k+1}Q_{k+1}y_k,$$

and that

$$x_n = Q_n x_n + (I - Q_n)x_n = \sum_{-\infty}^{\infty} G_{n,k+1}y_k.$$

The existence of a global bounded solution follows by verifying that $\{x_n\}_{n\in\mathbb{Z}}$, given by (9.4), is bounded and satisfies (9.3). To see that it is bounded note that

$$\begin{aligned}
\|x_n\|_X &\leqslant \sum_{-\infty}^{n-1} \|L_{n,k+1}(I - Q_{k+1})y_k\|_X + \sum_{n}^{\infty} \|L_{n,k+1}Q_{k+1}y_k\|_X \\
&\leqslant M\left(\sum_{-\infty}^{n-1} e^{-(n-k-1)\omega} + \sum_{n}^{\infty} e^{-(k+1-n)\omega}\right) \sup_{n\in\mathbb{Z}} \|y_n\|_X \\
&= M\,\frac{1+e^{-\omega}}{1-e^{-\omega}} \sup_{n\in\mathbb{Z}} \|y_n\|_X.
\end{aligned}$$

To see that $\{x_n\}_{n\in\mathbb{Z}}$ is a solution of (9.3), note that

$$\begin{aligned}
x_{n+1} &= \sum_{k=-\infty}^{\infty} G_{n+1,k+1}y_k = \sum_{-\infty}^{n} L_{n+1,k+1}(I - Q_{k+1})y_k - \sum_{n+1}^{\infty} L_{n+1,k+1}Q_{k+1}y_k \\
&= \sum_{-\infty}^{n-1} L_{n+1,k+1}(I - Q_{k+1})y_k - \sum_{n}^{\infty} L_{n+1,k+1}Q_{k+1}y_k + y_n \\
&= L_n\left(\sum_{-\infty}^{n-1} L_{n,k+1}(I - Q_{k+1})y_k - \sum_{n}^{\infty} L_{n,k+1}Q_{k+1}y_k\right) + y_n \\
&= L_n x_n + y_n.
\end{aligned}$$

This completes the proof. □

It follows directly from Theorem 9.4 that the projections associated with the exponential dichotomy are uniquely determined.

COROLLARY 9.5. *If $\{L_n\}_{n\in\mathbb{Z}} \subset \mathcal{L}(X)$ has discrete exponential dichotomy, then the associate projections $\{Q_n\}_{n\in\mathbb{Z}}$ are uniquely determined.*

PROOF. Assume that $\{Q_n^{(i)}\}_{n\in\mathbb{Z}}$, $i = 1, 2$, are two families of projections associated with the exponential dichotomy of $\{L_n\}_{n\in\mathbb{Z}}$. Let $\{f_n\}_{n\in\mathbb{Z}}$ with $f_n = 0$ for all $n \neq m-1$ and $f_{m-1} = x$. Thus, if $\{x_n\}_{n\in\mathbb{Z}}$ is the unique bounded solution of $x_{n+1} = L_n x_n + f_n$, from Theorem 9.4 we have

$$x_n = \sum_{-\infty}^{\infty} G^{(i)}_{n,k+1} f_k = G^{(i)}_{n,m} x, \quad i = 1, 2,$$

where

$$G^{(i)}_{n,m} = \begin{cases} L_{n,m}(I - Q^{(i)}_m), & n \geqslant m, \\ -L_{n,m}Q^{(i)}_m, & n < m, \end{cases}$$

for $i = 1, 2$. In particular, $x_m = G^{(i)}_{m,m}x = (I - Q^{(i)}_m)x$ and $Q^{(1)}_m x = Q^{(2)}_m x$ for each $x \in X$. □

The converse of Theorem 9.4 also holds (for a proof, see [**78**, Theorem 7.6]), that is:

THEOREM 9.6. *If $\{L_n\}_{n\in\mathbb{Z}} \subset \mathcal{L}(X)$, the following are equivalent:*
(i) *$\{L_n\}_{n\in\mathbb{Z}}$ has discrete exponential dichotomy;*
(ii) *for each bounded sequence $\{f_n\}_{n\in\mathbb{Z}}$ in X, there exists a unique bounded solution $\{x_n\}_{n\in\mathbb{Z}}$ of $x_{n+1} = L_n x_n + f_n$, $n \in \mathbb{Z}$.*

The following results play a crucial role in what follows. Their proofs can also be found in [**78**].

THEOREM 9.7. *Assume that $\{L^1_n\}_{n\in\mathbb{Z}} \subset \mathcal{L}(X)$ has discrete exponential dichotomy, with constant $M \geqslant 1$ and exponent $\omega > 0$. Given $M_1 > M$ and $0 < \omega_1 < \omega$, there exists $\epsilon > 0$ (depending only on M, M_1, ω and ω_1) such that, if $\{L^2_n\}_{n\in\mathbb{Z}} \subset \mathcal{L}(X)$ is such that $\sup_{n\in\mathbb{Z}} \|L^1_n - L^2_n\|_{\mathcal{L}(X)} < \epsilon$, then $\{L^2_n\}_{n\in\mathbb{Z}}$ has discrete exponential dichotomy, with constant M_1 and exponent ω_1.*

THEOREM 9.8. *Assume that $\{L^{(k)}_n\}_{n\in\mathbb{Z}}$ has discrete exponential dichotomy, with constant M and exponent $\omega > 0$, and associated projections $\{Q^{(k)}_n\}_{n\in\mathbb{Z}}$, $k = 1, 2$, respectively. If $\|L^{(1)}_n - L^{(2)}_n\|_{\mathcal{L}(X)} \leqslant \epsilon$ for all $n \in \mathbb{Z}$, then*

$$\|Q^{(1)}_n - Q^{(2)}_n\|_{\mathscr{L}(X)} \leqslant \frac{2M^2}{1 - e^{-\omega}}\,\epsilon.$$

DEFINITION 9.9. Let $\mathcal{S} = \{S_{n,m} \colon n \geqslant m\}$ be a C^1-discrete evolution process associated to a sequence $\{S_n\}_{n\in\mathbb{Z}}$. Let $\{y_n\}_{n\in\mathbb{Z}}$ be a global solution of $\mathcal{S}$, and L_n the derivative of S_n in y_n, that is, $L_n = S'_n(y_n)$, for each $n \in \mathbb{Z}$. We say that $\{y_n\}_{n\in\mathbb{Z}}$ is a **hyperbolic global solution** of $\mathcal{S}$ if $\{L_n\}_{n\in\mathbb{Z}}$ has discrete exponential dichotomy.

Considering a hyperbolic global solution $\{y_n\}_{n\in\mathbb{Z}}$ as a function $y\colon \mathbb{Z} \to X$ with $y(n) = y_n$ for each $n \in \mathbb{Z}$, we will also say that the global solution y of $\mathcal{S}$ is **hyperbolic**.

REMARK 9.10. Let $\mathcal{S}$ be a C^1-discrete evolution process associated to the sequence $\{S_n\}_{n\in\mathbb{Z}} \subset C^1(X)$. Consider the discrete initial value problem

$$\begin{cases} x_{n+1} = S_n x_n, \ n \geqslant m, \\ x_m = x. \end{cases}$$

Given a hyperbolic global solution y of $\mathcal{S}$, by writing $z_n = x_n - y_n$ for each $n \in \mathbb{Z}$, we have

$$\begin{cases} z_{n+1} = L_n z_n + S_n(z_n + y_n) - S_n y_n - L_n z_n, \ n \geqslant m, \\ z_m = x - y_m, \end{cases}$$

and define $T_n z = S_n(z + y_n) - S_n y_n - L_n z$ for all $z \in X$ and $n \geqslant m$. This form of the discrete initial value problem is more suitable for studying the neighborhood of the hyperbolic global solution y.

3. Unstable manifolds

3.1. Global unstable manifolds as graphs. We are now ready to study the behavior of the solutions of a discrete evolution process $\mathcal{S} = \{S_{n,m} : n \geqslant m\} \subset C(X)$ in the neighborhood of a hyperbolic global solution ζ of $\mathcal{S}$. More precisely, we will study the *unstable and stable manifolds* of ζ.

Assume that the discrete evolution process $\mathcal{S}$ is associated to the sequence $\{S_n\}_{n\in\mathbb{Z}} \subset C(X)$, and that each S_n is a perturbation of a bounded map $\tilde{L}_n \in \mathcal{L}(X)$ by a continuously differentiable map $T_n : X \to X$, that is, $S_n = \tilde{L}_n + T_n$, for each $n \in \mathbb{Z}$. Let $\tilde{\mathscr{L}} = \{\tilde{L}_{n,m} : n \geqslant m\}$ be the linear discrete evolution process associated to the sequence $\{\tilde{L}_n\}_{n\in\mathbb{Z}} \subset \mathcal{L}(X)$. Then, if

$$\begin{cases} y_{n+1} = S_n y_n = \tilde{L}_n y_n + T_n y_n, n \geqslant m, \\ y_m = x, \end{cases} \tag{9.5}$$

we have

$$S_{n,m}x = \tilde{L}_{n,m}x + \sum_{m}^{n-1} \tilde{L}_{n,k+1}T_k(S_{k,m}x) \quad \text{for } n \geqslant m. \tag{9.6}$$

Assume that ζ is a hyperbolic global solution of $\mathcal{S}$. Associated to ζ, we consider the linear discrete evolution process $\mathscr{L} = \{L_{n,m} : n \geqslant m\}$ given by the sequence of bounded linear operators $\{L_n\}_{n\in\mathbb{Z}}$, where $L_n = \tilde{L}_n + D_xT_n(\zeta_n)$ for each $n \in \mathbb{Z}$. Therefore $\{\ell_n = L_{n,m}x : n \geqslant m\}$ is the solution of $\ell_{n+1} = L_n\,\ell_n$, $\ell_m = x$ and

$$L_{n,m} = \tilde{L}_{n,m} + \sum_{m}^{n-1} \tilde{L}_{n,k+1}D_xT_k(\zeta_k)L_{k,m}. \tag{9.7}$$

From the hyperbolicity of ζ it follows that $\mathscr{L}$ has discrete exponential dichotomy, with constant $M \geqslant 1$ and exponent $\omega > 0$.

Definition 9.11. The **unstable set** of a hyperbolic global solution ζ of (9.6) is the set

$$\begin{aligned} W^u(\zeta) = \{\, (m,\eta) \in \mathbb{Z}\times X : \ &\text{there is a global solution } x\colon \mathbb{Z}\to X \text{ of (9.6)} \\ &\text{satisfying } x_m = \eta \text{ such that } \lim_{n\to-\infty} \|x_n - \zeta_n\|_X = 0\}. \end{aligned}$$

The **stable set** of a global hyperbolic solution ζ of (9.6) is the set

$$\begin{aligned} W^s(\zeta) = \{\, (m,\eta) \in \mathbb{Z}\times X : \ &\text{the solution } \{x_n\}_{n\geqslant m} \text{ of (9.6)} \\ &\text{satisfying } x_m = \eta \text{ is such that } \lim_{n\to\infty} \|x_n - \zeta_n\|_X = 0\}. \end{aligned}$$

We will show that the unstable and stable sets of the global hyperbolic solution ζ are given, respectively, by the graphs of the maps

$$\mathbb{Z}\times X \ni (n,x) \mapsto \varSigma_n^u(x) \in X \quad \text{and} \quad \mathbb{Z}\times X \ni (n,x) \mapsto \varSigma_n^s(x) \in X,$$

where $\varSigma_n^u(x) = \varSigma_n^u(Q_nx) \in \ker(Q_n)$ and $\varSigma_n^s(x) = \varSigma_n^s((I-Q_n)x) \in R(Q_n)$, for $n \in \mathbb{Z}$ and $x \in X$.

The points in the unstable and stable sets, respectively, are of the form

$$(n, Q_nx + \varSigma_n^u(Q_nx)),\ (n, (I-Q_n)x + \varSigma_n^s((I-Q_n)x)) \in \mathbb{Z}\times X,$$

with $(n,x) \in \mathbb{Z}\times X$ and x small.

We know that $\{y_n\}_{n\in\mathbb{Z}}$ is a solution of (9.6) if and only if $\{y_n\}_{n\in\mathbb{Z}} = \{x_n\}_{n\in\mathbb{Z}} + \{\zeta_n\}_{n\in\mathbb{Z}}$ where $\{x_n\}_{n\in\mathbb{Z}}$ satisfies

$$x_n = L_{n,m}x_m + \sum_m^{n-1} L_{n,k+1}F_k(x_k), \tag{9.8}$$

with $F_k(x) := T_k(x_k + \zeta_k) - T_k(\zeta_k) - D_x T_k(\zeta_k)x_k$. Hence, if $\{x_n\}_{n\in\mathbb{Z}}$ is a solution of (9.8), writing $x_n^+ = Q_n x_n$ and $x_n^- = x_n - x_n^+$, we have

$$\begin{cases} x_n^+ = L_{n,m}x_m^+ + \displaystyle\sum_m^{n-1} L_{n,k+1}H_k(x_k^+, x_k^-), \\ x_n^- = L_{n,m}x_m^- + \displaystyle\sum_m^{n-1} L_{n,k+1}G_k(x_k^+, x_k^-), \end{cases} \tag{9.9}$$

where

$$H_n(x_n^+, x_n^-) = Q_{n+1}[T_n(x_n^+ + x_n^- + \zeta_n) - T_n(\zeta_n) - D_x T_n(\zeta_n)(x_n^+ + x_n^-)] \quad \text{and}$$
$$G_n(x_n^+, x_n^-) = (I - Q_{n+1})[T_n(x_n^+ + x_n^- + \zeta_n) - T_n(\zeta_n) - D_x T_n(\zeta_n)(x_n^+ + x_n^-)].$$

Since in $(n, 0, 0)$ the functions H and G are zero with zero derivative (with respect to x_n^+ and x_n^-), from the continuous differentiability of H and G, uniformly with respect to n we obtain that, given $\rho > 0$ there exists $\delta > 0$ such that, if $\|x\|_X = \|x^+ + x^-\|_X < \delta$ and $\|\tilde{x}\|_X = \|\tilde{x}^+ + \tilde{x}^-\|_X < \delta$, then

$$\begin{aligned} &\|H_n(x^+, x^-)\|_X \leqslant \rho, \qquad \|G_n(x^+, x^-)\|_X \leqslant \rho, \\ &\|H_n(x^+, x^-) - H_n(\tilde{x}^+, \tilde{x}^-)\|_X \leqslant \rho(\|x^+ - \tilde{x}^+\|_X + \|x^- - \tilde{x}^-\|_X), \\ &\|G_n(x^+, x^-) - G_n(\tilde{x}^+, \tilde{x}^-)\|_X \leqslant \rho(\|x^+ - \tilde{x}^+\|_X + \|x^- - \tilde{x}^-\|_X). \end{aligned} \tag{9.10}$$

REMARK 9.12. It is possible to extend H and G outside the ball of radius δ in such a way that hypotheses (9.10) hold for all $x^+ \in X^+$, $x^- \in X^-$. In fact, given a Lipschitz continuous function g defined in the ball of radius δ of $X = V \times Z$ and taking values in W, where V, Z, W are Banach spaces, define $g_\delta \colon X \to W$ as

$$g_\delta(v, z) = \begin{cases} g(v, z), & \|(v, z)\|_X \leqslant \delta, \\ g\left(\dfrac{\delta\, v}{\|(v, z)\|_X}, \dfrac{\delta\, z}{\|(v, z)\|_X}\right), & \|(v, z)\|_X > \delta. \end{cases}$$

The extension g_δ is globally Lipschitz with the same Lipschitz constant as g.

The above considerations suggest that we first obtain the result about the stable and unstable manifolds assuming that the hypotheses (9.10) hold for all $z = (x^+, x^-) \in X$ and for some $\rho > 0$ suitably small. Also, assuming that (9.10) hold for all $z = (x^+, x^-) \in X$, we will prove the continuous differentiability of the stable and unstable manifolds and its continuous differentiability with respect to parameters. Finally, we localize the arguments of the case when H, G satisfy (9.10) only for $\|x\|_X = \|x^+ + x^-\|_X < \delta$ with $\delta > 0$ suitably small.

Hence, suppose also that H and G satisfy (9.10) for all $x_n^+ \in Q_n X$, $x_n^- \in (I - Q_n)X$ and for a certain $\rho > 0$, that will be defined below through (9.12). Let $W^u(0,0)$ be the unstable set of the equilibria $(0,0)$ of (9.9). We show that there is a bounded Lipschitz continuous function $\Sigma_n^{*,u} \colon Q_n X \to (I - Q_n)X$ such that

$$W^u(0,0) = \{(n, x^+, x^-) \colon x^- = \Sigma_n^{*,u}(x^+),\ x^+ \in Q_n X\}.$$

REMARK 9.13. Note that we are looking for a function $\Sigma_n^{*,u}$ such that, if $m \in \mathbb{Z}$ and $(\zeta, \Sigma_m^{*,u}(\zeta)) \in X$, the solution of (9.9) with $x_m^+ = \zeta$ and $x_m^- = \Sigma_m^{*,u}(\zeta)$ is such that x_n is in the graph of $\Sigma_n^{*,u}$ for all $n \in \mathbb{Z}$. Hence, $x_n^- = \Sigma_n^{*,u}(x_n^+)$ for all $n \in \mathbb{Z}$ and (9.9) becomes

$$\begin{cases} x_n^+ = L_{n,m}x_m^+ + \sum_m^{n-1} L_{n,k+1}H_k(x_k^+, \Sigma_k^{*,u}(x_k^+)), \\ x_n^- = L_{n,m}x_m^- + \sum_m^{n-1} L_{n,k+1}G_k(x_k^+, \Sigma_k^{*,u}(x_k^+)). \end{cases} \tag{9.11}$$

Besides that, the solution (x_n^+, x_n^-) must converge to zero as $n \to -\infty$ (in particular, it must remain bounded as $n \to -\infty$). Since

$$x_n^- = L_{n,m}(I - Q_m)x_m^- + \sum_m^{n-1} L_{n,k+1}(I - Q_{k+1})G_k(x_k^+, \Sigma_k^{*,u}(x_k^+)),$$

making $m \to -\infty$ we have

$$x_n^- = \Sigma_n^{*,u}(x_n^+) = \sum_{-\infty}^{n-1} L_{n,k+1}(I - Q_{k+1})G_k(x_k^+, \Sigma_k^{*,u}(x_k^+))$$

and, in particular, we obtain

$$\Sigma_m^{*,u}(\zeta) = \Sigma_m^{*,u}(x_m^+) = x_m^- = \sum_{-\infty}^{m-1} L_{m,k+1}(I - Q_{k+1})G_k(x_k^+, \Sigma_k^{*,u}(x_k^+)).$$

Before we proceed with the proof that the unstable manifold is given as the graph of a Lipschitz continuous map, let us show two discrete versions of the *Gronwall Inequality*.

LEMMA 9.14 (Upward Gronwall Lemma). *For $m \in \mathbb{Z}$ and $t \in (0, \infty)$, let $\mathbb{Z}_m^+ = \{k \in \mathbb{Z}\colon k \geqslant m\}$ and $r, \psi\colon \mathbb{Z}_m^+ \to [0, \infty)$ be such that*

$$\psi(k) \leqslant r(k) + t\sum_{j=m}^{k-1} \psi(j) \quad \textit{for all } k \in \mathbb{Z}_m^+.$$

Then for all $k \in \mathbb{Z}_m^+$ we obtain

$$\psi(k) \leqslant r(k) + t\sum_{j=m}^{k-1} r(j)(1+t)^{k-j-1},$$

and if r is nondecreasing, then $\psi(k) \leqslant r(k)(1+t)^{k-m}$.

PROOF. Define $v(j) = \sum_{i=m}^{j-1} \psi(i)$. Then $v(j+1) - v(j) = \psi(j) \leqslant r(j) + tv(j)$. Thus

$$v(j+1) - (1+t)v(j) \leqslant r(j),$$

and multiplying this expression by $(1+t)^{-j-1}$ we obtain

$$(1+t)^{-j-1}v(j+1) - (1+t)^{-j}v(j) \leqslant r(j)(1+t)^{-j-1}.$$

Adding from $j = m$ to $j = k-1$ and noting that $v(m) = 0$ we have

$$(1-t)^{-k}v(k) \leqslant \sum_{j=m}^{k-1} r(j)(1+t)^{-j-1},$$

and consequently

$$\psi(k) \leqslant r(k) + t \sum_{j=m}^{k-1} r(j)(1+t)^{k-j-1}.$$

The rest of the proof is immediate. □

LEMMA 9.15 (Downward Gronwall Lemma). *For $m \in \mathbb{Z}$ and $q \in (0,1)$, let $\mathbb{Z}_m^- = \{k \in \mathbb{Z} \colon k \leqslant m\}$ and $p, \phi \colon \mathbb{Z}_m^- \to [0,\infty)$ be such that*

$$\phi(k) \leqslant p(k) + q \sum_{j=k}^{m-1} \phi(j) \quad \text{for all } k \in \mathbb{Z}_m^-.$$

Then for all $k \in \mathbb{Z}_m^-$ we obtain

$$\phi(k) \leqslant p(k) + q \sum_{j=k}^{m-1} p(j)(1-q)^{k-j-1},$$

and if p is nonincreasing, then $\phi(k) \leqslant p(k)(1-q)^{k-m}$.

PROOF. Define $v(j) = \sum_{i=j}^{m-1} \phi(i)$. Then $v(j) - v(j+1) = \phi(j) \leqslant p(j) + qv(j)$. Hence

$$(1-q)v(j) - v(j+1) \leqslant p(j),$$

and multiplying this expression by $(1-q)^{-j-1}$ we obtain

$$(1-q)^{-j}v(j) - (1-q)^{-j-1}v(j+1) \leqslant p(j)(1-q)^{-j-1}.$$

Adding from $j = k$ to $j = m-1$ and noting that $v(m) = 0$ we have

$$(1-q)^{-k}v(k) \leqslant \sum_{j=k}^{m-1} p(j)(1-q)^{-j-1},$$

and consequently

$$\phi(k) \leqslant p(k) + q \sum_{j=k}^{m-1} p(j)(1-q)^{k-j-1}.$$

The rest of the proof is immediate. □

To show the existence of the function $\Sigma_m^{*,u}(\cdot)$ we use the *Banach Contraction Principle*. To that end, fix $D > 0$, $L > 0$, $0 < \vartheta < 1$ and choose $\rho > 0$ such that

$$\begin{aligned}
&0 < \nu = -\log(1 - \rho M(1+L+M)e^{-\omega}) < \frac{\omega}{2},\\
&0 < \nu' = \log(1 + \rho M(1+L+M)e^{\omega}) < \frac{\omega}{2},\\
&0 < \gamma = \log\left(1 + \rho M\left[e^{\omega-\nu} + \frac{\rho M(1+L+M)(1+M)}{1-e^{-(2\omega-\nu)}}\right]\right) < \frac{\omega}{2},\\
&\frac{\rho M(1+L+M)e^{-\omega}}{1-e^{-\omega}} < 1, \quad \frac{\rho M^2(1+L+M)e^{\nu}}{1-e^{-(\omega-\nu)}} \leqslant L, \quad \frac{\rho M}{1-e^{-\omega}} \leqslant D,\\
&\frac{\rho M}{1-e^{-\omega}}\left[1 + \frac{\rho M(1+L+M)}{1-e^{-(\omega-\nu)}}\right] \leqslant \vartheta < 1.
\end{aligned} \tag{9.12}$$

DEFINITION 9.16. Denote by $\mathcal{LB}(D,L)$ the complete metric space of all continuous functions $\Sigma\colon \mathbb{Z}\times X\to X$ satisfying

$$\begin{cases} \Sigma_m(x)=\Sigma_m(Q_mx)\in\ker(Q_m) \text{ for all } (m,x)\in\mathbb{Z}\times X,\\ \sup\limits_{(m,x)\in\mathbb{Z}\times X}\|\Sigma_m(x)\|_X\leqslant D,\\ \|\Sigma_m(x)-\Sigma_m(\tilde{x})\|_X\leqslant L\|x-\tilde{x}\|_X, \text{ for all } (m,x),(m,\tilde{x})\in\mathbb{Z}\times X, \end{cases} \tag{9.13}$$

where the distance between $\Sigma,\tilde{\Sigma}\in\mathcal{LB}(D,L)$ is defined by

$$|||\,\Sigma-\tilde{\Sigma}\,|||=\sup_{(m,x)\in\mathbb{Z}\times X}\|\Sigma_m(x)-\tilde{\Sigma}_m(x)\|_X.$$

THEOREM 9.17. *Assume that the conditions in* (9.12) *are satisfied. Then there exists* $\Sigma^{*,u}\in\mathcal{LB}(D,L)$ *such that the unstable manifold* $W^u(0,0)$ *of* (9.9) *is given by*

$$W^u(0,0)=\{(m,w)\in\mathbb{Z}\times X\colon w=(Q_mw,\Sigma^{*,u}_m(Q_mw))\}. \tag{9.14}$$

In addition, if $\{(x_n^+,x_n^-)\colon n\geqslant m_0\}$ *is a solution of* (9.9)*, then*

$$\|x_n^- -\Sigma^{*,u}_n(x_n^+)\|_X\leqslant Me^{-(\frac{\omega}{2}-\gamma)(n-m_0)}\|x_{m_0}^- -\Sigma^{*,u}_{m_0}(x_{m_0}^+)\|_X,\quad n\geqslant m_0.$$

PROOF. We split the proof into the *existence of a graph in the unstable manifold*, the *exponential attraction* and the *unstable manifold is in the graph*.

Existence of a graph in the unstable manifold.

For $m\in\mathbb{Z}$, $\zeta\in X$ and $\Sigma\in\mathcal{LB}(D,L)$, denote by $x_n^+=\psi(n,m,\zeta,\Sigma)$ the solution of

$$\begin{cases} x_n^+=L_{n,m}x_m^+-\sum\limits_{k=n}^{m-1}L_{n,k+1}H_k(x_k^+,\Sigma_k(x_k^+)),\ n<m,\\ x_m^+=Q_m\zeta\in Q_mX. \end{cases} \tag{9.15}$$

For $\Sigma\in\mathcal{LB}(D,L)$ and $(m,\zeta)\in\mathbb{Z}\times X$, we define

$$\Phi(\Sigma)(m,\zeta)=\sum_{-\infty}^{m-1}L_{m,k+1}(I-Q_{k+1})G_k(x_k^+,\Sigma_k(x_k^+)). \tag{9.16}$$

We show that, for $\rho>0$ satisfying (9.12), the map Φ takes $\mathcal{LB}(D,L)$ into itself and is a contraction. Consequently it has a unique fixed point in $\mathcal{LB}(D,L)$. First note that, since $\{L_{n,m}\colon n\geqslant m\}$ has discrete exponential dichotomy, with constant M and exponent $\omega>0$, it follows from (9.10), that

$$\|\Phi(\Sigma)(m,\zeta)\|_X\leqslant\sum_{-\infty}^{m-1}\rho Me^{-\omega(m-k-1)}=\frac{\rho M}{1-e^{-\omega}}, \tag{9.17}$$

and from (9.12) we have $\sup\limits_{(m,\zeta)\in\mathbb{Z}\times X}\|\Phi(\Sigma)(m,\zeta)\|_X\leqslant D.$

Next, suppose that Σ and $\tilde{\Sigma}$ are functions satisfying (9.13), $\zeta,\tilde{\zeta}\in X$ and denote $x_n^+=\psi(n,m,\zeta,\Sigma)$ and $\tilde{x}_n^+=\psi(n,m,\tilde{\zeta},\tilde{\Sigma})$. Then

$$x_n^+-\tilde{x}_n^+=L_{n,m}Q_m(\zeta-\tilde{\zeta})-\sum_{k=n}^{m-1}L_{n,k+1}Q_{k+1}[H_k(x_k^+,\Sigma_k(x_k^+))-H_k(\tilde{x}_k^+,\tilde{\Sigma}_k(\tilde{x}_k^+))],$$

and from (9.2) and (9.10) we obtain

$$\begin{aligned}
&\|x_n^+ - \tilde{x}_n^+\|_X \\
&\leqslant Me^{\omega(n-m)}\|\zeta - \tilde{\zeta}\|_X + M\sum_{k=n}^{m-1} e^{\omega(n-k-1)}\|H_k(x_k^+, \Sigma_k(x_k^+)) - H_k(\tilde{x}_k^+, \tilde{\Sigma}_k(\tilde{x}_k^+))\|_X \\
&\leqslant Me^{\omega(n-m)}\|\zeta - \tilde{\zeta}\|_X \\
&\qquad + \rho M\sum_{k=n}^{m-1} e^{\omega(n-k-1)}\Big(\|\Sigma_k(\tilde{x}_k^+) - \tilde{\Sigma}_k(\tilde{x}_k^+)\|_X + (1+L)\|x_k^+ - \tilde{x}_k^+\|_X\Big) \\
&\leqslant Me^{\omega(n-m)}\|\zeta - \tilde{\zeta}\|_X + \rho M\sum_{k=n}^{m-1} e^{\omega(n-k-1)}\Big((1+L)\|x_k^+ - \tilde{x}_k^+\|_X + |||\,\Sigma - \tilde{\Sigma}\,|||\Big) \\
&\leqslant Me^{\omega(n-m)}\|\zeta - \tilde{\zeta}\|_X + \rho M\;|||\,\Sigma - \tilde{\Sigma}\,|||\sum_{k=n}^{m-1} e^{\omega(n-k-1)} \\
&\qquad + \rho M(1+L)\sum_{k=n}^{m-1} e^{\omega(n-k-1)}\|x_k^+ - \tilde{x}_k^+\|_X.
\end{aligned}$$

If $\phi_n = e^{-\omega(n-m)}\|x_n^+ - \tilde{x}_n^+\|_X$, then

$$\phi_n \leqslant M\|\zeta - \tilde{\zeta}\|_X + \rho M\sum_{k=n}^{m-1} e^{\omega(m-k-1)}\;|||\,\Sigma - \tilde{\Sigma}\,|||\; + \rho M(1+L)e^{-\omega}\sum_{k=n}^{m-1}\phi_k,$$

and by the Downward Gronwall Lemma (Lemma 9.15) we have

$$\begin{aligned}
\|x_n^+ - \tilde{x}_n^+\|_X &\leqslant \Big(M\|\zeta - \tilde{\zeta}\|_X e^{\omega(n-m)} + \rho M\sum_{k=n}^{m-1} e^{\omega(n-k-1)}\;|||\,\Sigma - \tilde{\Sigma}\,|||\Big) \\
&\qquad\cdot(1 - \rho M(1+L)e^{-\omega})^{n-m} \\
&\leqslant \Big(M\|\zeta - \tilde{\zeta}\|_X + \rho M\frac{e^{-\omega}}{1-e^{-\omega}}\;|||\,\Sigma - \tilde{\Sigma}\,|||\Big) \\
&\qquad\cdot(1 - \rho M(1+L)e^{-\omega})^{n-m}.
\end{aligned} \tag{9.18}$$

Hence

$$\begin{aligned}
&\|\Phi(\Sigma)(m,\zeta) - \Phi(\tilde{\Sigma})(m,\tilde{\zeta})\|_X \\
&\qquad\leqslant M\sum_{-\infty}^{m-1} e^{-\omega(m-k-1)}\|G_k(x_k^+, \Sigma_k(x_k^+)) - G_k(\tilde{x}_k^+, \tilde{\Sigma}_k(\tilde{x}_k^+))\|_X \\
&\qquad\leqslant \rho M\sum_{-\infty}^{m-1} e^{-\omega(m-k-1)}\left(\|\Sigma_k(x_k^+) - \tilde{\Sigma}_k(\tilde{x}_k^+)\|_X + \|x_k^+ - \tilde{x}_k^+\|_X\right) \\
&\qquad\leqslant \rho M\sum_{-\infty}^{m-1} e^{-\omega(m-k-1)}\left[(1+L)\|x_k^+ - \tilde{x}_k^+\|_X + |||\,\Sigma - \tilde{\Sigma}\,|||\right].
\end{aligned}$$

Noting that (9.12) implies $\nu < \omega$, and using (9.18), we obtain

$$\|\Phi(\Sigma)(m,\zeta) - \Phi(\tilde{\Sigma})(m,\tilde{\zeta})\|_X \leqslant \rho M \sum_{-\infty}^{m-1} e^{-\omega(m-k-1)}\Big[(1+L)\|x_k^+ - \tilde{x}_k^+\|_X + |||\, \Sigma - \tilde{\Sigma}\, |||\Big]$$

$$\leqslant \rho M \sum_{-\infty}^{m-1} e^{-\omega(m-k-1)}\Big\{(1+L)\Big[M\|\zeta - \tilde{\zeta}\|_X + \rho M \frac{e^{-\omega}}{1-e^{-\omega}}\, |||\, \Sigma - \tilde{\Sigma}\, |||\Big] e^{\nu(m-k)} + |||\, \Sigma - \tilde{\Sigma}\, |||\Big\},$$

and consequently,

(9.19)
$$\|\Phi(\Sigma)(m,\zeta) - \Phi(\tilde{\Sigma})(m,\tilde{\zeta})\|_X \leqslant \frac{\rho M}{1-e^{-\omega}}\left[1 + \frac{\rho M(1+L)e^{-(\omega-\nu)}}{1-e^{-(\omega-\nu)}}\right] |||\, \Sigma - \tilde{\Sigma}\, ||| + \frac{\rho M^2(1+L)e^{\nu}}{1-e^{-(\omega-\nu)}}\|\zeta - \tilde{\zeta}\|_X$$
$$\leqslant I_\Sigma\, |||\, \Sigma - \tilde{\Sigma}\, ||| + I_\zeta\|\zeta - \tilde{\zeta}\|_X,$$

where

$$I_\Sigma = \frac{\rho M}{1-e^{-\omega}}\left[1 + \frac{\rho M(1+L)e^{-(\omega-\nu)}}{1-e^{-(\omega-\nu)}}\right] \quad \text{and} \quad I_\zeta = \frac{\rho M^2(1+L)e^{\nu}}{1-e^{-(\omega-\nu)}}.$$

From (9.12) we have $I_\Sigma \leqslant \frac{\rho M}{1-e^{-\omega}}(1 + Le^{-\omega}M) \leqslant \vartheta$ and $I_\zeta \leqslant L$. Hence we obtain

$$\|\Phi(\Sigma)(m,\zeta) - \Phi(\tilde{\Sigma})(m,\tilde{\zeta})\|_X \leqslant L\|\zeta - \tilde{\zeta}\|_X + \vartheta\, |||\, \Sigma - \tilde{\Sigma}\, |||. \tag{9.20}$$

Therefore, inequality (9.20) with $\Sigma = \tilde{\Sigma}$, together with (9.17), implies that Φ takes $\mathcal{LB}(D,L)$ into itself. Besides that, the estimate (9.20) with $\zeta = \tilde{\zeta}$ shows that Φ is a contraction. Hence, it has a unique fixed point $\Sigma^{*,u} = \Phi(\Sigma^{*,u})$ in $\mathcal{LB}(D,L)$.

Exponential attraction.

In what follows we show the existence of constants $M \geqslant 1$ and $\omega > \gamma > 0$ such that, if $\{(x_n^+, x_n^-) \colon n \geqslant m_0\}$ is a solution of (9.9), then

$$\|x_n^- - \Sigma_n^{*,u}(x_n^+)\|_X \leqslant M e^{-(\frac{\omega}{2}-\gamma)(n-m_0)}\|x_{m_0}^- - \Sigma_{m_0}^{*,u}(x_{m_0}^+)\|_X, \quad n \geqslant m_0. \tag{9.21}$$

In particular, if $\{(x_n^+, x_n^-) \colon n \in \mathbb{Z}\}$ is a global solution of (9.9) which remains bounded as $n \to -\infty$, making $m_0 \to -\infty$ in (9.21) we obtain $x_n^- = \Sigma_n^{*,u}(x_n^+)$ for each $n \in \mathbb{Z}$. This shows that any global solution which remains bounded as $n \to -\infty$ must be in the graph of $\Sigma^{*,u}$. From this, it also follows that $\Sigma_n^{*,u}(0) = 0$, since $(0,0)$ is a stationary solution of (9.9).

Let $\zeta_n = x_n^- - \Sigma_n^{*,u}(x_n^+)$ and let $\{y_{k,n}^+ \colon k \leqslant n\}$ be a solution of

$$y_k^+ = L_{k,n}x_n^+ - \sum_{j=k}^{n-1} L_{k,j+1}H_j(y_j^+, \Sigma_j^{*,u}(y_j^+)),\ k < n,\ y_{n,n}^+ = x_n^+.$$

Thus

$$\|y_{k,n}^+ - x_k^+\|_X = \Big\| \sum_{j=k}^{n-1} L_{k,j+1}[H_j(y_{j,n}^+, \Sigma_j^{*,u}(y_{j,n}^+)) - H_j(x_j^+, x_j^-)] \Big\|_X$$

$$\leqslant \rho M \sum_{j=k}^{n-1} e^{\omega(k-j-1)} \left[(1+L)\|y_{j,n}^+ - x_j^+\|_X + \|\zeta_j\|_X\right].$$

If $\psi_k = e^{\omega(n-k)}\|y_{k,n}^+ - x_k^+\|_X$, then

$$\psi_k \leqslant \rho M \sum_{j=k}^{n-1} e^{\omega(n-j-1)}\|\zeta_j\|_X + \rho M(1+L)e^{-\omega}\sum_{j=k}^{n-1}\psi_j, \quad k \leqslant n,$$

and using the Downward Gronwall Lemma we obtain

$$\|y_{k,n}^+ - x_k^+\|_X \leqslant \rho M \sum_{j=k}^{n-1} e^{\omega(k-j-1)}\|\zeta_j\|_X \; e^{\nu(n-k)}, \quad k \leqslant n. \tag{9.22}$$

If $k \leqslant m_0 \leqslant n$, then

$$\|y_{k,n}^+ - y_{k,m_0}^+\|_X = \|L_{k,m_0}Q_{m_0}(y_{m_0,n}^+ - x_{m_0}^+)\|_X$$

$$+ \Big\| \sum_{j=k}^{m_0-1} L_{k,j+1}Q_{j+1}\Big(H_j(y_{j,n}^+, \Sigma_j^{*,u}(y_{j,n}^+)) - H_j(y_{j,m_0}^+, \Sigma_j^{*,u}(y_{j,m_0}^+))\Big)\Big\|_X$$

$$\leqslant \rho M^2 e^{\omega(k-m_0)} \sum_{j=m_0}^{n-1} e^{\omega(m_0-j-1)}\|\zeta_j\|_X e^{\nu(n-m_0)}$$

$$+ \rho M(1+L)e^{-\omega} \sum_{j=k}^{m_0-1} e^{\omega(k-j)}\|y_{j,n}^+ - y_{j,m_0}^+\|_X.$$

Therefore, if $\phi_k = e^{-\omega k}\|y_{k,n}^+ - y_{k,m_0}^+\|_X$, we have

$$\phi_k \leqslant \rho M^2 \sum_{j=m_0}^{n-1} e^{-\omega(j+1)}\|\zeta_j\|_X e^{\nu(n-m_0)} + \rho M(1+L)e^{-\omega}\sum_{j=k}^{m_0-1}\phi_j,$$

and, using the Downward Gronwall Lemma, we obtain

$$\begin{aligned} \|y_{k,n}^+ - y_{k,m_0}^+\|_X &\leqslant \rho M^2 \sum_{j=m_0}^{n-1} e^{\omega(k-j-1)}\|\zeta_j\|_X \; e^{\nu(n-m_0)}e^{\nu(m_0-k)} \\ &\leqslant \rho M^2 \sum_{j=m_0}^{n-1} e^{\omega(n-j-1)}\|\zeta_j\|_X \; e^{-(\omega-\nu)(n-k)}. \end{aligned} \tag{9.23}$$

We will use (9.23) to estimate ζ_n. Note that

$$\begin{aligned} &\zeta_n - L_{n,m_0}(I - Q_{m_0})\zeta_{m_0} \\ &= x_n^- - \Sigma_n^{*,u}(x_n^+) - L_{n,m_0}(I - Q_{m_0})[x_{m_0}^- - \Sigma_{m_0}^{*,u}(x_{m_0}^+)] \\ &= \sum_{k=m_0}^{n-1} L_{n,k+1}(I - Q_{k+1})G_k(x_k^+, x_k^-) - \Sigma_n^{*,u}(x_n^+) + L_{n,m_0}(I - Q_{m_0})\Sigma_{m_0}^{*,u}(x_{m_0}^+) \end{aligned}$$

$$= \sum_{k=m_0}^{n-1} L_{n,k+1}(I - Q_{k+1})[G_k(x_k^+, x_k^-) - G_k(y_{k,n}^+, \Sigma_k^{*,u}(y_{k,n}^+))]$$

$$- \sum_{-\infty}^{m_0-1} L_{n,k+1}(I-Q_{k+1})[G_k(y_{k,n}^+, \Sigma_k^{*,u}(y_{k,n}^+))-G_k(y_{k,m_0}^+, \Sigma_k^{*,u}(y_{k,m_0}^+))].$$

Hence, using (9.22) and (9.23), we obtain

$$\begin{aligned}
&\|\zeta_n - L_{n,m_0}(I - Q_{m_0})\zeta_{m_0}\|_X \\
&\leqslant \rho M \sum_{k=m_0}^{n-1} e^{-\omega(n-k-1)} \left[\|x_k^+ - y_{k,n}^+\|_X + \|x_k^- - \Sigma_k^{*,u}(y_{k,n}^+)\|_X\right] \\
&\quad + \rho M(1+L) \sum_{-\infty}^{m_0-1} e^{-\omega(n-k-1)} \|y_{k,n}^+ - y_{k,m_0}^+\|_X \\
&\leqslant \rho M \sum_{k=m_0}^{n-1} e^{-\omega(n-k-1)} \left[(1+L)\|x_k^+ - y_{k,n}^+\|_X + \|x_k^- - \Sigma_k^{*,u}(x_k^+)\|_X\right] \\
&\quad + \rho M(1+L) \sum_{-\infty}^{m_0-1} e^{-\omega(n-k-1)} \|y_{k,n}^+ - y_{k,m_0}^+\|_X \\
&\leqslant \rho M \sum_{k=m_0}^{n-1} e^{-\omega(n-k-1)} \|\zeta_k\|_X \\
&\quad + \rho^2 M^2 (1+L) e^{\nu} \sum_{k=m_0}^{n-1} e^{-\omega(n-k-1)} \sum_{j=k}^{n-1} e^{\omega(k-j-1)} \|\zeta_j\|_X \, e^{\nu(n-k-1)} \\
&\quad + \rho^2 M^3 (1+L) e^{-(\omega-\nu)} \sum_{-\infty}^{m_0-1} e^{-\omega(n-k-1)} \sum_{j=m_0}^{n-1} e^{\omega(n-j-1)} \|\zeta_j\|_X e^{-(\omega-\nu)(n-k-1)} \\
&\leqslant \rho M \sum_{k=m_0}^{n-1} e^{-\omega(n-k-1)} \|\zeta_k\|_X \\
&\quad + \rho^2 M^2 (1+L) e^{\nu} \sum_{k=m_0}^{n-1} e^{-(\omega-\nu)(n-k-1)} \sum_{j=k}^{n-1} e^{\omega(k-j-1)} \|\zeta_j\|_X \\
&\quad + \rho^2 M^3 (1+L) e^{\nu} \sum_{-\infty}^{m_0-1} e^{-(\omega-\nu)(n-k-1)} \sum_{j=m_0}^{n-1} e^{\omega(k-j-1)} \|\zeta_j\|_X \\
&\leqslant \rho M \sum_{k=m_0}^{n-1} e^{-\omega(n-k-1)} \|\zeta_k\|_X \\
&\quad + \rho^2 M^2 (1+L) e^{-(\omega-\nu)} \sum_{j=m_0}^{n-1} \|\zeta_j\|_X e^{-(\omega-\nu)(n-j-1)} \sum_{k=m_0}^{j} e^{(\omega-\nu)(k-j)} e^{\omega(k-j)} \\
&\quad + \rho^2 M^3 (1+L) e^{-(\omega-\nu)} \sum_{j=m_0}^{n-1} \|\zeta_j\|_X e^{-(\omega-\nu)(n-j-1)} \sum_{k=-\infty}^{m_0-1} e^{(\omega-\nu)(k-j)} e^{\omega(k-j)}
\end{aligned}$$

$$\leqslant \rho M \sum_{k=m_0}^{n-1} e^{-\omega(n-k-1)} \|\zeta_k\|_X$$

$$+ \rho^2 M^2 (1+L)(1+M) \frac{e^{-(\omega-\nu)}}{1-e^{-(2\omega-\nu)}} \sum_{k=m_0}^{n-1} \|\zeta_k\|_X e^{-(\omega-\nu)(n-k-1)}$$

$$\leqslant \rho M \left[1 + \rho M(1+L)(1+M) \frac{e^{-(\omega-\nu)}}{1-e^{-(2\omega-\nu)}}\right] \sum_{k=m_0}^{n-1} \|\zeta_k\|_X e^{-(\omega-\nu)(n-k-1)},$$

and, if $\phi_n = e^{(\omega-\nu)(n-m_0)} \|\zeta_n\|_X$, $n \geqslant m_0$, we have

$$\phi_n \leqslant M \|\zeta_{m_0}\|_X e^{-\nu(n-m_0)} + \rho M \left[e^{\omega-\nu} + \frac{\rho M(1+L)(1+M)}{1-e^{-(2\omega-\nu)}}\right] \sum_{k=m_0}^{n-1} \phi_k.$$

Due to the Upward Gronwall Lemma we obtain

$$\|\zeta_n\|_X \leqslant \frac{2M}{1-e^{-(\nu+\gamma)}} \|\zeta_{m_0}\|_X \, e^{-(\frac{\omega}{2}-\gamma)(n-m_0)},$$

where γ is given in (9.12) and $0 < \gamma < \frac{\omega}{2}$. This proves (9.21) and also that

$$W^u(0,0) \subset \{(m,w) \in \mathbb{Z} \times X : w = (Q_m w, \Sigma_m^{*,u}(Q_m w))\}.$$

Unstable manifold is in the graph.

We now show that $\{(m,w) \in \mathbb{Z} \times X : w = (Q_m w, \Sigma_m^{*,u}(Q_m w))\} \subset W^u(0,0)$. Consider $x_1^+ \in Q_m X$ and the solution x_n^{+*}, $n \leqslant m$, of the problem

$$x_n^+ = L_{n,m} x_m^+ - \sum_{k=n}^{m-1} L_{n,k+1} H_k(x_k^+, \Sigma_k^{*,u}(x_k^+)), \quad x_m^+ = x_1^+.$$

That defines the sequence $\{(x_n^{+*}, \Sigma_n^{*,u}(x_n^{+*})) : n \leqslant m\}$. Recalling (9.16), we can verify that

$$\Sigma_n^{*,u}(x_n^{+*}) = \sum_{-\infty}^{n-1} L_{n,k+1}(I - Q_{k+1}) G_k(x_k^{+*}, \Sigma_k^{*,u}(x_k^{+*})), \quad n \leqslant m.$$

Thus $\Sigma_n^{*,u}(x_n^{+*})$ solves

$$x_n^- = L_{n,k} x_k^- + \sum_{j=k}^{n-1} L_{n,j+1} G_j(x_j^{+*}, \Sigma_j^{*,u}(x_j^{+*})), \ k \leqslant n \leqslant m,$$

and we conclude that $\{(x_n^{+*}, \Sigma_n^{*,u}(x_n^{+*})) : n \leqslant m\}$ is a solution of (9.9) through $(x_1^+, \Sigma_m^{*,u}(x_1^+))$ at the discrete time m, with $\Sigma_n^{*,u}(x_n^{+*}) \to 0$ as $n \to -\infty$. Since $\Sigma_n^{*,u}(0) = 0$, the reasoning that takes us to (9.18) can be used to ensure that

$$\|x_n^+\|_X \leqslant M \|x_m^+\|_X e^{(\omega-\nu)(n-m)}.$$

As a consequence $x_n^{+*} \to 0$ as $n \to -\infty$ and the proof is complete. □

3.2. Local unstable manifolds as graphs. In this section we will employ the results of Section 3 to obtain the existence and continuity of local stable manifolds when the conditions (9.10) are satisfied only locally.

Let $T\colon \mathbb{Z}\times X \to X$ be a continuously differentiable function such that $T_n(0) = 0$ and $D_x T_n(0) = 0 \in \mathcal{L}(X)$. Assume that the linear discrete evolution process $\mathscr{L} = \{L_{n,m}\colon n \geqslant m\}$ has discrete exponential dichotomy, with constant $M \geqslant 1$ and exponent $\omega > 0$. Consider the discrete initial value problem (9.8).

PROPOSITION 9.18. *Assume that the above conditions are satisfied. Then, for each suitably small $\delta > 0$, there are δ_1, δ_2, $0 < \delta_1 < \delta < \delta_2$, and a function $\Sigma^{*,u}\colon \mathbb{Z} \times X \to X$ such that*

$$\{(n,x) \in \mathbb{Z} \times X\colon x = (Q_n x, \Sigma_n^{*,u}(Q_n x)), \|x\| < \delta_1\} \subset W_\delta^u(0,0)$$
$$\subset \{(n,x) \in \mathbb{Z} \times X\colon x = (Q_n x, \Sigma_n^{*,u}(Q_n x)), \|x\| < \delta_2\}.$$

In addition, if $\{(x_n^+, x_n^-)\colon n \geqslant m_0\}$ is a solution of (9.9) with $(x_{m_0}^+, x_{m_0}^-) \in B_\delta(0,0)$, then

$$\|x_n^- - \Sigma_n^{*,u}(x_n^+)\|_X \leqslant M e^{-(\frac{\omega}{2}-\gamma)(n-m_0)} \|x_{m_0}^- - \Sigma_{m_0}^{*,u}(x_{m_0}^+)\|_X, \quad n \geqslant m_0, \tag{9.24}$$

as long as $(x_n^+, x_n^-) \in B_\delta(0,0)$.

PROOF. According to Remark 9.12 and Theorem 9.17, we need only to ensure that, given $\delta > 0$ there is a $0 < \delta' \leqslant \delta$ such that any solution (x_n^+, x_n^-) in the unstable manifold that satisfies $\|x_m^+\|_X + \|x_m^-\|_X < \delta'$ is such that $\|x_n^+\|_X + \|x_n^-\|_X < \delta$, for all $n \leqslant m$. This follows easily from the fact that x_n^+ is a solution of

$$x_n^+ = L_{n,m} x_m^+ - \sum_m^{n-1} L_{n,k+1} H_k(x_k^+, \Sigma_n^{*,u}(x_n^+)), \quad n \leqslant m.$$

As before

$$\|x_n^+\|_X \leqslant M \|x_m^+\|_X e^{(\omega-\nu)(n-m)} \tag{9.25}$$

and

$$\|x_n^-\|_X = \|\Sigma^*(x_n^+)\|_X \leqslant ML \|x_m^+\|_X e^{(\omega-\nu)(n-m)}.$$

The proof now follows easily. □

4. Stable manifolds

4.1. Global stable manifolds as graphs. Again we assume that S_n is such that H and G satisfy (9.10) for all $x_n^+ \in Q_n X$, $x_n^- \in (I - Q_n)X$ and for a certain $\rho > 0$, that verifies (9.12). Let $W^s(0,0)$ be the stable manifold of the equilibrium solution $(0,0)$ of (9.9). As in Section 3, we show the existence of a bounded Lipschitz continuous function $\Sigma^{*,s}\colon \mathbb{Z}\times X \to X$ such that $\Sigma_n^{*,s}(x) = \Sigma^{*,s}((I - Q_n)x) \in R(Q_n)$ for all $x \in X$ and such that

$$W^s(n,0,0) = \{(n, x^+, x^-)\colon x^+ = \Sigma_n^{*,s}(x^-),\ x^- \in (I - Q_n)X\}.$$

REMARK 9.19. In this case, we are looking for a function $\Sigma^{*,s}\colon \mathbb{Z} \times X \to X$ with the property that, for $m \in \mathbb{Z}$ and $\zeta \in (I - Q_m)X$, the solution $\{x_n\}_{n \geqslant m}$ of (9.9) with $x_m^+ = \Sigma_m^{*,s}(\zeta)$, $x_m^- = \zeta$, is such that x_n is in the graph of $\Sigma_n^{*,s}$ for all

$n \geqslant m$. That is, $x_n^+ = \Sigma_n^{*,s}(x_n^-)$ for all $n \geqslant m$. Besides that, $(x_n^+, x_n^-) \to (0,0)$ as $n \to \infty$ (in particular, it remains bounded as $n \to \infty$). Since, for $n \leqslant \ell$, we have

$$x_n^+ = L_{n,\ell} Q_\ell x_\ell^+ - \sum_{k=n}^{\ell-1} L_{n,k+1} Q_{k+1} H_k(\Sigma_k^{*,s}(x_k^-), x_k^-),$$

making $\ell \to \infty$ we obtain

$$x_n^+ = \Sigma_n^{*,s}(x_n^-) = -\sum_{k=n}^{\infty} L_{n,k+1} Q_{k+1} H_k(\Sigma_k^{*,s}(x_k^-), x_k^-).$$

Fix $D > 0$, $L > 0$, $0 < \vartheta < 1$ and choose $\rho > 0$ such that (9.12) is satisfied.

DEFINITION 9.20. Given $\eta > 0$, let $\mathcal{LB}^s(D, L)$ be the complete metric space of functions $\Sigma \colon \mathbb{Z} \times X \to X$ such that

$$(9.26) \qquad \begin{cases} \Sigma_m(x) = \Sigma_m((I - Q_m)x) \in Q_m X, \text{ for all } (m, x) \in \mathbb{Z} \times X, \\ \sup\limits_{(m,x) \in \mathbb{Z} \times X} \|\Sigma_m(x)\|_X \leqslant D, \\ \|\Sigma_m(x) - \Sigma_m(\tilde{x})\|_X \leqslant L \|x - \tilde{x}\|_X, \text{ for all } (m, x), (m, \tilde{x}) \in \mathbb{Z} \times X, \end{cases}$$

where the distance between Σ and $\tilde{\Sigma}$ in $\mathcal{LB}^s(D, L)$ is defined as

$$|\!|\!| \Sigma - \tilde{\Sigma} |\!|\!| = \sup_{(m,x) \in \mathbb{Z} \times X} \|\Sigma_m(x) - \tilde{\Sigma}_m(x)\|_X.$$

THEOREM 9.21. *Assume that the above conditions are satisfied. Then, there exists $\Sigma^{*,s} \in \mathcal{LB}^s(D, L)$, such that the stable manifold $W^s(0,0)$ of (9.9) is given by*

$$(9.27) \qquad W^s(0,0) = \{(m, w) \in \mathbb{Z} \times X \colon w = (\Sigma_m^{*,s}((I - Q_m)w), (I - Q_m)w)\}.$$

PROOF. For $m \in \mathbb{Z}$, $\zeta \in X$ and $\Sigma \in \mathcal{LB}^s(D, L)$, denote by $x_n^- = \psi(n, m, \zeta, \Sigma)$ the solution of

$$(9.28) \qquad \begin{cases} x_n^- = L_{n,m}(I - Q_m)\zeta + \sum\limits_{m}^{n-1} L_{n,k+1} G_k(\Sigma_k(x_k^-), x_k^-), \ n > m, \\ x_m^- = (I - Q_m)\zeta. \end{cases}$$

For $\Sigma \in \mathcal{LB}^s(D, L)$ and $(m, \zeta) \in \mathbb{Z} \times X$ we define

$$(9.29) \qquad \Phi(\Sigma)(m, \zeta) = -\sum_{k=m}^{\infty} L_{m,k+1} Q_{k+1} H_k(\Sigma_k(x_k^-), x_k^-).$$

Following the same reasoning as in Theorem 9.17, for $\rho > 0$ satisfying (9.12), it can be proved that the map Φ takes $\mathcal{LB}^s(D, L)$ into itself and it is a contraction. Hence, it has a unique fixed point in $\mathcal{LB}^s(D, L)$.

We first note that, from (9.10), we have

$$(9.30) \qquad \|\Phi(\Sigma)(m, \zeta)\|_X \leqslant \rho M \sum_{m}^{\infty} e^{\omega(m-k-1)} = \frac{\rho M e^{-\omega}}{1 - e^{-\omega}},$$

and from (9.12) we have $\sup\limits_{(m,x) \in \mathbb{Z} \times X} \|\Phi(\Sigma)(m, x)\|_X \leqslant D$.

Next, if $\Sigma, \tilde{\Sigma} \in \mathcal{LB}^s(D,L)$ and $\zeta, \tilde{\zeta} \in X$, denoting $x_n^- = \psi(n,m,\zeta,\Sigma)$ and $\tilde{x}_n^- = \psi(n,m,\tilde{\zeta},\tilde{\Sigma})$, we have

$$x_n^- - \tilde{x}_n^- = L_{n,m}(I-Q_m)(\zeta-\tilde{\zeta}) + \sum_m^{n-1} L_{n,k+1}(I-Q_{k+1})[G_k(\Sigma_k(x_k^-),x_k^-) - G_k(\tilde{\Sigma}_k(\tilde{x}_k^-),\tilde{x}_k^-)],$$

for which we can prove that

$$\|x_n^- - \tilde{x}_n^-\|_X \leqslant M\|\zeta-\tilde{\zeta}\|_X e^{-\omega\,(n-m)} + \rho M \sum_m^{n-1} e^{-\omega(n-k-1)}[(1+L)\|x_k^- - \tilde{x}_k^-\| + |||\Sigma - \tilde{\Sigma}\,|||].$$

Hence, if $\phi_n = e^{\omega\,(n-m)}\|x_n^- - \tilde{x}_n^-\|_X$, we have

$$\phi_n \leqslant \rho M \sum_m^{n-1} e^{-\omega(m-k-1)}\ |||\,\Sigma - \tilde{\Sigma}\,||| + M\|\zeta-\tilde{\zeta}\|_X + \rho M(1+L)e^{\omega}\sum_m^{n-1}\phi_k,$$

and from the Upward Gronwall Lemma (Lemma 9.14), if $\nu' = \log(1+\rho M(1+L)e^{\omega})$, we obtain

$$\phi_n \leqslant \Big(\rho M \sum_m^{n-1} e^{-\omega(m-k-1)}\ |||\,\Sigma - \tilde{\Sigma}\,|||\ + M\|\zeta-\tilde{\zeta}\|_X\Big)e^{\nu'(n-m)}.$$

Thus,
(9.31)

$$\begin{aligned}\|x_n^- - \tilde{x}_n^-\|_X &\leqslant \Big(M\|\zeta-\tilde{\zeta}\|_X e^{-\omega\,(n-m)} + \rho M\sum_m^{n-1} e^{-\omega(n-k-1)}\ |||\,\Sigma-\tilde{\Sigma}\,|||\Big)e^{\nu'(n-m)}\\ &\leqslant \Big(M\|\zeta-\tilde{\zeta}\|_X + \frac{\rho M}{1-e^{-\omega}}\ |||\,\Sigma-\tilde{\Sigma}\,|||\Big)e^{\nu'(n-m)},\end{aligned}$$

and we arrive at

$$\|\Phi(\Sigma)(m,\zeta) - \Phi(\tilde{\Sigma})(m,\tilde{\zeta})\|_X \leqslant I_\Sigma\ |||\,\Sigma-\tilde{\Sigma}\,||| + I_\zeta\|\zeta-\tilde{\zeta}\|_X, \tag{9.32}$$

where

$$I_\Sigma = \frac{\rho M e^{-\omega}}{1-e^{-\omega}}\left[1 + \frac{\rho M(1+L)}{1-e^{-(\omega-\nu')}}\right] \quad \text{and} \quad I_\zeta = \frac{\rho M^2(1+L)e^{-\omega}}{1-e^{-(\omega-\nu')}}.$$

It follows from (9.12) and (9.32) that $I_\Sigma \leqslant \vartheta$, $I_\zeta \leqslant L$ and hence

$$\|\Phi(\Sigma)(m,\zeta) - \Phi(m,\tilde{\Sigma})(\tilde{\zeta})\|_X \leqslant L\|\zeta-\tilde{\zeta}\|_X + \vartheta\ |||\,\Sigma-\tilde{\Sigma}\,|||. \tag{9.33}$$

Together, (9.30) and (9.33) with $\Sigma = \tilde{\Sigma}$ imply that Φ takes $\mathcal{LB}^s(D,L)$ into itself. From (9.12) and (9.33) with $\zeta = \tilde{\zeta}$ we conclude that Φ is a contraction. Therefore Φ has a unique fixed point $\Sigma^{*,s} = \Phi(\Sigma^{*,s})$ in $\mathcal{LB}^s(D,L)$.

Now, if $\{(x_n^+, x_n^-)\colon n \geqslant m\}$ is a solution of (9.9) which remains bounded as $n \to \infty$, it can be proved that there exist constants $M \geqslant 1$ and $\gamma > 0$ such that

$$\|x_n^+ - \Sigma_n^{*,s}(x_n^-)\|_X \leqslant Me^{\gamma(n-m_0)}\|x_{m_0}^+ - \Sigma_{m_0}^{*,s}(x_{m_0}^-)\|_X, \quad m_0 \geqslant n \geqslant m. \tag{9.34}$$

Making $m_0 \to \infty$ in (9.34) we obtain $x_n^+ = \Sigma_n^{*,s}(x_n^-)$ for each $n \geqslant m$. That also ensures that $\Sigma_n^{*,s}(0) = 0$, since $(0,0)$ is a stationary solution of (9.9).

Now, let us show that $\{(m,w)\in\mathbb{Z}\times X\colon w=(\Sigma_m^{*,u}((I-Q_m)w),(I-Q_m)w)\}\subset W^s(0,0)$. We consider $x_0^-\in(I-Q_m)X$ and the solution $\{x_n^{-,*}\colon n\geqslant m\}$ of the discrete initial value problem

$$\begin{cases} x_n^- = L_{n,m}x_0^- + \sum_m^{n-1} L_{n,k+1}G_k(\Sigma_k^{*,s}(x_k^-),x_k^-), \ n>m,\\ x_m^- = x_0^-. \end{cases}$$

That defines a sequence $\{(\Sigma_n^{*,s}(x_n^{-,*}),x_n^{-,*})\colon n\geqslant m\}$. Recalling (9.29) we can verify that

$$\Sigma_n^{*,s}(x_n^{-,*}) = -\sum_m^{\infty} L_{n,k+1}Q_{k+1}H_k(\Sigma_k^{*,s}(x_k^{-,*}),x_k^{-,*}), \quad n\geqslant m.$$

Hence $\{\Sigma_n^{*,s}(x_n^{-,*})\colon n\geqslant m\}$ solves

$$x_n^+ = L_{n,m}x_m^+ + \sum_m^{n-1} L_{n,k+1}H_k(\Sigma_k^{*,s}(x_k^{-,*}),x_k^{-,*}), \ n\geqslant m,$$

and we conclude that $\{(\Sigma_n^{*,s}(x_n^{-,*}),x_n^{-,*})\colon n\geqslant m\}$ is the solution of (9.9), through $(\Sigma_m^{*,s}(x_0^-),x_0^-)$ at the discrete time m, with $\Sigma_n^{*,s}(x_n^{-,*})\to 0$ as $n\to\infty$. Since $\Sigma_n^{*,s}(0)=0$, the reasoning that takes us to (9.31) can also be used to ensure that

$$\|x_n^{-,*}\|_X \leqslant M\|x_0^-\|_X e^{-(\omega-\nu')(n-m)}.$$

Consequently, $x_n^{-,*}\to 0$ as $n\to\infty$. $\square$

4.2. Local stable manifolds as graphs. In this section we will employ the results of Section 4.1 to obtain the existence and continuity of local stable manifolds when the conditions (9.10) are satisfied only locally.

Let $T\colon \mathbb{Z}\times X\to X$ be a continuously differentiable function such that $T_n(0)=0$ and $D_xT_n(0)=0\in\mathcal{L}(X)$. Assume that the linear discrete evolution process $\mathscr{L}=\{L_{n,m}\colon n\geqslant m\}$ has discrete exponential dichotomy, with constant $M\geqslant 1$ and exponent $\omega>0$. Consider the discrete initial value problem (9.8).

PROPOSITION 9.22. *Assume that the above conditions are satisfied. Then, for any suitably small $\delta>0$ there are δ_1,δ_2, $0<\delta_1<\delta<\delta_2$, and a function $\Sigma^{*,s}\colon\mathbb{Z}\times X\to X$ such that*

$$\begin{aligned}\{(n,x)\in\mathbb{Z}\times X\colon x=(\Sigma_n^{*,s}((I-Q_n)x),(I-Q_n)x),\|x\|<\delta_1\}\subset W_\delta^s(0,0)\\ \subset\{(n,x)\in\mathbb{Z}\times X\colon x=(\Sigma_n^{*,s}((I-Q_n)x),(I-Q_n)x),\|x\|<\delta_2\}.\end{aligned}$$

PROOF. According to Remark 9.19 and Theorem 9.21, we only need to ensure that, given $\delta>0$ there exists $0<\delta'\leqslant\delta$ such that each solution (x_n^+,x_n^-) in the stable manifold that satisfies $\|x_m^+\|_X+\|x_m^-\|_X<\delta'$ is such that $\|x_n^+\|_X+\|x_n^-\|_X<\delta$, for all $n\geqslant m$. That is easily seen since x_n^+ is the solution of

$$\begin{cases} x_n^- = L_{n,m}\zeta + \sum_m^{n-1} L_{n,k+1}G_k(\Sigma_k^{*,s}(x_k^-),x_k^-), \ n>m,\\ x_m^- \in (I-Q_m)X, \end{cases}$$

and, as before, $\|x_n^-\|_X\leqslant M\|x_m^-\|_X e^{-(\omega-\nu')(n-m)}$. Therefore

$$\|x_n^+\|_X = \|\Sigma_n^{*,s}(x_n^-)\|_X \leqslant ML\|x_m^-\|_X e^{-(\omega-\nu)(n-m)}$$

and the proof follows easily. □

5. Global hyperbolic solutions

If $\{S_n\}_{n\in\mathbb{Z}} \subset C^1(X)$, $\mathcal{S} = \{S_{n,m}\colon n \geqslant m\}$ is its associated discrete evolution process and ζ is a global solution of $\mathcal{S}$, let $L_n = D_x S_n(\zeta_n)$ and $T_n z = S_n(z+\zeta_n) - S_n(\zeta_n) - L_n z$ for each $n \in \mathbb{Z}$. Associated to ζ consider the linear discrete evolution process $\mathscr{L} = \{L_{n,m}\colon n \geqslant m\} \subset \mathcal{L}(X)$ associated with the sequence $\{L_n\}_{n\in\mathbb{Z}}$.

DEFINITION 9.23. We say that $\zeta\colon \mathbb{Z} \to X$ is a **hyperbolic global solution** for the discrete evolution process $\mathcal{S}$ if it is bounded and the linear discrete evolution process $\mathscr{L}$ has discrete exponential dichotomy.

In what follows we characterize the global hyperbolic solutions for the discrete evolution process $\mathcal{S}$. Assume that ζ is a hyperbolic global solution of $\mathcal{S}$, that is, $\mathscr{L}$ has discrete exponential dichotomy, with constant $M \geqslant 1$ and exponent $\omega > 0$.

If $\sup_{n\in\mathbb{Z}} \|\zeta_n\|_X < K < \infty$, suppose that

$$\sup_{\|x\|\leqslant K} \{\|T_n(x)\|_X + \|D_x T_n(x)\|_{\mathcal{L}(X)}\} < \infty, \tag{9.35}$$

and that

$$\rho(\delta) = \sup_{\substack{\|x\|,\|y\|\leqslant\delta \\ x\neq y}} \sup_{n\in\mathbb{Z}} \frac{\|T_n(x) - T_n(y)\|_X}{\|x-y\|_X} \to 0 \quad \text{as } \delta \to 0. \tag{9.36}$$

Let $\phi\colon \mathbb{Z} \to X$ be a bounded global solution of $x_{n+1} = S_n x_n$ and define $\eta = \phi - \zeta$. Then, the function $\eta\colon \mathbb{Z} \to X$ is a bounded global solution of $x_{n+1} = L_n x_n + T_n x_n$. Then

$$\eta_n = L_{n,m}\eta_m + \sum_{m}^{n-1} L_{n,k+1} T_k(\eta_k).$$

It is clear from Theorem 9.4 that

$$\eta_n = \sum_{-\infty}^{\infty} G_{n,k+1} T_k(\eta_k),$$

where

$$G_{n,m} = \begin{cases} L_{n,m}(I - Q_m), & n \geqslant m, \\ -L_{n,m}Q_m, & n < m. \end{cases}$$

With this characterization of bounded global solutions, it is not difficult to prove that ζ is isolated in $\ell^\infty(\mathbb{Z}, X)$. In fact, if $\phi\colon \mathbb{Z} \to X$ is a bounded global solution of $x_{n+1} = S_n x_n$ satisfying $\sup_{n\in\mathbb{Z}} \|\phi_n - \zeta_n\|_X \leqslant \delta$, then

$$\sup_{n\in\mathbb{Z}} \|\phi_n - \zeta_n\|_X \leqslant M\frac{1+e^{-\omega}}{1-e^{-\omega}}\rho(\delta) \sup_{n\in\mathbb{Z}} \|\phi_n - \zeta_n\|_X,$$

and making δ suitably small we conclude that $\phi_n = \zeta_n$, for all $n \in \mathbb{Z}$.

Now we will deal with the problem of *permanence* of hyperbolic global solutions under small perturbation. To that end, let $\{S_n^1\}_{n\in\mathbb{Z}}, \{S_n^2\}_{n\in\mathbb{Z}} \subset C^1(X)$ and $\zeta\colon \mathbb{Z} \to X$ be a hyperbolic global solution of $\mathcal{S}_1 = \{S_{n,m}^1\colon n \geqslant m\}$. Hence, the linear discrete evolution process $\mathscr{L}_1 = \{L_{n,m}^1\colon n \geqslant m\}$ given by $L_{n,m}^1 = L_{n-1}^1 \circ \cdots \circ L_m^1$,

$L_n^1 = D_x S_n^1(\zeta_n)$, $n \in \mathbb{Z}$, has discrete dichotomy, with constant $M \geqslant 1$ and exponent $\omega > 0$. Then

$$\eta_n = \sum_{-\infty}^{\infty} G_{n,k+1}^1 T_k^1(\eta_k),$$

where $G_{n,m}^1 \in \mathcal{L}(X)$ and T_n are given by

$$G_{n,m}^1 = \begin{cases} L_{n,m}^1(I - Q_m), & n \geqslant m, \\ -L_{n,m}^1 Q_m, & n < m, \end{cases}$$

and

$$T_n^1(z) = S_n^1(\zeta_n + z) - S_n^1(\zeta_n) - L_n^1 z, \quad z \in X.$$

THEOREM 9.24. *Assume that* $\{T_n^1\}_{n\in\mathbb{Z}}$ *satisfies conditions* (9.36) *and* (9.35), *and choose* $\delta > 0$ *such that*

$$M\frac{1+e^{-\omega}}{1-e^{-\omega}}\rho(\delta)\delta < \frac{\delta}{4}. \tag{9.37}$$

If, for $K = \sup_{n\in\mathbb{Z}} \|\zeta_n\| + \delta$, $\{S_n^1\}_{n\in\mathbb{Z}}$ *and* $\{S_n^2\}_{n\in\mathbb{Z}}$ *satisfy*

$$\sup_{\|x\|\leqslant K} \|S_n^1(x) - S_n^2(x)\|_X + \|D_x S_n^1(x) - D_x S_n^2(x)\|_{\mathscr{L}(X)} < \delta\frac{1-e^{-\omega}}{2M(1+e^{-\omega})}, \tag{9.38}$$

then $\mathcal{S}_2 = \{S_{n,m}^2 \colon n \geqslant m\}$ *has a unique hyperbolic global solution* $\xi\colon \mathbb{Z} \to X$ *such that*

$$\sup_{n\in\mathbb{Z}} \|\zeta_n - \xi_n\|_X < \delta.$$

PROOF. If $y\colon \mathbb{Z} \to X$ is a bounded global solution of $\mathcal{S}_2$, then

$$\phi_n = L_{n,m}^1\phi_m + \sum_{m}^{n-1} L_{n,k+1}^1 \left[T_k^1(\phi_k) + S_k^2(\phi_k + \zeta_k) - S_k^1(\phi_k + \zeta_k)\right], \tag{9.39}$$

where $\phi_n = y_n - \zeta_n$, $n \in \mathbb{Z}$. This implies that any bounded global solution $y\colon \mathbb{Z} \to X$ of $\mathcal{S}_2$ must satisfy $y_n = \zeta_n + \phi_n$, $n \in \mathbb{Z}$, where $\phi\colon \mathbb{Z} \to X$ is a fixed point of

$$\mathcal{T}(\phi)_n = \sum_{-\infty}^{\infty} G_{n,k+1}^1 \tilde{T}_k(\phi_k),$$

where $\tilde{T}_k(\phi_k) = T_k^1(\phi_k) + S_k^2(\phi_k + \zeta_k) - S_k^1(\phi_k + \zeta_k)$. We show that $\mathcal{T}$ has a unique fixed point in

$$B_\delta = \{\phi\colon \mathbb{Z} \to X \text{ such that } \sup_{n\in\mathbb{Z}} \|\phi_n\|_X \leqslant \delta\}$$

for sufficiently small $\delta > 0$. In fact, from (9.36), (9.37) and (9.38), we have

$$\begin{aligned}
\|\mathcal{T}(\phi)_n\|_X &\leqslant M\sum_{-\infty}^{\infty} e^{-\omega|n-k-1|}\|\tilde{T}_k(\phi_k)\|_X \\
&\leqslant M\frac{1+e^{-\omega}}{1-e^{-\omega}} \sup_{n\in\mathbb{Z}} \|S_n^2(y_n) - S_n^1(y_n)\|_X \\
&+ M\frac{1+e^{-\omega}}{1-e^{-\omega}} \sup_{\|z\|\leqslant\delta} \sup_{n\in\mathbb{Z}} \frac{\|S_n^1(\zeta_n + z) - S_n^1(\zeta_n) - L_n^1 z\|_X}{\|z\|_X}\delta \\
&\leqslant \frac{\delta}{2} + M\frac{1+e^{-\omega}}{1-e^{-\omega}}\rho(\delta)\delta < \delta,
\end{aligned}$$

which implies that $\mathcal{T}$ takes B_δ into itself. Using (9.38) it is not difficult to see, for δ sufficiently small, that

$$\|\mathcal{T}(\phi)_n - \mathcal{T}(\psi)_n\|_X \leqslant \frac{1+\delta}{2} \sup_{n\in\mathbb{Z}} \|\phi_n - \psi_n\|_X,$$

showing that $\zeta\colon \mathbb{Z} \to X$ is the only global solution of $\mathcal{S}_2$ that remains in a δ-neighborhood around itself.

Let $L_n^2 = D_x S_n^2(\eta_n)$. Since $\xi\colon \mathbb{Z} \to X$ is uniformly close to $\zeta\colon \mathbb{Z} \to X$, it follows from (9.38) and from Theorem 9.7 that, for δ suitably small, $L_{n,m}^2 = L_{n-1}^2 \circ \cdots \circ L_m^2$ has discrete exponential dichotomy and ξ is hyperbolic. □

6. Stable and unstable manifolds under perturbation

In this section we prove the continuity of the unstable manifolds relative to perturbations. Before that, we consider the continuity of the projections associated to the exponential dichotomy. Let $\mathscr{L}_1 = \{L_{n,m}^{(1)}\colon n \geqslant m\}$ be a linear discrete evolution process that has discrete exponential dichotomy, with constant $M \geqslant 1$, exponent $\omega > 0$ and projections $\{Q_n^{(1)}\}_{n\in\mathbb{Z}}$. If $\{B_n\}_{n\in\mathbb{Z}} \subset \mathcal{L}(X)$ we define $\{L_{n,m}^{(2)}\colon n \geqslant m\}$ by

$$L_{n,m}^{(2)} = L_{n,m}^{(1)} + \sum_{k=m}^{n-1} L_{n,k+1}^{(1)} B_k L_{k,m}^{(2)}, \tag{9.40}$$

that is, $\{L_{n,m}^{(2)}\colon n \geqslant m\}$ is the evolution process given by $y_{n+1} = (L_n^{(1)} + B_n)y_n$. Using Theorem 9.7, given $M_1 > M$ and $\omega_1 < \omega$ there exists $\epsilon > 0$ such that, if

$$\sup_{n\in\mathbb{Z}} \|B_n\|_{\mathcal{L}(X)} < \epsilon,$$

then $\{L_{n,m}^{(2)}\colon n \geqslant m\}$ has discrete exponential dichotomy, with constant $M_1 \geqslant 1$, exponent $w_1 > 0$ and projections $\{Q_n^{(2)}\}_{n\in\mathbb{Z}}$.

From Theorem 9.8 we have the following.

Lemma 9.25. *Let $\{Q_n^{(i)}\}_{n\in\mathbb{Z}}$ be the associated projections to the discrete exponential dichotomy of $\{L_{n,m}^{(i)}\colon n \geqslant m\}$, $i = 1, 2$. Then we have*

$$\sup_{k\in\mathbb{Z}} \|Q_k^{(1)} - Q_k^{(2)}\|_{\mathcal{L}(X)} \leqslant \frac{2M_1^2}{1 - e^{-\omega_1}} \sup_{k\in\mathbb{Z}} \|B_k\|_{\mathcal{L}(X)}. \tag{9.41}$$

6.1. Continuity of unstable and stable manifolds in the C^0 topology. Let $\zeta^{(1)}\colon \mathbb{Z} \to X$ be a hyperbolic global solution of the C^1 discrete evolution process $\{S_{n,m}^{(1)}\colon n \geqslant m\}$ and let $L_n^{(1)} = D_x S_n^{(1)}(\zeta_n^{(1)})$, where $S_n^{(1)} = S_{n+1,n}^{(1)}$. Then, $\{L_{n,m}^{(1)}\colon n \geqslant m\}$ has discrete exponential dichotomy, with constant $M_1 \geqslant 1$ and exponent $\omega_1 > 0$. Given $M_2 > M_1$ and $0 < \omega_2 < \omega_1$, if

$$d(\{S_n^{(1)}\}, \{S_n^{(2)}\}) = \sup_{n\in\mathbb{Z}} \sup_{x\in X} \|S_n^{(1)}(x) - S_n^{(2)}(x)\|_X + \|D_x S_n^{(1)}(x) - D_x S_n^{(2)}(x)\|_{\mathcal{L}(X)} \tag{9.42}$$

is suitably small, $\{S_{n,m}^{(2)}\colon n \geqslant m\}$ has a hyperbolic global solution $\zeta^{(2)}$ near $\zeta^{(1)}$ and $\{L_{n,m}^{(2)}\colon n \geqslant m\}$ has discrete exponential dichotomy, with constant $M_2 \geqslant 1$ and exponent $\omega_2 > 0$, where $L_n^{(2)} = D_x S_n^{(2)}(\zeta_n^{(2)})$, $n \in \mathbb{Z}$. Therefore $\{L_{n,m}^{(i)}\colon n \geqslant m\}$

has discrete exponential dichotomy with constant $M_i \geqslant 1$, exponent $\omega_i > 0$ and projections $\{Q_n^{(i)}\}_{n\in\mathbb{Z}}$.

We know that $y_n^{(i)}$ is a solution of $\{S_{n,m}^{(i)} : n \geqslant m\}$, $i = 1, 2$, if and only if $y_n^{(i)} = x_n^{(i)} + \zeta_n^{(i)}$ where $x_n^{(i)}$ satisfies

$$x_n^{(i)} = L_{n,m}^{(i)} x_m^{(i)} + \sum_m^{n-1} L_{n,k+1}^{(i)} [S_k^{(i)}(x_k^{(i)} + \zeta_k^{(i)}) - S_k^{(i)}(\zeta_k^{(i)}) - L_k^{(i)} x_k^{(i)}]. \tag{9.43}$$

We can decompose a solution $x_n^{(i)}$ of (9.43) as $x_n^{+,i} = Q_n^{(i)}(x_n^{(i)})$ and $x_n^{-,i} = (I - Q_n^{(i)})(x_n^{(i)})$. Then

$$\begin{cases} x_n^{+,i} = L_{n,m}^{(i)} x_m^{+,i} + \displaystyle\sum_m^{n-1} L_{n,k+1}^{(i)} H_k(x_k^{+,i}, x_k^{-,i}), \\ x_n^{-,i} = L_{n,m}^{(i)} x_m^{-,i} + \displaystyle\sum_m^{n-1} L_{n,k+1}^{(i)} G_k(x_k^{+,i}, x_k^{-,i}), \end{cases} \tag{9.44}$$

where

(9.45)

$$\begin{cases} h_n^{(i)}(x_n^{+,i} + x_n^{-,i} + \zeta_n^{(i)}) = S_n^{(i)}(x_n^{+,i} + x_n^{-,i} + \zeta_n^{(i)}) - S_n^{(i)}(\zeta_n^{(i)}) - L_k^{(i)}(x_n^{+,i} + x_n^{-,i}), \\ H_n^{(i)}(x_n^{+,i}, x_n^{-,i}) = Q_{n+1}^{(i)} h_n^{(i)}(x_n^{+,i} + x_n^{-,i} + \zeta_n^{(i)}), \\ G_n^{(i)}(x_n^{+,i}, x_n^{-,i}) = (I - Q_{n+1}^{(i)}) h_n^{(i)}(x_n^{+,i} + x_n^{-,i} + \zeta_n^{(i)}). \end{cases}$$

Let $D > 0$, $L > 0$, $0 < j < 1$, $\rho > 0$ such that (9.12) holds and assume that (9.10) holds for all $x_n^{+,i} \in Q_n^{(i)} X$ and $x_n^{-,i} \in (I - Q_n^{(i)})X$.

THEOREM 9.26. *For $i = 1, 2$, assume that the above conditions are satisfied and (9.12) holds, hence there exists a function $\Sigma_i^{*,u} : \mathbb{Z} \times X \to X$, such that $\Sigma_i^{*,u}(n, x) = \Sigma_i^{*,u}(n, Q_n^{(i)} x) \in (I - Q_n^{(i)})X$ and the unstable manifold $W_i^u(0, 0)$ of the equilibrium solution $(0, 0)$ of (9.44) is given by*

$$W_i^u(0, 0) = \{(m, w) \in \mathbb{Z} \times X : w = (Q_m^{(i)} w, \Sigma_i^{*,u}(m, Q_m^{(i)} w))\};$$

also, for any $\zeta \in X_i^+ = Q_m^{(i)} X$,

$$\Sigma_i^{*,u}(m, \zeta) = \sum_{-\infty}^{m-1} L_{m,k+1}^{(i)} (I - Q_{k+1}^{(i)}) G_k^{(i)}(x_k^{+,i}, \Sigma_i^{*,u}(k, x_k^{+,i})).$$

Furthermore, if $\{(x_n^{+,i}, x_n^{-,i}) : n \geqslant m_0\}$ is a solution of (9.44) then, for $n \geqslant m_0$,

$$\|x_n^{-,i} - \Sigma_i^{*,u}(n, x_n^{+,i})\|_X \leqslant M e^{-(\omega/2-\gamma)(n-m_0)} \|x_{m_0}^{-,i} - \Sigma_i^{*,u}(m_0, x_{m_0}^{+,i})\|_X. \tag{9.46}$$

If in addition to the assumptions above we have

$$\frac{\rho M}{1 - e^{-\omega}} \left[1 + \frac{\rho M(1 + L)}{1 - e^{-(2\omega-\nu)}}\right] \leqslant \frac{1}{2},$$

where $M = \max\{M_1, M_2\}$ and $\omega = \min\{\omega_1, \omega_2\}$, then, for each $r > 0$,

$$|||\, \Sigma_2^{*,u} - \Sigma_1^{*,u} \,|||_r \to 0 \quad \textit{as} \quad d(\{S_n^{(1)}\}, \{S_n^{(2)}\}) \to 0,$$

and

$$|||\, \Sigma_2^{*,u} - \Sigma_1^{*,u} \,|||_r = \sup_{k\leqslant m} \sup_{\|x\|_X \leqslant r} \|\Sigma_2^{*,u}(k, x) - \Sigma_1^{*,u}(k, x)\|_X.$$

PROOF. If $x \in X$ with $\|x\|_X \leqslant r$, then

(9.47)

$$\begin{aligned}
&\Sigma_2^{*,u}(m,x) - \Sigma_1^{*,u}(m,x) = \Sigma_2^{*,u}(m,Q_m^{(2)}x) - \Sigma_1^{*,u}(m,Q_m^{(1)}x) \\
&= \sum_{k=-\infty}^{m-1} [L_{n,k+1}^{(2)}(I - Q_{k+1}^{(2)}) - L_{n,k+1}^{(1)}(I - Q_{k+1}^{(1)})] h_k^{(2)}(x_k^{+,2}, \Sigma_2^{*,u}(k, x_k^{+,2})) \\
&\quad + \sum_{k=-\infty}^{m-1} L_{n,k+1}^{(1)}(I - Q_{k+1}^{(1)})[h_k^{(2)}(x_k^{+,2}, \Sigma_2^{*,u}(k,x_k^{+,2})) - h_k^{(2)}(x_k^{+,1}, \Sigma_1^{*,u}(k,x_k^{+,1}))] \\
&\quad + \sum_{k=-\infty}^{m-1} L_{n,k+1}^{(1)}(I - Q_{k+1}^{(1)})[h_k^{(2)}(x_k^{+,1}, \Sigma_1^{*,u}(k,x_k^{+,1})) - h_k^{(1)}(x_k^{+,1}, \Sigma_1^{*,u}(k,x_k^{+,1}))] \\
&= I_1 + I_2 + I_3.
\end{aligned}$$

Given $\epsilon > 0$, choose $\ell \in \mathbb{N}$ such that

$$\Big\| \sum_{k=-\infty}^{m-\ell-1} [L_{m,k+1}^{(2)}(I - Q_{k+1}^{(2)}) - L_{m,k+1}^{(1)}(I - Q_{k+1}^{(1)})] h_k^{(2)}(x_k^{+,2}, \Sigma_2^{*,u}(k, x_k^{+,2})) \Big\|_X < \frac{\epsilon}{6}.$$

Recalling that

$$\sup_{k\in\mathbb{Z}} \|Q_k^{(1)} - Q_k^{(2)}\|_{\mathcal{L}(X)} \leqslant \frac{2M^2}{1-e^{-\omega}} \sup_{k\in\mathbb{Z}} \|L_k^{(1)} - L_k^{(2)}\|_{\mathcal{L}(X)}$$

and that

$$L_{m,k+1}^{(1)}(I - Q_{k+1}^{(2)}) - L_{m,k+1}^{(2)}(I - Q_{k+1}^{(2)}) = \sum_{i=k+1}^{m-1} L_{m,i+1}^{(1)}(L_i^{(1)} - L_i^{(2)}) L_{i,k+1}^{(2)}(I - Q_{k+1}^{(2)}),$$

choosing $d(\{S_n^{(1)}\}, \{S_n^{(2)}\})$ small we have

$$\begin{aligned}
&\Big\| \sum_{k=m-\ell}^{m-1} [L_{m,k+1}^{(2)}(I - Q_{k+1}^{(2)}) - L_{m,k+1}^{(1)}(I - Q_{k+1}^{(1)})] h_k^{(2)}(x_k^{+,2}, \Sigma_2^{*,u}(k, x_k^{+,2})) \Big\| \\
&\leqslant \rho \sum_{k=m-\ell}^{m-1} \|L_{m,k+1}^{(1)}(I - Q_{k+1}^{(2)}) - L_{m,k+1}^{(2)}(I - Q_{k+1}^{(2)})\| \\
&\quad + \rho \sum_{k=m-\ell}^{m-1} \|L_{m,k+1}^{(1)}(Q_{k+1}^{(2)} - Q_{k+1}^{(1)})\| \\
&\leqslant \rho \sum_{k=m-\ell}^{m-1} \sum_{i=k+1}^{m-1} \|L_{m,i+1}^{(1)}\| \|(L_i^{(1)} - L_i^{(2)})\| \|L_{i,k+1}^{(2)}(I - Q_{k+1}^{(2)})\| \\
&\quad + \rho \sup_{k\in\mathbb{Z}} \|Q_k^{(1)} - Q_k^{(2)}\| \sum_{k=m-\ell}^{m-1} \|L_{m,k+1}^{(1)}\| \\
&\leqslant \rho \sup_{k\in\mathbb{Z}} \|L_k^{(1)} - L_k^{(2)}\| \sup_{k\in\mathbb{Z}} \|L_k^{(1)}\|^{\ell-1} \sum_{k=m-\ell}^{m-1} \sum_{i=k+1}^{m-1} e^{-\omega(i-k-1)} \\
&\quad + \rho \sup_{k\in\mathbb{Z}} \|L_k^{(1)}\|^{\ell} \frac{2M^2\ell}{1-e^{-\omega}} \sup_{k\in\mathbb{Z}} \|L_k^{(1)} - L_k^{(2)}\|
\end{aligned}$$

$$\leqslant \frac{\rho\ell(2M^2+1)}{1-e^{-\omega}}(1+\sup_{k\in\mathbb{Z}}\|L_k^{(1)}\|^{\ell})d(\{S_n^{(1)}\},\{S_n^{(2)}\}) < \frac{\epsilon}{6},$$

and $\|I_1\|_X \leqslant \frac{\epsilon}{3}$. Also choosing $d(\{S_n^{(1)}\},\{S_n^{(2)}\})$ small we obtain $\|I_3\|_X \leqslant \frac{\epsilon}{3}$.

Next we estimate I_2. Recall that

$$x_n^{+,i} = L_{n,m}^{(i)}Q_m^i x - \sum_{k=n}^{m-1} L_{n,k+1}^{(i)}Q_{k+1}^i h_k^{(i)}(x_k^{+,i},\Sigma_i^{*,u}(k,x_k^{+,i})),$$

from which, in a similar way as in (9.18), for some $0<\nu<\omega=\min\{\omega_1,\omega_2\}$, we obtain

$$\|x_n^{+,i}\|_X \leqslant Me^{(\omega-\nu)(n-m)}\|Q_m^{(i)}x\|_X. \tag{9.48}$$

Since

$$\|I_2(\epsilon)\|_X \leqslant \rho M\sum_{-\infty}^{m-1} e^{-\omega(m-k-1)}\Big[\|x_k^{+,2}-x_k^{+,1}\|+\|\Sigma_2^{*,u}(k,x_k^{+,2})-\Sigma_1^{*,u}(k,x_k^{+,1})\|\Big]$$

$$\leqslant \rho M\sum_{-\infty}^{m-1} e^{-\omega(m-k-1)}\Big[(1+L)\|x_k^{+,2}-x_k^{+,1}\|+\|\Sigma_2^{*,u}(k,Q_k^{(2)}x_k^{+,1})-\Sigma_1^{*,u}(k,x_k^{+,1})\|\Big]$$

$$\leqslant \rho M(1+L)\sum_{-\infty}^{m-1} e^{-\omega(m-k-1)}\|x_k^{+,2}-x_k^{+,1}\| + \frac{\rho M}{1-e^{-\omega}}\, |||\, \Sigma_2^{*,u}-\Sigma_1^{*,u}\, |||_r,$$

where

$$|||\, \Sigma_2^{*,u}-\Sigma_1^{*,u}\, |||_r = \sup_{k\leqslant m}\sup_{\|x\|_X\leqslant r}\|\Sigma_2^{*,u}(k,x)-\Sigma_1^{*,u}(k,x)\|_X.$$

Replacing this in (9.47) we obtain

$$\begin{aligned}\big\|\Sigma_2^{*,u}(m,x)-\Sigma_1^{*,u}(m,x)\big\|_X &\leqslant o(1)+\frac{\rho M}{1-e^{-\omega}}\, |||\, \Sigma_2^{*,u}-\Sigma_1^{*,u}\, |||_r \\ &\quad + \rho M(1+L)\sum_{-\infty}^{m-1} e^{-\omega(m-k-1)}\|x_k^{+,2}-x_k^{+,1}\|_X.\end{aligned}\tag{9.49}$$

Also we have

$$\begin{aligned}\|x_n^{+,2}-x_n^{+,1}\|_X &\leqslant \big\|L_{n,m}^{(2)}Q_m^{(2)}x - L_{n,m}^{(1)}Q_m^{(1)}x\big\|_X \\ &\quad + \Big\|\sum_n^{m-1}[L_{n,k+1}^{(2)}Q_{k+1}^{(2)}h_k^{(2)}(x_k^{+,2},\Sigma_2^{*,u}(k,x_k^{+,2})) \\ &\qquad\qquad - L_{n,k+1}^{(1)}Q_{k+1}^{(1)}h_k^{(1)}(x_k^{+,1},\Sigma_1^{*,u}(k,x_k^{+,1}))]\Big\|_X \\ &\leqslant \big\|L_{n,m}^{(2)}Q_m^{(2)}x - L_{n,m}^{(1)}Q_m^{(1)}x\big\|_X \\ &\quad + \|\sum_n^{m-1}[L_{n,k+1}^{(2)}Q_{k+1}^{(2)} - L_{n,k+1}^{(1)}Q_{k+1}^{(1)}]h_k^{(1)}(x_k^{+,1},\Sigma_1^{*,u}(k,x_k^{+,1}))\|_X \\ &\quad + \|\sum_n^{m-1} L_{n,k+1}^{(2)}Q_{k+1}^{(2)}\Big[h_k^{(2)}(x_k^{+,1},\Sigma_1^{*,u}(k,x_k^{+,1})) - h_k^{(1)}(x_k^{+,1},\Sigma_1^{*,u}(k,x_k^{+,1}))\Big]\|_X \\ &\quad + \|\sum_n^{m-1} L_{n,k+1}^{(2)}Q_{k+1}^{(2)}\Big[h_k^{(2)}(x_k^{+,2},\Sigma_2^{*,u}(k,x_k^{+,2})) - h_k^{(2)}(x_k^{+,1},\Sigma_1^{*,u}(k,x_k^{+,1}))\Big]\|_X\end{aligned}$$

$$\leqslant o(1) + \rho M \sum_{k=n}^{m-1} e^{\omega(n-k-1)} \Big[(1+L)\|x_k^{+,2} - x_k^{+,1}\|_X + \|\Sigma_2^{*,u}(k, Q_k^{(2)} x_k^{+,1}) - \Sigma_1^{*,u}(k, x_k^{+,1})\|_X \Big]$$

$$\leqslant o(1) + \frac{\rho M}{1-e^{-\omega}} |||\, \Sigma_2^{*,u} - \Sigma_1^{*,u} \,|||_r + \rho M(1+L) \sum_{k=n}^{m-1} e^{\omega(n-k-1)} \|x_k^{+,2} - x_k^{+,1}\|_X,$$

and the Downward Gronwall Inequality implies that

$$\|x_n^{+,2} - x_n^{+,1}\|_X \leqslant \big(o(1) + \frac{\rho M}{1-e^{-\omega}} |||\, \Sigma_2^{*,u} - \Sigma_1^{*,u} \,|||_r\big) e^{(\omega-\nu)(n-m)}. \tag{9.50}$$

Applying (9.50) in (9.49) we obtain that

$$\begin{aligned} \|\Sigma_2^{*,u}(m,x) - \Sigma_1^{*,u}(m,x)\|_X &\leqslant o(1) + \frac{\rho M}{1-e^{-\omega}} |||\, \Sigma_2^{*,u} - \Sigma_1^{*,u} \,|||_r \\ &+ \rho M(1+L) \sum_{-\infty}^{m} e^{-(2\omega-\nu)(m-k)} \Big[o(1) + \frac{\rho M}{1-e^{-\omega}} |||\, \Sigma_2^{*,u} - \Sigma_1^{*,u} \,|||_r \Big] \\ &\leqslant o(1) + \frac{\rho M}{1-e^{-\omega}} \Big[1 + \frac{\rho M(1+L)}{1-e^{-(2\omega-\nu)}} \Big] |||\, \Sigma_2^{*,u} - \Sigma_1^{*,u} \,|||_r \\ &= o(1) + \frac{1}{2} |||\, \Sigma_2^{*,u} - \Sigma_1^{*,u} \,|||_r. \end{aligned} \tag{9.51}$$

It follows from (9.51) that $|||\, \Sigma_2^{*,u} - \Sigma_1^{*,u} \,|||_r \to 0$ as $\epsilon \to 0$, and the proof is complete. □

Analogously, we prove the continuity of the stable manifolds in the C^0-topology.

6.2. Unstable and stable manifolds are C^1 and $C^{1+\delta}$. Assume that $h \in C^1(X)$ in (9.45) is such that

$$\sup_{n\in\mathbb{Z}} \|h'(n,x) - h'(n,y)\|_{\mathcal{L}(X)} \leqslant w_{h'}(\|x-y\|_X),$$

with $w_{h'}\colon \mathbb{R}^+ \to \mathbb{R}^+$ increasing and such that $\sup_{t\in\mathbb{R}^+} w_{h'}(t) \leqslant 2\delta$ and $\lim_{t\to 0^+} w_{h'}(t) = 0$. Let $\Sigma \in \mathcal{LB}(D,L) \cap C^1(X)$. Then

$$\sup_{(n,x)\in\mathbb{Z}\times X} \|\Sigma_n'(x)\|_{\mathcal{L}(X)} \leqslant L. \tag{9.52}$$

Assume that

$$\sup_{n\in\mathbb{Z}} \|\Sigma_n'(x) - \Sigma_n'(y)\|_{\mathcal{L}(X)} \leqslant w_{\Sigma'}(\|x-y\|) \leqslant 3\bar{w}_{h'}(\|x-y\|), \tag{9.53}$$

where $w_{\Sigma'}\colon \mathbb{R}^+ \to \mathbb{R}^+$ is increasing and such that $\sup_{t\in\mathbb{R}^+} w_{\Sigma'}(t) \leqslant 2L$, $\lim_{t\to 0^+} w_{\Sigma'}(t) = 0$ and

$$\bar{w}_{h'}(t) = \sum_{-\infty}^{m-1} M^{2+\delta}(1+L)^{1+\delta} w_{h'}(M(1+L)(1-\rho M(1+L)e^{-\omega})^{k-m}\, t) e^{\omega''(k-m)}.$$

Let $y_k = y_k(m,\zeta,\Sigma)$, $k \leqslant m$, be the solution of

$$y_k(\zeta) = L_{k,m} Q_m \zeta - \sum_{i=k}^{m-1} L_{k,i+1} H_i(y_i(\zeta), \Sigma_i(y_i(\zeta))), \quad k < m.$$

Then $y'_k = \frac{\partial}{\partial\zeta} y_k(m,\zeta,\Sigma)$, for $k < m$, satisfies

$$y'_k(\zeta) = L_{k,m}Q_m - \sum_{i=k}^{m-1} L_{k,i+1}(H_i)_x(y_i(\zeta), \Sigma_i(y_i(\zeta)))(y'_i(\zeta), \Sigma'_i(y_i(\zeta))y'_i(\zeta)).$$

Thus

$$e^{-\omega k}\|y'_k(\zeta)\| \leqslant Me^{-\omega m} + M\rho(1+L)e^{-\omega} \sum_{i=k}^{m-1} e^{-\omega i}\|y'_i(\zeta)\|, \quad k < m,$$

and, from Lemma 9.15, we obtain

$$\|y'_k(\zeta)\| \leqslant Me^{\omega'(k-m)} \tag{9.54}$$

where $\omega' = \omega - \log(1 - M\rho(1+L)e^{-\omega})^{-1}$.

From Theorem 9.17, $W^u(0,0) = \{(m,\zeta) \in \mathbb{Z} \times X \colon \zeta = (Q_m\zeta, \Sigma_m(\zeta)\}$ where, for all $m \in \mathbb{Z}$ and $\zeta \in X$, $\Sigma_m \colon X \to X$ satisfies $\Sigma_m(\zeta) = \Sigma_m(Q_m\zeta)$, $\Sigma_m(0) = 0$, Lip $\Sigma_m \leqslant L$ and is given by

$$\begin{cases} x_n^+ = L_{n,m}Q_m\zeta - \displaystyle\sum_{k=n}^{m-1} L_{n,k+1}H_k(x_k^+, \Sigma_k(x_k^+)), \; n < m, \\ \Sigma_m(\zeta) = \displaystyle\sum_{-\infty}^{m-1} L_{m,k+1}(I - Q_{k+1})G_k(x_k^+, \Sigma_k(x_k^+)). \end{cases}$$

From now on, to avoid notation pollution, we will omit the indexes $\mathcal{L}(X)$ and X on the norms. Now note that

$$\begin{aligned} \|y'_k(\zeta) - y'_k(\bar\zeta)\| &\leqslant \| \sum_{i=k}^{m-1} L_{k,i+1} \left[(H_i)_x - \overline{(H_i)}_x\right] (y'_i(\zeta), \Sigma'_i(y_i(\zeta))y'_i(\zeta))\| \\ &\quad + \| \sum_{i=k}^{m-1} L_{k,i+1}\overline{(H_i)}_x \left[(y'_i(\zeta), \Sigma'_i(y_i(\zeta))y'_i(\zeta)) - (y'_i(\bar\zeta), \Sigma'_i(y_i(\bar\zeta))y'_i(\bar\zeta))\right] \| \\ &\leqslant \sum_{i=k}^{m-1} M(1+L)^1 e^{\omega(k-i-1)} w_{h'}((1+L)\|y_i(\zeta) - y_i(\bar\zeta)\|)\|y'_i(\zeta)\| \\ &\quad + \sum_{i=k}^{m-1} Me^{\omega(k-i-1)} \|\overline{(H_i)}_x\| \Big(w_{\Sigma'}(\|y_i(\zeta) - y_i(\bar\zeta)\|)\|y'(\zeta)\| \\ &\qquad\qquad + (1+L)\|y'_i(\zeta) - y'_i(\bar\zeta)\| \Big) \\ &\leqslant \sum_{i=k}^{m-1} M(1+L)^{1+\delta} e^{\omega(k-i-1)} \Big(w_{h'}((1+L)\|y_i(\zeta) - y_i(\bar\zeta)\|) \\ &\qquad\qquad + \rho\, w_{\Sigma'}(\|y_i(\zeta) - y_i(\bar\zeta)\|) \Big) \|y'_i(\zeta)\| \\ &\quad + \sum_{i=k}^{m-1} \rho M(1+L)^2 e^{\omega(k-i-1)} \|y'_i(\zeta) - y'_i(\bar\zeta)\| \\ &\leqslant \sum_{i=k}^{m-1} M^{2+\delta}(1+L)^{1+\delta} e^{\omega(k-i-1)} (w_{h'} + \rho w_{\Sigma'})(A_i) e^{\omega'(i-m)} \end{aligned}$$

$$
\begin{aligned}
&+ \sum_{i=k}^{m-1} \rho M(1+L)^2 e^{-\omega} e^{\omega(k-i)} \|y_i'(\zeta) - y_i'(\bar\zeta)\| \\
&\leqslant \sum_{i=k}^{m-1} M^{2+\delta}(1+L)^{1+\delta} e^{\omega(k-i-1)} (w_{h'} + \rho w_{\Sigma'})(A_i)^{i-m} e^{\omega''(i-m)} \\
&+ \rho ML(1+L)^2 e^{-\omega} \sum_{i=k}^{m-1} e^{\omega''(k-i)} \|y_i'(\zeta) - y_i'(\bar\zeta)\|,
\end{aligned}
$$

where $(H_i)_x = (H_i)_x(y_i(\zeta), \Sigma_i(y_i(\zeta)))$, $\overline{(H_i)_x} = (H_i)_x(y_i(\bar\zeta), \Sigma_i(y_i(\bar\zeta)))$, $A_i := M(1+L)\|\zeta - \bar\zeta\| e^{\omega(i-m)}(1 - \rho M(1+L)e^{-\omega})^{i-m}$, we have used (9.54) and (9.18) and $\omega'' = \omega' - \delta \log{(1 - M\rho(1+L)e^{-\omega})^{-1}}$. From the Downward Gronwall Lemma (Lemma 9.15) and for ρ sufficiently small

$$
\begin{aligned}
&\|y_k'(\zeta) - y_k'(\bar\zeta)\| \\
&\leqslant \sum_{i=k}^{m-1} M^{2+\delta}(1+L)^{1+\delta} e^{\omega(k-i-1)} (w_{h'} + \rho w_{\Sigma'})(A_i) e^{\omega'(i-m)} (1 - \rho M(1+L)e^{-\omega})^{k-m} \\
&= \sum_{i=k}^{m-1} M^{2+\delta}(1+L)^{1+\delta} e^{\omega(k-i-1)} (w_{h'} + \rho w_{\Sigma'})(A_i) e^{\omega'(i-m)} e^{(\omega'-\omega)(k-i)+(\omega''-\omega')(i-m)} \\
&\leqslant e^{-\omega} \sum_{i=k}^{m-1} M^{2+\delta}(1+L)^{1+\delta} e^{\omega'(k-i)} (w_{h'} + \rho w_{\Sigma'})(A_i) e^{\omega''(i-m)} \\
&= w_{y'}(\|\zeta - \bar\zeta\|) \leqslant e^{-\omega} \bar w_{h'}(\|\zeta - \bar\zeta\|) + \frac{3\rho M^{2+\delta}(1+L)^{1+\delta}}{1 - e^{-\omega''}} \bar w_{h'}(M(1+L)\|\zeta - \bar\zeta\|) \\
&\leqslant 2\bar w_{h'}(\|\zeta - \bar\zeta\|).
\end{aligned}
$$

Recall that

$$
\Phi(\Sigma)(m,\zeta) = \sum_{-\infty}^{m-1} L_{m,k+1}(I - Q_{k+1}) G_k(y_k(\zeta), \Sigma_k(y_k(\zeta))). \tag{9.55}
$$

Hence, if $\Phi(\Sigma)'(m,\zeta) = \frac{\partial}{\partial\zeta}\Phi(\Sigma)(m,\zeta)$,

(9.56)

$$
\Phi(\Sigma)'(m,\zeta) = \sum_{-\infty}^{m-1} L_{m,k+1}(I - Q_{k+1})(G_k)_x(y_k(\zeta), \Sigma_k(y_k(\zeta)))(y_k'(\zeta), \Sigma_k'(y_k(\zeta))y_k'(\zeta))
$$

and it is easy to see that, making ρ suitably small, $\|\frac{\partial}{\partial\zeta}\Phi(\Sigma)(m,\zeta)\| \leqslant L$. To see that

$$
\|\Phi(\Sigma)'(m,\zeta) - \Phi(\Sigma)'(m,\bar\zeta)\| \leqslant w_{\Phi'}(\|\zeta - \bar\zeta\|)
$$

we use the estimate for $\|y_k'(\zeta) - y_k'(\bar\zeta)\|$ in (9.56). Now

$$
\begin{aligned}
&\|\Phi(\Sigma)'(m,\zeta) - \Phi(\Sigma)'(m,\bar\zeta)\| \\
&\leqslant \sum_{-\infty}^{m-1} M e^{-\omega(m-k-1)} \|(G_k)_x(y_k(\zeta), \Sigma_k(y_k(\zeta))) - (G_k)_x(y_k(\bar\zeta), \Sigma_k(y_k(\bar\zeta)))\| \\
&\qquad \cdot \|(y_k'(\zeta), \Sigma_k'(y_k(\zeta))y_k'(\zeta))\|
\end{aligned}
$$

$$
\begin{aligned}
&+\sum_{-\infty}^{m-1} Me^{-\omega(m-k-1)}\|(G_k)_x(y_k(\bar\zeta),\Sigma_k(y_k(\bar\zeta)))\| \\
&\qquad\qquad\cdot\|(y_k'(\zeta),\Sigma_k'(y_k(\zeta))y_k'(\zeta))-(y_k'(\bar\zeta),\Sigma_k'(y_k(\bar\zeta))y_k'(\bar\zeta)\| \\
&\leqslant M(1+L)^{1+\delta}\sum_{-\infty}^{m-1} e^{-\omega(m-k-1)}w_{h'}((1+L)\|y_k(\zeta)-y_k(\bar\zeta)\|)\|y_k'(\zeta)\| \\
&\quad+\rho M(1+L)\sum_{-\infty}^{m-1} e^{-\omega(m-k-1)}\|y_k'(\zeta)-y_k'(\bar\zeta)\| \\
&\quad+\rho M\sum_{-\infty}^{m-1} e^{-\omega(m-k-1)}w_{\Sigma'}(\|y_k(\zeta)-y_k(\bar\zeta)\|)\|y_k'(\zeta)\| \\
&\leqslant M^{2+\delta}(1+L)^{1+\delta}\sum_{-\infty}^{m-1} e^{-\omega(m-k-1)}(w_{h'}\rho w_{\Sigma'})(A_i)e^{\omega'(k-m)} \\
&\quad+\frac{\rho M(1+L)}{1-e^{-\omega}}w_{y'}(\|\zeta-\bar\zeta\|) \\
&\leqslant 3\bar w_{h'}(\|\zeta-\bar\zeta\|),
\end{aligned}
$$

with $A_i = M(1+L)\|\zeta-\bar\zeta\|e^{\omega(i-m)}(1-\rho M(1+L)e^{-\omega})^{i-m}$, as before. From that we obtain that the fixed point Σ^* of Φ must satisfy (4.5). In fact, we first take Σ differentiable and satisfying (9.52) and (9.53) and notice that $\Phi^j(\Sigma)$ also is differentiable and satisfies (9.52) and (9.53). It follows that

$$
\begin{aligned}
&\|\Phi^j(\Sigma)_n(\zeta+h)-\Phi^j(\Sigma)_n(\zeta)-\Phi^j(\Sigma)_n(\zeta_0+h)+\Phi^j(\Sigma)_n(\zeta_0)\| \\
&\qquad=\left\|\int_0^1[\Phi^j(\Sigma)_n'(\zeta+sh)-\Phi^j(\Sigma)_n'(\zeta_0+sh)]hds\right\| \\
&\qquad\leqslant 6\bar w_{h'}(\|\zeta-\zeta_0\|)\|h\| = o(\|h\|),
\end{aligned}
$$

as $(\zeta,h)\to(\zeta_0,0)$ and $\Phi^j(\Sigma)$ satisfies (4.5). Noticing that $\Sigma^* = \lim_{j\to\infty}\Phi^j(\Sigma)$ we have that it must satisfy (4.5) and therefore it is continuously differentiable.

Now, we assume that $h\in C^{1+\delta}(X)$ in (9.45), that is,

$$
\begin{cases}
\|(H_n)_x(x)-(H_n)_x(y)\|_{\mathcal{L}(X)}\leqslant C_{H'}\|x-y\|^\delta, \\
\|(G_n)_x(x)-(G_n)_x(y)\|_{\mathcal{L}(X)}\leqslant C_{G'}\|x-y\|^\delta,
\end{cases}
$$

and let $\Sigma\in\mathcal{LB}(D,L)\cap C^{1+\delta}(X)$. Then

$$
\sup_{x\in X}\|\Sigma_n'(t,x)\|_{\mathcal{L}(X)}\leqslant L,
$$

and

$$
\|\Sigma_n'(t,x)-\Sigma_n'(t,y)\|_{\mathcal{L}(X)}\leqslant C_{\Sigma'}\|x-y\|^\delta.
$$

Let $y_k = y_k(m,\zeta,\Sigma)$, $k\leqslant m$, be the solution of

$$
y_k(\zeta) = L_{k,m}Q_m\zeta - \sum_{i=k}^{m-1} L_{k,i+1}H_i(y_i(\zeta),\Sigma_i(y_i(\zeta))),\quad k<m.
$$

Then $y_k' = \frac{\partial}{\partial\zeta} y_k(m,\zeta,\Sigma)$, for $k < m$, satisfies

$$y_k'(\zeta) = L_{k,m}Q_m - \sum_{i=k}^{m-1} L_{k,i+1}(H_i)_x(y_i(\zeta), \Sigma_i(y_i(\zeta)))(y_i'(\zeta), \Sigma_i'(y_i(\zeta))y_i'(\zeta)).$$

Thus

$$e^{-\omega k}\|y_k'(\zeta)\| \leqslant Me^{-\omega m} + M\rho(1+L)e^{-\omega}\sum_{i=k}^{m-1} e^{-\omega i}\|y_i'(\zeta)\|, \quad k < m,$$

and from the Downward Gronwall Lemma

$$\|y_k'(\zeta)\| \leqslant Me^{\omega'(k-m)}, \tag{9.57}$$

where $\omega' = \omega - \log(1 - M\rho(1+L)e^{-\omega})^{-1}$. Also

$$\begin{aligned}
\|y_k'(\zeta) - y_k'(\bar\zeta)\| &\leqslant \|\sum_{i=k}^{m-1} L_{k,i+1}\left[(H_i)_x - \overline{(H_i)}_x\right](y_i'(\zeta), \Sigma_i'(y_i(\zeta))y_i'(\zeta))\| \\
&\quad + \|\sum_{i=k}^{m-1} L_{k,i+1}\overline{(H_i)}_x\left[(y_i'(\zeta), \Sigma_i'(y_i(\zeta))y_i'(\zeta)) - (y_i'(\bar\zeta), \Sigma_i'(y_i(\bar\zeta))y_i'(\bar\zeta))\right]\| \\
&\leqslant \sum_{i=k}^{m-1} MC_{H'}(1+L)^{1+\delta}e^{\omega(k-i-1)}\|y_i(\zeta) - y_i(\bar\zeta)\|^\delta\|y_i'(\zeta)\| \\
&\quad + \sum_{i=k}^{m-1} Me^{\omega(k-i-1)}\|\overline{(H_i)}_x\|\Big(C_{\Sigma'}\|y_i(\zeta) - y_i(\bar\zeta)\|^\delta\|y'(\zeta)\| \\
&\qquad\qquad + (1+L)\|y_i'(\zeta) - y_i'(\bar\zeta)\|\Big) \\
&\leqslant \sum_{i=k}^{m-1} M(1+L)^{1+\delta}e^{\omega(k-i-1)}(C_{H'} + C_{\Sigma'}\|\overline{(H_i)}_x\|)\|y_i(\zeta) - y_i(\bar\zeta)\|^\delta\|y_i'(\zeta)\| \\
&\quad + \sum_{i=k}^{m-1}\|\overline{(H_i)}_x\|M(1+L)e^{\omega(k-i-1)}\|y_i'(\zeta) - y_i'(\bar\zeta)\| \\
&\leqslant \sum_{i=k}^{m-1} M^{2+\delta}(1+L)^{1+\delta}e^{\omega(k-i-1)}(C_{H'} + C_{\Sigma'}\rho)(1 - \rho M(1+L)e^{-\omega})^{\delta(i-m)} \\
&\qquad\qquad \cdot e^{\omega'(i-m)}\|\zeta - \tilde\zeta\|_X^\delta \\
&\quad + \sum_{i=k}^{m-1}\rho M(1+L)e^{-\omega}e^{\omega(k-i)}\|y_i'(\zeta) - y_i'(\bar\zeta)\| \\
&\leqslant \sum_{i=k}^{m-1} M^{2+\delta}(1+L)^{1+\delta}e^{\omega(k-i-1)}(C_{H'} + C_{\Sigma'}\rho)e^{\omega''(i-m)}\|\zeta - \tilde\zeta\|_X^\delta \\
&\quad + \sum_{i=k}^{m-1}\rho M(1+L)e^{-\omega}e^{\omega(k-i)}\|y_i'(\zeta) - y_i'(\bar\zeta)\| \\
&\leqslant M^{2+\delta}(1+L)^{1+\delta}e^{-\omega''}(C_{H'} + C_{\Sigma'}\rho)\sum_{i=k}^{m-1} e^{(\omega-\omega'')(k-i-1)}\, e^{\omega''(k-m)}\|\zeta - \tilde\zeta\|_X^\delta
\end{aligned}$$

$$+\sum_{i=k}^{m-1} \rho M(1+L)e^{-\omega}e^{\omega''(k-i)}\|y_i'(\zeta)-y_i'(\bar\zeta)\|$$

$$\leqslant M^{2+\delta}(1+L)^{1+\delta}\frac{e^{-\omega}}{1-e^{-(\omega-\omega'')}}(C_{H'}+C_{\Sigma'}\rho)\; e^{\omega''(k-m)}\|\zeta-\tilde\zeta\|_X^\delta$$

$$+\rho M(1+L)e^{-\omega}\sum_{i=k}^{m-1} e^{\omega''(k-i)}\|y_i'(\zeta)-y_i'(\bar\zeta)\|,$$

where $(H_i)_x = (H_i)_x(y_i(\zeta),\Sigma_i(y_i(\zeta)))$, $\overline{(H_i)}_x = (H_i)_x(y_i(\bar\zeta),\Sigma_i(y_i(\bar\zeta)))$, and we have used (9.57), (9.18) and $\omega'' = \omega' - \delta\log(1-M\rho(1+L)e^{-\omega})^{-1}$. From the Downward Gronwall Lemma (Lemma 9.15) we obtain

$$\begin{aligned}\|y_k'(\zeta)-y_k'(\bar\zeta)\| &\leqslant M^{2+\delta}(1+L)^{1+\delta}\frac{(C_{H'}+C_{\Sigma'}\rho)e^{-\omega}}{1-e^{-(\omega-\omega'')}}e^{\omega''(k-m)}\\ &\qquad\cdot(1-\rho M(1+L)e^{-\omega})^{k-m}\|\zeta-\tilde\zeta\|^\delta\\ &\leqslant M^{2+\delta}(1+L)^{1+\delta}\frac{(C_{H'}+C_{\Sigma'}\rho)e^{-\omega}}{1-e^{-(\omega-\omega'')}}e^{\omega'''(k-m)}\|\zeta-\tilde\zeta\|^\delta,\end{aligned}\tag{9.58}$$

where $\omega''' = \omega'' - \log(1-\rho M(1+L)e^{-\omega})^{-1}$. Recall that

$$\Phi(\Sigma)(m,\zeta) = \sum_{-\infty}^{m-1} L_{m,k+1}(I-Q_{k+1})G_k(y_k(\zeta),\Sigma_k(y_k(\zeta))).\tag{9.59}$$

Hence, if $\Phi(\Sigma)'(m,\zeta) = \frac{\partial}{\partial\zeta}\Phi(\Sigma)(m,\zeta)$, we have

$$\begin{aligned}\Phi(\Sigma)'(m,\zeta) = \sum_{-\infty}^{m-1} L_{m,k+1}(I-Q_{k+1})&(G_k)_x(y_k(\zeta),\Sigma_k(y_k(\zeta)))\\ &\cdot(y_k'(\zeta),(\Sigma_k)_x(y_k(\zeta))y_k'(\zeta)),\end{aligned}\tag{9.60}$$

and it is easy to see that, making ρ suitably small, $\|\frac{\partial}{\partial\zeta}\Phi(\Sigma)(m,\zeta)\| \leqslant L$. To see that

$$\|\Phi(\Sigma)'(m,\zeta) - \Phi(\Sigma)'(m,\bar\zeta)\| \leqslant C\|\zeta-\bar\zeta\|^\delta,$$

we use (9.58) in (9.60). Finally

$$\begin{aligned}&\|\Phi(\Sigma)'(m,\zeta)-\Phi(\Sigma)'(m,\bar\zeta)\|\\ &\leqslant \sum_{-\infty}^{m-1} Me^{-\omega(m-k-1)}\|(G_k)_x(y_k(\zeta),\Sigma_k(y_k(\zeta))) - (G_k)_x(y_k(\bar\zeta),\Sigma_k(y_k(\bar\zeta)))\|\\ &\qquad\cdot\|(y_k'(\zeta),(\Sigma_k)_x(y_k(\zeta))y_k'(\zeta))\|\\ &\quad+\sum_{-\infty}^{m-1} Me^{-\omega(m-k-1)}\|(G_k)_x(y_k(\bar\zeta),\Sigma_k(y_k(\bar\zeta)))\|\\ &\qquad\cdot\|(y_k'(\zeta),(\Sigma_k)_x(y_k(\zeta))y_k'(\zeta)) - (y_k'(\bar\zeta),(\Sigma_k)_x(y_k(\bar\zeta))y_k'(\bar\zeta))\|\\ &\leqslant MC_{G'}(1+L)^{1+\delta}\sum_{-\infty}^{m-1} e^{-\omega(m-k-1)}\|y_k(\zeta)-y_k(\bar\zeta)\|^\delta\|y_k'(\zeta)\|\\ &\quad+\rho M(1+L)\sum_{-\infty}^{m-1} e^{-\omega(m-k-1)}\|y_k'(\zeta)-y_k'(\bar\zeta)\|\end{aligned}$$

$$
\begin{aligned}
&+\rho M \sum_{-\infty}^{m-1} e^{-\omega(m-k-1)} C_{\Sigma'} \|y_k(\zeta)-y_k(\bar{\zeta})\|^{\delta} \|y_k'(\zeta)\| \\
&\leqslant M^{2+\delta}(C_{G'}+\rho C_{\Sigma'})(1+L)^{1+\delta}\|\zeta-\tilde{\zeta}\|^{\delta} \\
&\qquad\qquad \cdot \sum_{-\infty}^{m-1} e^{-\omega(m-k-1)}(1-\rho M(1+L)e^{-\omega})^{\delta(k-m)} e^{\omega'(k-m)} \\
&+\rho M^{3+\delta}(1+L)^{2+\delta} \frac{(C_{H'}+C_{\Sigma'}\rho)e^{-(\omega-\omega''')}}{(1-e^{-(\omega-\omega'')})^2} \|\zeta-\tilde{\zeta}\|^{\delta} \\
&\leqslant (C_{G'}+\rho C_{\Sigma'}) M^{2+\delta}(1+L)^{1+\delta} \frac{e^{-\omega''}}{1-e^{-(\omega-\omega'')}} \|\zeta-\tilde{\zeta}\|^{\delta} \\
&+\rho M^{3+\delta}(1+L)^{2+\delta} \frac{(C_{H'}+C_{\Sigma'}\rho)e^{-(\omega-\omega''')}}{(1-e^{-(\omega-\omega'')})^2} \|\zeta-\tilde{\zeta}\|^{\delta},
\end{aligned}
$$

and choosing C_G sufficiently small we can make the Hölder constant for $\Phi(\Sigma)'(m,\zeta)$ equal to $C_{\Sigma'}$, and proves that the unstable manifolds are $C^{1+\delta}$. Analogously, the result follows for the stable manifolds.

6.3. Continuity of unstable and stable manifolds in the C^1 topology. In this subsection, our goal is to see that the fixed point $\Sigma^{*,u}$ of Φ behaves continuously in the C^1 topology, with respect to C^1 perturbations, that is, both $\Sigma^{*,u}$ and $(\Sigma^{*,u})'$ behave continuously.

The continuity of $\Sigma^{*,u}$ was already proven in Subsection 6.1. Now, to see that $(\Sigma^{*,u})'$ behaves continuously with respect to C^1 perturbations, we note that all terms under the summation sign of

$$
\begin{aligned}
&(\Sigma^{*,u})'_m(\zeta) = \\
&\sum_{k=-\infty}^{m-1} L_{m,k+1}(I-Q_{k+1})(G_k)_x(y_k(\zeta),\Sigma_k^{*,u}(y_k(\zeta)))(y_k'(\zeta),(\Sigma^{*,u})_k'(y_k(\zeta))y_k'(\zeta))
\end{aligned}
$$

behave continuously with respect to C^1 perturbations $\{S_n^\eta\}_{n\in\mathbb{Z}}$ of $\{S_n\}_{n\in\mathbb{Z}}$. That is, $\{L_{m,n}\colon m\geqslant n\}$, $\{Q_n\}_{n\in\mathbf{Z}}$, $\{G_n\}_{n\in\mathbf{Z}}$, $\{\Sigma_n^{*,u}\}_{n\in\mathbf{Z}}$, $\{y_n(\zeta)\}_{n\in\mathbf{Z}}$ and $\{y_n'(\zeta)\}_{n\in\mathbf{Z}}$ behave continuously with respect to C^1 perturbations $\{S_n^\eta\}_{n\in\mathbb{Z}}$ of $\{S_n\}_{n\in\mathbb{Z}}$. Using arguments as in Subsection 6.1, we obtain the desired result. In fact, we have for $\Sigma^\eta=\Sigma^{*,u,\eta}$ and $\Sigma=\Sigma^{*,u,0}$, that

$$
\begin{aligned}
&\Sigma_m^\eta(\zeta+h)-\Sigma_m^\eta(\zeta)-\Sigma_m(\zeta+h)+\Sigma_m(\zeta) \\
&=\sum_{k=-\infty}^{m-1} L_{m,k+1}^\eta(I-Q_{k+1}^\eta)[G_k^\eta(y_k^\eta(\zeta+h),\Sigma_k^\eta(y_k^\eta(\zeta+h)))-G_k^\eta(y_k^\eta(\zeta),\Sigma_k^\eta(y_k^\eta(\zeta)))] \\
&\quad-\sum_{k=-\infty}^{m-1} L_{m,k+1}(I-Q_{k+1})[G_k(y_k(\zeta+h),\Sigma_k^\eta(y_k(\zeta+h)))-G_k(y_k^\eta(\zeta),\Sigma_k^\eta(y_k^\eta(\zeta)))] \\
&=\sum_{k=-\infty}^{m-1} L_{m,k+1}^\eta(I-Q_{k+1}^\eta)(G_k^\eta)_x(\iota(\eta)) \\
&\qquad\qquad \cdot (y_k^\eta(\zeta+h)-y_k^\eta(\zeta),\Sigma_k^\eta(y_k^\eta(\zeta+h))-\Sigma_k^\eta(y_k^\eta(\zeta)))
\end{aligned}
$$

$$
\begin{aligned}
&- \sum_{k=-\infty}^{m-1} L_{m,k+1}(I - Q_{k+1})(G_k)_x(\iota(0)) \\
&\qquad\qquad \cdot (y_k(\zeta+h) - y_k(\zeta), \Sigma_k(y_k(\zeta+h)) - \Sigma_k(y_k(\zeta))) \\
&= \sum_{k=-\infty}^{m-1} [L^{\eta}_{m,k+1}(I - Q^{\eta}_{k+1}) - L_{m,k+1}(I - Q_{k+1})(G^{\eta}_k)_x(\iota(\eta)) \\
&\qquad\qquad \cdot [(y^{\eta}_k(\zeta+h) - y^{\eta}_k(\zeta), \Sigma^{\eta}_k(y^{\eta}_k(\zeta+h)) - \Sigma^{\eta}_k(y^{\eta}_k(\zeta)))] \\
&+ \sum_{k=-\infty}^{m-1} L^{\eta_0}_{m,k+1}(I - Q^{\eta_0}_{k+1})\,[(G^{\eta}_k)_x(\iota(\eta)) - (G^{\eta_0}_k)_x(\iota(\eta_0))] \\
&\qquad\qquad \cdot [(y^{\eta}_k(\zeta+h) - y^{\eta}_k(\zeta), \Sigma^{\eta}_k(y^{\eta}_k(\zeta+h)) - \Sigma^{\eta}_k(y^{\eta}_k(\zeta)))] \\
&+ \sum_{k=-\infty}^{m-1} L_{m,k+1}(I - Q_{k+1})(G_k)_x(\iota(0))[(y^{\eta}_k(\zeta+h) - y^{\eta}_k(\zeta) - y_k(\zeta+h) + y_k(\zeta)] \\
&+ \sum_{k=-\infty}^{m-1} L_{m,k+1}(I - Q_{k+1})(G_k)_x(\iota(0)) \\
&\qquad\qquad \cdot [\Sigma^{\eta}_k(y^{\eta}_k(\zeta+h)) - \Sigma^{\eta}(y^{\eta}_k(\zeta)) - \Sigma_k(y_k(\zeta+h)) + \Sigma(y_k(\zeta)))],
\end{aligned}
$$

where $\iota(\eta)$ is a point in the segment $[(y^{\eta}_k(\zeta+h), \Sigma^{\eta}_k(y^{\eta}_k(\zeta+h))), (y^{\eta}_k(\zeta), \Sigma^{\eta}_k(y^{\eta}_k(\zeta)))]$.

The first three terms in the last equality are $o(\|h\|)$ as $(\eta, h) \to (0,0)$. The last term can be absorbed in the left-hand side after taking the supremum in m and for all ζ and $\zeta + h$ in a suitable small ball B such that all their negative orbits under any of the evolution processes associated to any η suitably near 0 stay in the ball B.

Since the maps Σ^{η} are differentiable, we conclude that their derivatives behave continuously with respect to η.

7. Permanence and continuity of hyperbolic global solutions

Now we consider the discrete evolution equation

$$
x_{n+1} = S_n x_n = L_n x_n + T_n x_n, \tag{9.61}
$$

where $\{L_n\}_{n\in\mathbb{Z}} \subset \mathcal{L}(X)$ and $\{T_n\}_{n\in\mathbb{Z}}$ is a sequence of nonlinear maps from X into itself. It is clear that given $\zeta \in X$ and $m \in \mathbb{Z}$, there exists only one sequence $\{x_n\}_{n\geqslant m}$ satisfying $x_m = \zeta$ and $x_{n+1} = L_n x_n + T_n x_n$, $n \geqslant m$. In some situations we can also obtain the existence of a sequence $\{x_n\}_{n\in\mathbb{Z}}$ for (9.61).

Let $T_n \colon X \to X$ be a family of nonlinear Lipschitz continuous maps such that

$$
\sup_{n\in\mathbb{Z}} \sup_{\substack{x,y\in X \\ x\neq y}} \frac{\|T_n x - T_n y\|_X}{\|x-y\|_X} \leqslant \rho < \infty \quad \text{and} \quad \sup_{n\in\mathbb{Z}} \sup_{x\in X} \|T_n x\|_X \leqslant N < \infty,
$$

and let $\{L_{n,m} \colon n \geqslant m\}$ be the linear discrete evolution process associated to $\{L_n\}_{n\in\mathbb{Z}} \subset \mathcal{L}(X)$ which has discrete exponential dichotomy, with constant $M \geqslant 1$, exponent $\omega > 0$ and projections $\{Q_n\}_{n\in\mathbb{Z}}$.

Assume also that there exists a map $\Sigma\colon \mathbb{Z}\times X \to X$ such that $\Sigma_n(x) = \Sigma_n(Q_n x)$ and $Q_n(\Sigma_n(x)) = 0$ for all $n \in \mathbb{Z}$ and $x \in X$. Assume also that

$$\sup_{n\in\mathbb{Z}} \sup_{\substack{x,y\in X\\ x\neq y}} \frac{\|\Sigma_n(x) - \Sigma_n(y)\|_X}{\|x-y\|_X} \leqslant L < \infty \quad \text{and} \quad \sup_{n\in\mathbb{Z}} \sup_{x\in X} \|\Sigma_n(x)\|_X \leqslant N < \infty.$$

Consider the discrete equation

$$x_{n+1} = L_n Q_n x_n + H_n(x_n),$$

where $H_n(x_n) = Q_{n+1}T_n(x_n + \Sigma_n(x_n))$; then $x_n = Q_n x_n$. If $m > n$, it is clear that

$$x_m = L_{m,n}x_n + \sum_{k=n}^{m-1} L_{m,k+1}H_k(x_k) \tag{9.62}$$

and since $L_{m,n}\colon Q_nX \to Q_mX$ is an isomorphism, we have (9.62) is equivalent to

$$x_n = L_{n,m}x_m - \sum_{k=n}^{m-1} L_{n,k+1}H_k(x_k). \tag{9.63}$$

THEOREM 9.27. *If*

$$\frac{\rho M(1+L+M)e^{-\omega}}{1-e^{-\omega}} < 1,$$

then for each $x \in X$ *and* $m \in \mathbb{Z}$ *there exists a unique global solution of* (9.62) $x\colon \mathbb{Z} \to X$ *satisfying* $x_m = Q_m x$ *and* $x_n = Q_n x_n$ *for all* $n \in \mathbb{Z}$.

PROOF. The definition $\{x_n\}_{n\geqslant m}$ is clear from (9.62). We will use (9.63) to obtain $\{x_n\}_{n<m}$. Let $\mathcal{Z} = \{\zeta = \{x_n\}_{-\infty\leqslant n\leqslant m}\colon x_m = Q_m x\}$ with the norm $\|\zeta\|_{\mathcal{Z}} = \sup_{-\infty\leqslant n\leqslant m} \|x_n\|_X$ and define $\Psi\colon \mathcal{Z} \to \mathcal{Z}$ by

$$(\Psi\zeta)_n = L_{n,m}Q_m x - \sum_{k=n}^{m-1} L_{n,k+1}H_k(Q_k x_k).$$

Let us show that Ψ is a contraction. To that end, for $n \leqslant m$, we note first that

$$\|(\Psi\zeta)_n\|_X \leqslant Me^{-\omega(m-n)}\|x\| + MN\sum_{k=n}^{m-1} e^{\omega(n-k-1)} \leqslant M\left(\|x\| + \frac{N(M+1)e^{-\omega}}{1-e^{-\omega}}\right),$$

and hence, $\Psi(\mathcal{Z}) \subset \mathcal{Z}$. Now, if $\zeta, \tilde{\zeta} \in \mathcal{Z}$, then

$$\begin{aligned}\|(\Psi\zeta)_n - (\Psi\tilde{\zeta})_n\|_X &\leqslant \rho(1+L+M)M\sum_{k=n}^{m-1} e^{\omega(n-k-1)}\|x_k - \tilde{x}_k\|_X\\ &\leqslant \frac{\rho(1+L+M)Me^{-\omega}}{1-e^{-\omega}}\|\zeta - \tilde{\zeta}\|_{\mathcal{Z}},\end{aligned}$$

which shows that Ψ is a contraction and, consequently, has a unique fixed point. $\square$

Notes. In this chapter we have extended the results of Chapter 4 and of [**141**] to treat the neighborhood of a hyperbolic global solution, giving conditions under which the unstable and stable manifolds are given as graphs of a Lipschitz or a smooth map, and proved that these manifolds depend continuously (in the C^0 or C^1 topology) with respect to perturbations. These ingredients are essential to prove the structural stability under non-autonomous perturbations. Part of this has been considered in [**78**] in the continuous setting, that is, the proof that the unstable and stable manifolds are given as graphs of Lipschitz maps and that they depend continuously (in the C^0 topology) with respect to perturbations. The smoothness of the local unstable manifolds and their continuous dependence relative to perturbations is done here. In [**19**] the authors study, in the autonomous framework, the regularity and continuity in $C^{1,\theta}$ topology of invariant manifolds, for dynamical systems associated to parabolic type equations, including also the possibility that the phase spaces are also parameter-dependent. We do not pursue the non-autonomous λ-lemma, as it is only used in the autonomous case in this book, but it would certainly be needed if we were to consider perturbations of non-autonomous Morse-Smale problems.

CHAPTER 10

Non-autonomous Morse-Smale Dynamical Systems

This chapter is devoted to the proof that, under suitable conditions, a *small* non-autonomous perturbation of a Morse-Smale semigroup is an evolution process which has a pullback attractor with the same structure of connections. We will also show this property in the skew-product framework.

Consider X a metric space, $\Lambda = [0,1]$ the space of parameters, and $\mathcal{S}_\lambda = \{S_\lambda(t,s) \colon (t,s) \in \mathcal{P}\}$ an evolution process in X for each $\lambda \in \Lambda$. In Theorem 8.14 we have proved that if $\mathcal{S}_{\lambda_0}$ is a dynamically gradient semigroup $\mathcal{T} = \{T(t) \colon t \geqslant 0\}$ (that is, $\mathcal{S}_{\lambda_0}$ is an autonomous evolution process and $T(t) = S_\lambda(t+\tau, \tau)$ for each $t \geqslant 0$ and $\tau \in \mathbb{R}$) with a global attractor $\mathcal{A}$ and a disjoint collection of *unitary* isolated invariant sets (equilibria) $\mathbf{E} = \{e_1^*, \cdots, e_n^*\}$, then the *small non-autonomous perturbation* $\mathcal{S}_\lambda$ (of $\mathcal{T}$), which has a pullback attractor $\hat{A}_\lambda = \{A_\lambda(t)\}_{t\in\mathbb{R}}$ with a disjoint collection of *unitary* isolated invariant families (isolated global solutions) $\mathbf{E}_\lambda = \{\xi_{1,\lambda}^*, \cdots, \xi_{n,\lambda}^*\}$, that approaches $\mathbf{E}$ as $\lambda \to \lambda_0$ and satisfy a uniform *topological hyperbolicity property* (see condition (d) of Theorem 8.14), is a dynamically gradient evolution process.

This is saying that any global solution of $\mathcal{S}_\lambda$ in the pullback attractor $\hat{A}_\lambda$ connects (from $-\infty$ to ∞) two of the isolated global solutions of $\mathbf{E}_\lambda$, and that there are no homoclinic structures in $\hat{A}_\lambda$.

The aim of this chapter is to show that, assuming that $\mathcal{T}$ is a Morse-Smale semigroup together with some additional technical conditions, the connections between equilibria in the global attractor $\mathcal{A}$ of $\mathcal{T}$ are also seen in the pullback attractor $\hat{A}_\lambda$ of $\mathcal{S}_\lambda$.

The crucial additional condition is the *transversality of unstable and stable manifolds along a connection for the unperturbed problem.* This condition ensures that a connection existing between two equilibria of the limiting autonomous Morse-Smale semigroup $\mathcal{T}$ will also appear, under smooth non-autonomous perturbation, in the associated nearby isolated global solutions in $\mathbf{E}_\lambda$ of the perturbed evolution process $\mathcal{S}_\lambda$. Also, having connections between two global isolated solutions $\xi_{i,\lambda}^*$ and $\xi_{j,\lambda}^*$ of $\mathbf{E}_\lambda$ for all λ suitably close to λ_0 ensures, according to Proposition 10.6, that there exists a sequence of connections between elements of $\mathbf{E}$ starting at e_i^* and ending at e_j^*. Now, as a consequence of the λ-lemma, we have proved in Proposition 5.9 that there is a direct connection between e_i^* and e_j^*.

This chapter is dedicated to precisely state the conditions ensuring that the above reasoning holds, that is, that the diagram of connections between elements of $\mathbf{E}_\lambda$ and $\mathbf{E}$ is the same.

A Morse-Smale semigroup (see Definition 5.2) is a reversible semigroup $\mathcal{T}$ with a global attractor $\mathcal{A}$, a finite set of hyperbolic fixed points and normally hyperbolic periodic solutions, all with finite-dimensional local unstable manifolds, and such that the intersection between the local unstable manifold and the stable manifold of two different critical elements - fixed points or periodic orbits - is transversal.

In Section 1 (see Corollary 5.12) we have proved that a reversible semigroup, with finitely many equilibria (all hyperbolic) and finitely many periodic orbits (all normally hyperbolic) having unstable and stable manifolds that intersect transversally along connections, is Morse-Smale if and only if it is dynamically gradient.

The results of Section 1 are synthesised next for easy reference.

THEOREM 10.1. *Let* $\mathcal{T} = \{T(t) : t \geqslant 0\}$ *be a semigroup with a global attractor* $\mathcal{A}$.

(a) *If* $\mathcal{T}$ *is a Morse-Smale semigroup with nonwandering set*

$$\mathbf{E} = \{e_1^*, \cdots, e_n^*, \xi_1^*(\mathbb{R}), \cdots, \xi_m^*(\mathbb{R})\},$$

where each e_i^*, $1 \leqslant 1 \leqslant n$, *is a hyperbolic equilibrium point and* $\xi_j^* : \mathbb{R} \to X$, $1 \leqslant j \leqslant m$, *is a normally hyperbolic periodic orbit, then* $\mathbf{E}$ *is a disjoint collection of isolated invariant sets and* $\mathcal{T}$ *is a generalized dynamically gradient semigroup with respect to* $\mathbf{E}$.

(b) *If* $\mathcal{T}$ *is a reversible dynamically gradient semigroup associated to the disjoint collection of isolated invariant sets*

$$\mathbf{E} = \{e_1^*, \cdots, e_n^*, \xi_1^*(\mathbb{R}), \cdots, \xi_m^*(\mathbb{R})\},$$

where each e_i^*, $1 \leqslant i \leqslant n$, *is a hyperbolic equilibrium point and* $\xi_j^* : \mathbb{R} \to X$, $1 \leqslant j \leqslant m$, *is a normally hyperbolic periodic orbit, with* $\dim W^u_{loc}(E) < \infty$ *whenever* $E \in \mathbf{E}$, *and* $W^u(E) \pitchfork W^s_{loc}(\tilde{E})$ *whenever* $E, \tilde{E} \in \mathbf{E}$, *then* $\mathcal{T}$ *is a Morse-Smale semigroup with nonwandering set* $\mathbf{E}$.

As we have mentioned before, we already know that a small non-autonomous perturbation of a dynamically gradient semigroup is a dynamically gradient evolution process (the so-called *topological structural stability*). With the concept of Morse-Smale semigroups at hand, we can go one step further and prove that *the structure of connections remains the same*, a property that here we will refer to as **geometrical structural stability**.

THEOREM 10.2. *Let* $\Lambda = [0,1]$ *and let* $\{\mathcal{S}_\lambda\}_{\lambda\in\Lambda}$ *be a family of evolution processes satisfying the following conditions:*

(a) $\mathcal{S}_\lambda$ *has a pullback attractor* $\hat{A}_\lambda$ *and is reversible for each* $\lambda \in \Lambda$;
(b) $\overline{\bigcup_{\lambda\in\Lambda}\bigcup_{t\in\mathbb{R}} A_\lambda(t)}$ *is compact;*
(c) *for each compact set* $K \subset \mathbb{R}^+ \times X$ *we have*

$$\sup_{s\in\mathbb{R}} \sup_{(t,x)\in K} \|S_\lambda(t+s,s)x - S_{\lambda_0}(t+s,s)x\|_X \to 0 \quad \text{as } \lambda \to \lambda_0;$$

(d) *for each compact set* $J \in \mathbb{R}^+$ *we have*

$$\sup_{s\in\mathbb{R}} \sup_{t\in J} \sup_{z\in A(s)} \|S'_\lambda(t+s,s)(z) - S'_{\lambda_0}(t+s,s)(z)\|_{\mathcal{L}(X)} \to 0 \quad \text{as } \lambda \to \lambda_0;$$

(e) $S_{\lambda_0}(t,s) = T(t-s)$ *for all* $t \geqslant s$, *where* $\mathcal{T} = \{T(t) : t \geqslant 0\}$ *is a Morse-Smale semigroup in* X, *with global attractor* $\mathcal{A}$ *and set of isolated stationary solutions* $\mathbf{E} = \{e_1^*, \cdots, e_n^*\}$.

Then

(i) $\mathcal{S}_\lambda$ *has a disjoint set of isolated hyperbolic global solutions*

$$\mathbf{E}_\lambda = \{\xi^*_{1,\lambda}, \cdots, \xi^*_{n,\lambda}\} \quad \textit{for each } \lambda \in \Lambda,$$

with

$$\lim_{\lambda \to \lambda_0} \max_{1 \leqslant i \leqslant n} \operatorname{dist}_H(\xi^*_{i,\lambda}(\mathbb{R}), z^*_i) = 0;$$

(ii) *there exist* $\delta > 0$ *and* $\epsilon_0 > 0$ *such that, if* $|\lambda - \lambda_0| < \epsilon_0$, $\xi_\lambda \colon \mathbb{R} \to X$ *is a global solution of* $\mathcal{S}_\lambda$ *in* $\hat{A}_\lambda$, $t_0 \in \mathbb{R}$ *and* $\operatorname{dist}_H(\xi_\lambda(t), \xi^*_{i,\lambda}(\mathbb{R})) \leqslant \delta$ *for all* $t \leqslant t_0$ $(t \geqslant t_0)$, *then* $\operatorname{dist}_H(\xi_\lambda(t), \xi^*_{i,\lambda}(t)) \to 0$ *as* $t \to \infty$ $(t \to \infty)$;

(iii) *there exists* $\epsilon_1 \in (0, \epsilon_0)$ *such that* $\mathcal{S}_\lambda$ *is a Morse-Smale evolution process (see Definition* 10.3*) for all* $|\lambda - \lambda_0| < \epsilon_1$ *and there exists a phase diagram isomorphism between* $\mathcal{S}_\lambda$ *and* $\mathcal{T}$.

The rest of the chapter will be dedicated to giving precise meaning to the term *Morse-Smale evolution process*, to the proof of Theorem 10.2, and to present some applications. We also write the associated results for skew-product semigroups and cocycles and their respective attractors (global and uniform).

The approach of this chapter is taken from our work, together with Geneviève Raugel [**31**]. In a finite-dimensional framework, the structural stability of Morse-Smale maps under small time-dependent perturbations has been proved by Franks in [**121**], and by Kryzhevich and Pliss in [**158**]. The study of non-autonomous perturbations of a Morse-Smale semigroup in the infinite-dimensional case has been done in [**31**], were we show that if the nonwandering set of a Morse-Smale semigroup (having a compact global attractor) is reduced to a finite number of hyperbolic equilibria, then the resulting evolution process is still of Morse-Smale type (in a sense to be specified in Definition 10.3) and there is a phase diagram isomorphism (see Definition 10.4) between the dynamical systems defined by the limiting semigroup and the perturbed evolution processes (see Theorem 10.2).

1. Morse-Smale evolution processes

In this section, given a Morse-Smale semigroup, we explore the structure's behavior of its global attractor under a small non-autonomous perturbation. More precisely, we verify that the *phase diagrams* (see Definition 10.4) between the Morse-Smale semigroup and the perturbed non-autonomous evolution processes are *isomorphic*, that is, the connections and their orientations between the isolated invariant sets of the semigroup and the isolated invariant families of the perturbed non-autonomous evolution process are the same. We will restrict ourselves, however, to the case of a Morse-Smale semigroup with nonwandering set containing only hyperbolic equilibria, that is, without normally hyperbolic periodic orbits.

Next we ask the reader to recall the notions of *dynamically gradient evolution processes*, given in Definition 8.10 (see Chapter 8 - Section 3), and of *dynamically gradient semigroups*, given in Definition 3.12 (see Chapter 3 - Section 2). With these definitions, we know that a small non-autonomous perturbation of a dynamically gradient semigroup is a dynamically gradient evolution process, that is, for a dynamically gradient semigroup, conditions (G1) and (G2) of Definition 3.12 are stable under non-autonomous perturbations. This is what we called *topological structural stability*, and it was proved in Theorem 8.14.

Moreover, inspired by Theorem 8.14, we define *Morse-Smale evolution processes*.

DEFINITION 10.3. Let $\mathcal{S}$ be an evolution process in a Banach space X with a pullback attractor $\hat{A}$. We say that $\mathcal{S}$ is a **Morse-Smale evolution process** if:

(i) $\mathcal{S}$ is a dynamically gradient evolution process with respect to a disjoint collection of isolated hyperbolic global solutions $\mathbf{E} = \{\xi_1, \cdots, \xi_n\}$;
(ii) $\mathcal{S}$ is a reversible evolution process, that is, $S(t,s)|_{A(s)}\colon A(s) \to A(t)$ is injective, $S(t,s)$ is differentiable in $A(s)$ and $S'(t,s)(z)\colon X \to X$ is an injective bounded linear operator, for each $t \geqslant s \in \mathbb{R}$;
(iii) $W^u_{loc}(\xi_i)$ is finite dimensional, for all $i = 1, \cdots, n$;
(iv) given $\xi_i, \xi_j \in \mathbf{E}$ such that $W^u(\xi_i)(t_0) \cap W^s_{loc}(\xi_j)(t_0) \neq \varnothing$ for some $t_0 \in \mathbb{R}$, there exists a point of transversal intersection in $W^u(\xi_i)(t_0) \cap W^s_{loc}(\xi_j)(t_0)$.

Before proving our main result, we need the definition of a *phase diagram isomorphism* between two evolution processes, and also some auxiliary results.

DEFINITION 10.4. Let $\mathcal{S}$ be a dynamically gradient evolution process in a metric space X associated with the disjoint collection of isolated invariant sets $\mathbf{E} = \{\xi_1, \cdots, \xi_n\}$, where $\xi_i\colon \mathbb{R} \to X$ is an isolated bounded global solution of $\mathcal{S}$ for each $1 \leqslant i \leqslant n$. The *phase diagram* of $\mathcal{S}$ is the disjoint collection $\mathbf{E} = \{\xi_1, \cdots, \xi_n\}$ together with all the bounded global solutions $\xi\colon \mathbb{R} \to X$ of $\mathcal{S}$.

DEFINITION 10.5. Let $\mathcal{R} = \{R(t,s)\colon (t,s) \in \mathcal{P}\}$ and $\mathcal{S} = \{S(t,s)\colon (t,s) \in \mathcal{P}\}$ be two dynamically gradient evolution processes in a metric space (X,d) associated with the collections of isolated invariant bounded solutions $\mathbf{E}_{\mathcal{R}} = \{\xi_1, \cdots, \xi_n\}$ and $\mathbf{E}_{\mathcal{S}} = \{\psi_1, \cdots, \psi_m\}$, respectively. We say that there exists a **phase diagram isomorphism** between $\mathcal{R}$ and $\mathcal{S}$ if there exists a bijection $\mathcal{B}\colon \mathbf{E}_{\mathcal{R}} \to \mathbf{E}_{\mathcal{S}}$ (in particular, $m = n$) satisfying the following: there exists a global solution ξ of $\mathcal{R}$ with

$$\lim_{t\to-\infty} d(\xi_i(t), \xi(t)) = 0 \quad \text{and} \quad \lim_{t\to\infty} d(\xi_j(t), \xi(t)) = 0,$$

if and only if there exists a global solution ψ of $\mathcal{S}$ with

$$\lim_{t\to-\infty} d(\mathcal{B}(\xi_i)(t), \psi(t)) = 0 \quad \text{and} \quad \lim_{t\to\infty} d(\mathcal{B}(\xi_j)(t), \psi(t)) = 0.$$

PROPOSITION 10.6. *Suppose that all the hypotheses of Theorem 8.14 are satisfied. Assume also that $\{\lambda_k\} \subset \Lambda \setminus \{\lambda_0\}$, $\lambda_k \to \lambda_0$ as $k \to \infty$ and, for each $k \in \mathbb{N}$, ξ_k is a global solution of $\mathcal{S}_k$ with*

$$\lim_{t\to-\infty} d(\xi_k(t), \xi^*_{i,\lambda_k}(t)) = 0 \text{ and } \lim_{t\to\infty} d(\xi_k(t), \xi^*_{j,\lambda_k}(t)) = 0.$$

Then there exist a finite sequence $\{\xi^l\}_{l=1}^m$ (with $m \leqslant n$) of global solutions of $\mathcal{T}$, and equilibria $\{e^l\}_{l=1}^m$ with $e^1 = e^*_i$, $e^m = e^*_j$ (for some $1 \leqslant i,j \leqslant n$) such that

$$e^l \overset{t\to-\infty}{\longleftarrow} \xi^l(t) \overset{t\to\infty}{\longrightarrow} e^{l+1}, \text{ for each } 1 \leqslant l < m.$$

PROOF. Taking subsequences, if necessary, we may assume that

$$\lim_{k\to\infty} d(\xi_k(t), \xi^1(t)) = 0, \text{ uniformly in compact subsets of } \mathbb{R}.$$

Since $\lim_{t\to-\infty} d(\xi_k(t), \xi^*_{i,\lambda_k}(t)) = 0$ and $\mathrm{dist}_H(\xi^*_{i,\lambda}(\mathbb{R}), e^*_i) \to 0$ as $k \to \infty$ we have $\lim_{t\to-\infty} d(\xi^1(t), e^*_i) = 0$, and we set $e^1 = e^*_i$. Now (G1) implies that there exists an equilibrium e^2 of $\mathcal{T}$ (which, from (G2), is different from e^1) such that $\lim_{t\to\infty} d(\xi^1(t), e^2) = 0$. If $e^2 = e^*_j$ the proof is complete. Otherwise there exists $t'_\ell > t_\ell$ such that $d(\xi_{k_\ell}(t), e^2) < \delta$ for all $t \in [t_\ell, t'_\ell)$ and $d(\xi_{k_\ell}(t'_\ell), e^2) = \delta$. Taking subsequences, if necessary, let $\xi^2(t) = \lim_{\ell\to\infty} \xi_{k_\ell}(t+t_\ell)$. Then, since $t'_\ell - t_\ell \to \infty$ as $\ell \to \infty$, $d(\xi^2(t), e^2) \leqslant \delta$ for all $t \leqslant 0$ and consequently $\lim_{t\to-\infty} d(\xi^2(t), e^2) = 0$. Furthermore, from (G1) and (G2), $\lim_{t\to\infty} d(\xi^2(t), e^3) = 0$ for some equilibrium e^3 of $\mathcal{T}$ (different from e^1 and e^2). This reasoning can be done as long as e^l is different from e_j^*, and since we have just a finite number of equilibrium points of $\mathcal{T}$, we conclude the result. □

Now we are able to present the proof of Theorem 10.2. We recall that the auxiliary results concerning perturbation of stable and unstable sets are included in Chapter 9 - Subsection 6.2.

PROOF OF THEOREM 10.2. From Theorem 8.14, we know that there exists $\epsilon_0 > 0$ such that $\mathcal{S}_\lambda$ is a dynamically gradient evolution process associated with the disjoint collection of isolated global solutions $\mathbf{E}_\lambda$ for $|\lambda - \lambda_0| < \epsilon_0$. Also ϵ_0 can be taken sufficiently small so that each global solution in $\mathbf{E}_\lambda$ is hyperbolic (see Theorem 9.24). Now, define $\mathcal{B}\colon \mathbf{E} \to \mathbf{E}_\lambda$ by

$$\mathcal{B}(e_i^*) = \xi_{i,\lambda}^*, \text{ for } i = 1, \cdots, n.$$

Then $\mathcal{B}$ is clearly a bijection, and we can assume by the openness of the transversality property that ϵ_0 is small enough so that, if $W^u(e_i^*) \cap W^s_{loc}(e_j^*) \neq \varnothing$, then $W^u(\xi_{i,\lambda}^*)(t) \cap W^s_{loc}(\xi_{j,\lambda})(t) \neq \varnothing$ for all $t \in \mathbb{R}$, and moreover, there is a point in which the intersection is transversal (this holds since, from the results in Section 6.2, the unstable and stable manifolds behave continuously in the C^1 topology).

This shows that if there exists a connection between e_i^* and e_j^*, then there exists a connection between $\xi_{i,\lambda}^*$ and $\xi_{j,\lambda}^*$. It remains to show the converse.

Assume that we have a sequence $\lambda_k \to \lambda_0$ and global solutions ξ_k of S_{λ_k} for each k such that

$$\lim_{t\to-\infty} \|\xi_{i,\lambda}^*(t) - \xi_k(t)\| = 0 \text{ and } \lim_{t\to\infty} \|\xi_{j,\lambda}^*(t) - \xi_k(t)\| = 0.$$

Then, by Proposition 10.6, there exist a finite sequence $\{\xi^l\}_{l=1}^m$ $(m \leqslant n)$ of global solutions of $\mathcal{T}$ and equilibria $\{e_l\}_{l=1}^m$ in $\mathbf{E}$ with $e_1 = e_i^*$, $e_m = e_j^*$ such that

$$e_l \overset{t\to-\infty}{\leftarrow} \varphi_l(t) \overset{t\to\infty}{\to} e_{l+1} \text{ for all } l = 1, \cdots, m-1.$$

But Proposition 5.9 ensures the existence of a global solution ξ of $\mathcal{T}$ such that

$$e_i^* \overset{t\to-\infty}{\leftarrow} \varphi(t) \overset{t\to\infty}{\to} e_j^*,$$

and completes the proof. □

2. Geometrical structural stability for skew-product semigroups

In this section we prove the geometrical structural stability for global attractors of skew-product semigroups. As a corollary, we also obtain the structural stability of uniform attractors for cocycles.

We begin by considering the initial value problem:

$$\begin{cases} \dot{u} = f_\lambda(t,u), \ t > \tau \geqslant 0, \\ u(\tau) = u_0 \in X, \end{cases} \tag{10.1}$$

where X is a Banach space and $f_\lambda\colon \mathbb{R}^+ \times D \subset \mathbb{R} \times X \to X$ is a map belonging to some complete metric space $(\mathcal{C}, \rho)$ for each $\lambda \in \Lambda = [0,1]$. We assume that for each $g \in \mathcal{C}$, $\tau \in \mathbb{R}$ and $u_0 \in X$ the problem:

$$\begin{cases} \dot{u} = g(t,u), \ t > \tau \geqslant 0, \\ u(\tau) = u_0 \in X \end{cases} \tag{10.2}$$

is globally well posed, that is:

(i) for each $u_0 \in X$, $\tau \in \mathbb{R}^+$ and $g \in \mathcal{C}$, there exists a unique function $[\tau, \infty) \ni t \mapsto u(t, \tau, g, u_0) \in X$ satisfying (10.2), in a suitable sense;

(ii) $\{(t, \tau, u_0) \in \mathbb{R}^+ \times \mathbb{R}^+ \times X : t \geqslant \tau\} \ni (t, \tau, u_0) \mapsto u(t, \tau, g, u_0) \in X$ is continuous.

We are interested in the case when f_{λ_0} is independent of time and $\rho(f_\lambda, f_{\lambda_0}) \to 0$ as $\lambda \to \lambda_0$. In particular, we are interested in the case where the dynamical system associated to (10.1), with $\lambda = \lambda_0$, is a Morse-Smale semigroup $\mathcal{T}_0 = \{T_0(t)\colon t \geqslant 0\}$ in the sense of Definition 5.2. For λ suitably close to λ_0, the dynamical system associated to (10.1) will be a non-autonomous perturbation of a Morse-Smale semigroup, and hence, a Morse-Smale evolution process (Definition 10.3). We prove that the structure of connections of a Morse-Smale semigroup remains the same under non-autonomous perturbations, that is, we have geometrical structural stability.

Define

$$H_\lambda = \overline{\{f_\lambda(s + \cdot, \cdot)\colon s \in \mathbb{R}^+\}}^{\,\rho} \quad \text{and} \quad M = \overline{\bigcup_{\lambda \in \Lambda} H_\lambda}^{\,\rho}.$$

Clearly $M \subset \mathcal{C}$ and we define the (driving) semigroup $\Theta = \{\Theta_t \colon t \geqslant 0\}$ in M by $\Theta_t h(\cdot, \cdot) = h(t + \cdot, \cdot)$ for each $h \in M$. For each $h \in M$ and $u_0 \in X$ we consider the initial value problem

$$\begin{cases} \dot{u} = h(t,u), t > 0, \\ u(0) = u_0 \in X \end{cases}$$

and let $\mathbb{R}^+ \ni t \mapsto \mathcal{K}(t,h)u_0 \in X$ be its only solution. With this, we have defined the cocycle

$$\mathbb{R}^+ \times M \ni (t,h) \mapsto \mathcal{K}(t,h) \in C(X),$$

and the skew-product semigroup $\Pi_M = \{\Pi_M(t)\colon t \geqslant 0\}$ in $\mathbb{X} = X \times M$ by

$$\Pi_M(t)(u,h) = (\mathcal{K}(t,h)u, \Theta_t h) \text{ for } t \geqslant 0.$$

We assume that Π_M has a global attractor $\mathbb{A}_M$ in $\mathbb{X}_M$, and as we know, this implies in particular that Θ has a global attractor $\mathscr{S}_M$ in M.

2.1. Skew-product semigroups and evolution processes. For each $h_\lambda \in H_\lambda$ and $u_0 \in X$ consider the initial value problem:

$$\begin{cases} \dot{u} = h_\lambda(t,u), \ t > 0, \\ u(0) = u_0 \in X \end{cases}$$

and let $\mathbb{R}^+ \ni t \mapsto \mathcal{K}(t, h_\lambda)u_0 \in X$ be its only solution. With this, we have defined the cocycle

$$\mathbb{R}^+ \times H_\lambda \ni (t, h_\lambda) \mapsto \mathcal{K}(t, h_\lambda) \in C(X),$$

and the skew-product semigroup $\Pi_\lambda = \{\Pi_\lambda(t)\colon t \geqslant 0\}$ in $\mathbb{X}_\lambda = X \times H_\lambda$ by

$$\Pi_\lambda(t)(u, h_\lambda) = (\mathcal{K}(t, h_\lambda)u, \Theta_t h_\lambda) \text{ for } t \geqslant 0.$$

If f_{λ_0} is time-independent, $H_{\lambda_0} = \{f_{\lambda_0}\}$, $\mathbb{X}_{\lambda_0} = X \times \{f_{\lambda_0}\}$, $\Theta_t f_{\lambda_0} = f_{\lambda_0}$ for each $t \geqslant 0$, and setting $T_0(t) = \mathcal{K}(t, f_{\lambda_0})$ for each $t \geqslant 0$, $\mathcal{T}_0 = \{T_0(t) \colon t \geqslant 0\}$ defines a semigroup in X and

$$\Pi_{\lambda_0}(t)(u, f_{\lambda_0}) = (\mathcal{K}(t, f_{\lambda_0})u, f_{\lambda_0}) = (T_0(t)u, f_{\lambda_0}) \text{ for } t \geqslant 0.$$

It is clear that the skew-product semigroup Π_λ has a global attractor $\mathbb{A}_\lambda$ in $\mathbb{X}_\lambda = X \times H_\lambda$ and, if Θ^λ denotes the restriction of the semigroup Θ to H_λ, it has a global attractor $\mathscr{S}_\lambda$ in H_λ.

Given a bounded global solution η_λ of Θ_λ through h_λ, we consider the problem:

$$\begin{cases} \dot{u} = \eta_\lambda(\tau)(t-\tau, u), \ t > \tau, \\ u(\tau) = u_0 \in X. \end{cases}$$

From the previous assumptions, this problem generates an evolution process $\mathcal{S}_{\eta_\lambda}$ given by $S_{\eta_\lambda}(t,\tau) = \mathcal{K}(t-\tau, \eta_\lambda(\tau))$ for each $t \geqslant \tau$, and $\mathcal{S}_{\eta_\lambda}$ has a pullback attractor $\hat{A}_{\eta_\lambda}$.

We have seen in Theorem 8.14 the robustness of the dynamically gradient structure of pullback attractors for evolution processes, i.e., there exists an $\epsilon > 0$ such that $\mathcal{S}_{\eta_\lambda}$ is a dynamically gradient evolution process, for all $\lambda \in \Lambda$ such that $|\lambda - \lambda_0| < \epsilon$, when $\mathcal{T}_0$ is a dynamically gradient semigroup. Now if $\epsilon > 0$ is as in Theorem 8.14 and, for $|\lambda - \lambda_0| < \epsilon$, we define

$$\mathbb{M}_{\lambda,i} = \{(\xi^*_{i,\lambda}(0), \eta_\lambda(0)) \colon \eta_\lambda \text{ is a bounded global solution of } \Theta^\lambda \text{ in } H_\lambda\},$$

the following result holds as a corollary of the proof of Theorem 8.14.

PROPOSITION 10.7. *Suppose the assumptions of Theorem* 8.14 *hold,* $\lambda_n \to \lambda_0$ *as* $n \to \infty$, *and there exist global solutions* $\pi_{\lambda_n} \colon \mathbb{R} \to \mathbb{A}_{\lambda_n}$ *of* Π_{λ_n} *in* $\mathbb{X}_{\lambda_n}$ *with*

$$\mathbb{M}_{\lambda_n,i} \overset{t\to-\infty}{\longleftarrow} \pi_{\lambda_n}(t) \overset{t\to\infty}{\longrightarrow} \mathbb{M}_{\lambda_n,j} \quad \textit{for some } 1 \leqslant i,j \leqslant n.$$

Then there exist global solutions $\pi_{0,k} \colon \mathbb{R} \to \mathbb{A}_0$ *of* Π_{λ_0} *in* $\mathbb{X}_{\lambda_0}$, *for* $1 \leqslant k \leqslant n_0 \leqslant n$ *such that*

$$\mathbb{M}_{0,i_{k-1}} \overset{t\to-\infty}{\longleftarrow} \pi_{0,k}(t) \overset{t\to\infty}{\longrightarrow} \mathbb{M}_{0,i_k} \quad \textit{for } k = 1, \cdots, n_0,$$

where $i_0 = i$ *and* $i_{n_0} = j$.

REMARK 10.8 (Abstract phase diagram commutativity result). Under the assumptions of Theorem 8.14, to ensure that the structure of connections between the perturbed and limiting attractors remains the same, we need to show the following:

(a) if there is an *indirect connection* between two isolated invariant sets $\mathbb{M}_{\lambda_0,i}$ and $\mathbb{M}_{\lambda_0,j}$ of the limiting problem, then there is a direct connection between $\mathbb{M}_{\lambda_0,i}$ and $\mathbb{M}_{\lambda_0,j}$ (which is achieved using the transversality property together with the λ-lemma);
(b) if there is a connection between two isolated invariant sets $\mathbb{M}_{\lambda_0,i}$ and $\mathbb{M}_{\lambda_0,j}$ of the limiting problem, then there will be a connection between the associated isolated invariant sets $\mathbb{M}_{\lambda,i}$ and $\mathbb{M}_{\lambda,j}$ of the perturbed problem (which is achieved using the transversality property).

2.2. Geometrical structural stability. Now we are ready to show the phase diagram commutativity under non-autonomous perturbations for Morse-Smale semigroups (see [**195**]).

THEOREM 10.9 (cf. [**31**]). *Assume that Π_{λ_0} is a C^1-reversible Morse-Smale semigroup with global attractor $\mathbb{A}_0$ on $X \times \{f_{\lambda_0}\}$ and that its nonwandering set consists of a finite number of hyperbolic equilibria $\Omega(\Pi_{\lambda_0}) = \{(z_1^*, f_{\lambda_0}), \cdots, (z_n^*, f_{\lambda_0})\}$. Assume also that:*

(a) *Π_λ has a global attractor $\mathbb{A}_\lambda$ in $\mathbb{X}_\lambda$ for each $\lambda \in \Lambda$, and $\mathbb{K} = \overline{\bigcup_{\lambda \in (0,1]} \mathbb{A}_\lambda}$ is compact;*

(b) *$\Pi_\lambda(t)(x, h_\lambda) \to \Pi_{\lambda_0}(t)(x, f_{\lambda_0})$ in $\mathbb{X}_M$ as $\lambda \to \lambda_0$, uniformly for $h_\lambda \in H_\lambda$ and for (t, x) in compact subsets of $\mathbb{R}^+ \times X$;*

(c) *for each compact set $J \in \mathbb{R}^+$ we have*

$$\|D\mathcal{K}(t, h_\lambda)(z) - D\mathcal{K}(t, f_{\lambda_0})(z)\|_{\mathcal{L}(X)} \to 0 \text{ as } \lambda \to \lambda_0,$$

uniformly for $t \in J$, $h_\lambda \in H_\lambda$ and $z \in \pi_X(\mathbb{K})$.

Then there exists $\epsilon_0 > 0$ such that for $|\lambda - \lambda_0| < \epsilon_0$ we have:

(i) *for any bounded global solution η_λ of Θ_λ in H_λ, the evolution process $\mathcal{S}_{\eta_\lambda}$ has a disjoint collection of isolated hyperbolic global solutions*

$$\{\xi_{1,\eta_\lambda}^*, \cdots, \xi_{n,\eta_\lambda}^*\}$$

with

$$\lim_{\lambda \to \lambda_0} \sup_{\eta_\lambda \in \mathscr{S}_\lambda} \max_{1 \leqslant i \leqslant n} \mathrm{dist}_H^X(\xi_{i,\eta_\lambda}^*(\mathbb{R}), z_i^*) = 0,$$

(ii) *for any bounded global solution η_λ of Θ_λ in H_λ, $W_{loc}^u(\xi_{i,\eta_\lambda}^*)$ is finite dimensional, and*

$$W^u(\xi_{i,\eta_\lambda}^*)(t_0) \pitchfork W_{loc}^s(\xi_{j,\eta_\lambda}^*)(t_0),$$

for some $t_0 \in \mathbb{R}$, for each $1 \leqslant i, j \leqslant n$;

(iii) *if $(\xi_\lambda, \eta_\lambda)$ is a bounded global solution of Π_λ, $t_0 \in \mathbb{R}$ and*

$$\mathrm{dist}_H^H(\xi_\lambda(t), \xi_{i,\eta_\lambda}^*(t)) \leqslant \delta \text{ for all } t \leqslant t_0 \text{ (resp. } t \geqslant t_0),$$

then

$$\mathrm{dist}_H^X(\xi_\lambda(t), \xi_{i,\eta_\lambda}^*(t)) \to 0 \text{ as } t \to -\infty \text{ (resp. } t \to \infty);$$

(iv) *Π_λ is dynamically gradient relative to the disjoint collection of isolated invariant sets $\{\mathbb{M}_{\lambda,1}, \cdots, \mathbb{M}_{\lambda,n}\}$, $\mathbb{M}_{\lambda,i} = \bigcup_{\eta_\lambda} \{(\xi_{i,\eta_\lambda}^*(0), \eta_\lambda(0))\}$, where the union is taken over all bounded global solutions of Θ_λ in H_λ;*

(v) *there exists a phase diagram isomorphism between $\mathcal{S}_\lambda$ and $\mathcal{T}_0$, and between Π_λ and Π_{λ_0}.*

This theorem can also be written in the context of uniform attractors for co-cycles, with the same hypotheses, using the notion of lifted invariant sets and projecting the isolated invariant sets on the first coordinate to obtain the lifted invariant isolated invariant sets. Before stating this result, we need the following definition.

DEFINITION 10.10. Let $(\mathcal{K}, \Theta)_{(X,\Sigma)}$ be a non-autonomous dynamical system in a Banach space X, with a uniform attractor $\mathcal{A}$ and such that the associated skew-product semigroup Π has a global attractor $\mathbb{A}$ in $\mathbb{X}_\Sigma$. We say that $(\mathcal{K}, \Theta)_{(X,\Sigma)}$ (or simply, $\mathcal{K}$) is a **Morse-Smale non-autonomous dynamical system** if for each bounded global solution η of Θ in Σ, the associated evolution process $\mathcal{S}_\eta$, given by $S_\eta(t, s) = \mathcal{K}(t - s, \eta(s))$ for each $t \geqslant s$, is a Morse-Smale evolution process with nonwandering set given as a finite number of hyperbolic global solutions $\mathbf{E}_\eta = \{\xi_{\eta,1}, \cdots, \xi_{\eta,n}\}$.

COROLLARY 10.11 (Structural stability for uniform attractors). *Let the hypotheses in Theorem* 10.9 *hold. In particular,* $(\mathcal{K}(t,h_\lambda),\Theta)_{(X,H_\lambda)}$ *is a continuously differentiable non-autonomous dynamical system, for each* $\lambda\in\Lambda$, *which we will denote briefly by* $\mathcal{K}_\lambda$. *Assume that:*

(a) $\mathcal{K}_\lambda$ *has a uniform attractor* $\mathcal{A}_\lambda$ *for each* $\lambda\in\Lambda$, *and* $\mathcal{G}=\overline{\bigcup_{\lambda\in\Lambda}\mathcal{A}_\lambda}$ *is compact;*

(b) $\mathcal{T}_0=\mathcal{K}_{\lambda_0}$ *is a Morse-Smale semigroup with its nonwandering set consisting of only hyperbolic equilibria* $\mathcal{E}_0=\{z_{0,1},\cdots,z_{0,n}\}$;

(c) *for each compact set* $J\subset\mathbb{R}^+$ *we have*

$$\|\mathcal{K}(t,h_\lambda)x-T_0(t)x\|_X\to 0,$$

and

$$\|\mathcal{K}'(t,h_\lambda)(x)-T_0'(t)(x)\|_{\mathcal{L}(X)}\to 0,$$

as $\lambda\to\lambda_0$, *uniformly for* $t\in J$, $h_\lambda\in H_\lambda$ *and* $x\in\mathcal{G}$;

Then there exists $\epsilon_0>0$ *such that for* $|\lambda-\lambda_0|<\epsilon_0$ *we have:*

(i) $\mathcal{K}_\lambda$ *has a disjoint family of isolated* $\mathcal{K}_\lambda$*-lifted invariant sets*

$$\mathcal{E}_\lambda=\{E_{\lambda,1},\cdots,E_{\lambda,n}\}$$

such that

$$\max_{1\leqslant i\leqslant n}\operatorname{dist}_H^X(E_{\lambda,i},E_{\lambda_0,i})\to 0,\ \text{as } \lambda\to\lambda_0;$$

(ii) $(\mathcal{K}(t,h_\lambda),\Theta)_{(X,H_\lambda)}$ *is a* Morse-Smale non-autonomous dynamical system *with* $\mathcal{E}_\lambda=\{E_{\lambda,1},\cdots,E_{\lambda,n}\}$ *as its nonwandering set.*

3. Applications

We now illustrate the theory developed in this chapter with some examples and models, applying Theorem 10.9 (or, equivalently, Corollary 10.11 for the uniform attractor).

3.1. Applications to partial differential equations. In this section we recall the two applications in Chapter 5 - Section 3: parabolic equations and a damped wave equation, for which the previous results in this chapter are satisfied under non-autonomous perturbation.
Application of Theorem 10.9 *to reaction-diffusion equations.*

Let Ω be a bounded smooth domain in $\mathbb{R}^n$, $n\geqslant 1$, and let $f_0\colon\overline{\Omega}\times\mathbb{R}\to\mathbb{R}$ be a C^k-function, $k\geqslant 2$. We consider the problem

$$(10.3)\qquad \begin{cases} u_t=\Delta u+f(x,u), & t>0,\ x\in\Omega,\\ u=0, & t>0,\ x\in\partial\Omega,\\ u=u_0, & t=0,\ x\in\Omega,\end{cases}$$

and we assume that all the conditions of Chapter 5 - Section 3.2 hold for (5.5).

We now fix a function $f\in\mathcal{G}^k_{MS}$, $k\geqslant 2$, and consider non-autonomous perturbations of (10.3), that is, we consider the problem:

$$(10.4)\qquad \begin{cases} u_t=\Delta u+f(t,x,u), & t>s,\ x\in\Omega,\\ u=0, & t>s,\ x\in\partial\Omega,\\ u=u_s, & t=s,\ x\in\Omega,\end{cases}$$

where $u_s \in X$ and $f\colon \mathbb{R}\times\overline{\Omega}\times\mathbb{R}\to\mathbb{R}$ is a suitably small non-autonomous perturbation of f_0, satisfying for all $(t,x,y)\in\mathbb{R}\times\Omega\times\mathbb{R}$ the dissipative conditions:

$$f(t,x,y)y < C_0 + \mu y^2 \quad \text{and} \quad F(t,x,y) < C_0 + \frac{1}{2}\mu y^2, \tag{10.5}$$

where $F(t,x,y) = \int_0^y f(t,x,s)ds$.

It follows that Theorem 10.9 applies for f a suitably C^2 (in the third variable) small perturbation of f_0 uniformly in $x\in\Omega$ and in $t\in\mathbb{R}$ and there is a phase diagram isomorphism between the pullback attractors (10.3) and the global attractor of (10.4).

Application of Theorem 10.9 *to the damped wave equation.*

Let us recall the application in Chapter 5 - Section 3.3. Consider Ω a bounded smooth domain in $\mathbb{R}^n$, $n=1,2,3$, $f\colon\overline{\Omega}\times\mathbb{R}\to\mathbb{R}$ a C^k-function, $k\geqslant 2$, a constant $\gamma>0$ and consider the following damped wave equation:

$$\begin{cases} u_{tt}+\gamma u_t-\Delta u = f_0(x,u), & t>0,\ x\in\Omega,\\ u(x,t)=0, & t>0,\ x\in\partial\Omega,\\ (u(x,0),u_t(x,0))=(u_0,v_0), \end{cases} \tag{10.6}$$

where $(u_0,v_0)\in Y$, $Y=H_0^1(\Omega)\times L^2(\Omega)$ if $n=1$, or $Y=(H^2(\Omega)\cap H_0^1(\Omega))\times H_0^1(\Omega)$ if $n=2,3$.

In Chapter 5 - Section 3.3 we presented results ensuring that (10.6) generates a Morse-Smale semigroup (for a generic set of nonlinearities f). Assuming that the semigroup associated to (10.6) is Morse-Smale, it follows from Theorem 10.9 that, for all suitably small non-autonomous perturbations f of f_0 (as in the previous example), the evolution process associated to

$$\begin{cases} u_{tt}+\gamma u_t-\Delta u = f(t,x,u), & t>0,\ x\in\Omega,\\ u(x,t)=0, & t>0,\ x\in\partial\Omega,\\ (u(x,0),u_t(x,0))=(u_0,v_0), \end{cases} \tag{10.7}$$

is a Morse-Smale evolution process, and there is a phase diagram isomorphism between the pullback attractors (10.7) and the global attractor of (10.6).

3.2. A non-autonomous perturbation of a hyperbolic periodic orbit. This example exhibits, in a particular case, the object we obtain when we make a non-autonomous perturbation of a periodic orbit (a fixed point in polar coordinates). So far, in the non-autonomous setting, we are not able to make a small non-autonomous perturbation of a hyperbolic periodic orbit in the general case, but we can give a full disclosure of what happens in this particular example.

Consider the autonomous equation (3.8) of Chapter 3. Note that $P_0=\{(0,0)\}$ is a hyperbolic point and $P_1=\{r=1\}$ and $P_2=\{r=2\}$ are normally hyperbolic periodic orbits. Also note that the intersections of $W^u(P_i)$ and $W^s_{loc}(P_j)$ are always transverse, for $0\leqslant i,j\leqslant 2$. Consider, for $\lambda\in\Lambda=[0,1]$, the perturbed problem:

$$\begin{cases} \dot r = -r(r-1)(r-2)+f_\lambda(t), & t\in\mathbb{R},\\ \dot\theta = 1 \end{cases} \tag{10.8}$$

where $f_{\lambda_0}\equiv 0$, $f_\lambda(t)\geqslant 0$, for all $\lambda\in\Lambda$ and $t\in\mathbb{R}$, and $\sup_{t\in\mathbb{R}} f_\lambda(t)\to 0$ as $\lambda\to\lambda_0$. Let $\mathcal{S}_\lambda=\{S_\lambda(t,s)\colon (t,s)\in\mathcal{P}\}$ be the evolution process generated by (10.8).

Since P_2 is a stable periodic orbit, each region $\{(x,y)\in\mathbb{R}^2\colon \alpha_1\leqslant x^2+y^2\leqslant\alpha_2\}$ where $1<\alpha_1<4<\alpha_2<\infty$ is positively $\mathcal{S}_\lambda$-invariant, provided that λ is

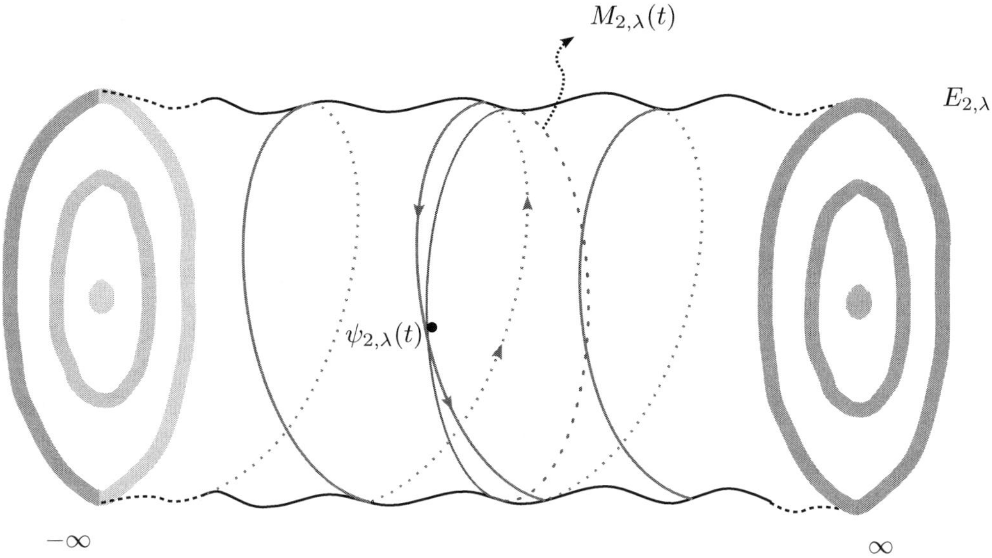

FIGURE 10.1. Solutions in the pullback attractor.

sufficiently close to 0 (and hence $\sup_{t\in\mathbb{R}} f_\lambda(t)$ is small). Therefore, if we consider the restriction of $\mathcal{S}_\lambda$ to a small annulus $\{(x,y)\in\mathbb{R}^2\colon \alpha_1 \leqslant x^2+y^2 \leqslant \alpha_2\}$, where $1<\alpha_1<4<\alpha_2<\infty$, it possesses a pullback attractor $\hat{M}_{2,\lambda}=\{M_{2,\lambda}(t)\}_{t\in\mathbb{R}}$.

An analogous reasoning can be used to prove the existence of pullback attractors $\hat{M}_{1,\lambda}=\{M_{1,\lambda}(t)\}_{t\in\mathbb{R}}$ and $\hat{M}_{0,\lambda}=\{M_{0,\lambda}(t)\}_{t\in\mathbb{R}}$ of the evolution process $\mathcal{S}_\lambda$ restricted to a small annulus around the circle of radius 1 and to a small disc around 0, respectively. In the case of the circle of radius 1, we can see that it is a repeller for the limiting problem, and the argument can be done by simply reversing time.

By the continuity with respect to the parameter, since the solution to the limiting problem (i.e. $\lambda=\lambda_0$) with $r=1,2$ is a circle, then we can see that the projection in the phase space of the solutions in $\{M_{i,\lambda}(t)\}_{t\in\mathbb{R}}$ must complete a whole lap, provided that λ is sufficiently close to λ_0. Thus the pullback and the uniform dynamics around P_2 (the dynamics around 0 and P_1 are analogous) can be described by Figure 10.1.

Since the limiting equation generates a generalized dynamically gradient semigroup with $\{0,P_1,P_2\}$ the family of disjoint isolated invariant sets, Theorem 8.14 implies that each evolution process $\mathcal{S}_\lambda$ is dynamically gradient with respect to the family $\{\hat{M}_{0,\lambda},\hat{M}_{1,\lambda},\hat{M}_{2,\lambda}\}$.

Now, using the results of [**75**] on perturbation of hyperbolic points, we can assure that there exist hyperbolic global solutions $\xi_{j,\lambda}\colon\mathbb{R}\to\mathbb{R}$ for $j=0,1,2$ of the first equation of (10.8), provided that λ is sufficiently close to λ_0, such that:

$$\sup_{t\in\mathbb{R}} |\xi_{\lambda,0}(t)| \to 0, \quad \sup_{t\in\mathbb{R}} |\xi_{\lambda,1}(t) - 1| \to 0 \text{ and } \sup_{t\in\mathbb{R}} |\xi_{\lambda,2}(t) - 2| \to 0, \tag{10.9}$$

as $\lambda \to \lambda_0$. Therefore, (10.8) has infinite normally hyperbolic global solutions, given by

$$\begin{aligned} P_{i,\lambda,k} \colon \mathbb{R} &\to \mathbb{R}^2 \\ t &\mapsto (\xi_{\lambda,i}(t)\cos(\theta+k), \xi_{\lambda,i}(t)\sin(\theta+k)), \end{aligned}$$

for each $i = 0, 1$ and 2 and each $k \in \mathbb{R}$.

Notes. The concept of Morse-Smale evolution process appears in [**31, 111**]. The structural stability of Morse-Smale maps under small non-autonomous perturbations has been proved by Franks and Kryzhevich and Pliss in [**121, 158**]. It is shown in [**111**] that the concept is robust under perturbation. In [**31**] it is also shown that there exists a phase diagram isomorphism between the limit Morse-Smale semigroup and the evolution process obtained under non-autonomous perturbation. The definition of lifted invariance allows us to define a Morse-Smale characterization for the uniform attractors, and its robustness under perturbation. The theory of local manifolds around a hyperbolic global solution of a given discrete evolution process can be found in [**78**] and in [**142**] for the continuous case whereas the continuous dependence with respect to C^1 norms in the non-autonomous case is done in [**31**].

Bibliography

[1] S. B. Angenent, *The Morse-Smale property for a semilinear parabolic equation*, J. Differential Equations **62** (1986), no. 3, 427–442, DOI 10.1016/0022-0396(86)90093-8. MR837763

[2] S. B. Angenent, *The zero set of a solution of a parabolic equation*, J. Reine Angew. Math. **390** (1988), 79–96, DOI 10.1515/crll.1988.390.79. MR953678

[3] Cung The Anh and Nguyen Duong Toan, *Existence and upper semicontinuity of uniform attractors in $H^1(\mathbb{R}^N)$ for nonautonomous nonclassical diffusion equations*, Ann. Polon. Math. **111** (2014), no. 3, 271–295, DOI 10.4064/ap111-3-5. MR3229425

[4] E. R. Aragão-Costa, T. Caraballo, A. N. Carvalho, and J. A. Langa, *Continuity of Lyapunov functions and of energy level for a generalized gradient semigroup*, Topol. Methods Nonlinear Anal. **39** (2012), no. 1, 57–82. MR2934334

[5] E. R. Aragão-Costa, T. Caraballo, A. N. Carvalho, and J. A. Langa, *Stability of gradient semigroups under perturbations*, Nonlinearity **24** (2011), no. 7, 2099–2117, DOI 10.1088/0951-7715/24/7/010. MR2805595

[6] ———, *Non-autonomous morse decomposition and lyapunov functions for gradient-like processes*, Transactions of the American Mathematical Society **10** (2013), 5277–5312. MR3074374

[7] José M. Arrieta and A. N. Carvalho, *Abstract parabolic problems with critical nonlinearities and applications to Navier-Stokes and heat equations*, Trans. Amer. Math. Soc. **352** (2000), no. 1, 285–310, DOI 10.1090/S0002-9947-99-02528-3. MR1694278

[8] ———, *Spectral convergence and nonlinear dynamics of reaction-diffusion equations under perturbations of the domain*, J. Differential Equations **199** (2004), no. 1, 143–178. MR2041515

[9] José M. Arrieta, A. N. Carvalho, and Jack K. Hale, *A damped hyperbolic equation with critical exponent*, Comm. Partial Differential Equations **17** (1992), no. 5-6, 841–866, DOI 10.1080/03605309208820866. MR1177295

[10] José M. Arrieta, A. N. Carvalho, J. A. Langa, and A. Rodriguez-Bernal, *Continuity of dynamical structures for nonautonomous evolution equations under singular perturbations*, J. Dynam. Differential Equations **24** (2012), no. 3, 427–481, DOI 10.1007/s10884-012-9269-y. MR2964788

[11] José M. Arrieta, A. N. Carvalho, and German Lozada-Cruz, *Dynamics in dumbbell domains. I. Continuity of the set of equilibria*, J. Differential Equations **231** (2006), no. 2, 551–597, DOI 10.1016/j.jde.2006.06.002. MR2287897

[12] José M. Arrieta, A. N. Carvalho, and German Lozada-Cruz, *Dynamics in dumbbell domains. II. The limiting problem*, J. Differential Equations **247** (2009), no. 1, 174–202, DOI 10.1016/j.jde.2009.03.014. MR2510133

[13] José M. Arrieta, A. N. Carvalho, and German Lozada-Cruz, *Dynamics in dumbbell domains. III. Continuity of attractors*, J. Differential Equations **247** (2009), no. 1, 225–259, DOI 10.1016/j.jde.2008.12.014. MR2510135

[14] José M. Arrieta, A. N. Carvalho, Marcone C. Pereira, and Ricardo P. Silva, *Semilinear parabolic problems in thin domains with a highly oscillatory boundary*, Nonlinear Anal. **74** (2011), no. 15, 5111–5132, DOI 10.1016/j.na.2011.05.006. MR2810693

[15] José M. Arrieta, A. N. Carvalho, and A. Rodriguez-Bernal, *Critical nonlinearities at the boundary* (English, with English and French summaries), C. R. Acad. Sci. Paris Sér. I Math. **327** (1998), no. 4, 353–358, DOI 10.1016/S0764-4442(99)80047-0. MR1649955

[16] José M. Arrieta, A. N. Carvalho, and A. Rodríguez-Bernal, *Parabolic problems with nonlinear boundary conditions and critical nonlinearities*, J. Differential Equations **156** (1999), no. 2, 376–406, DOI 10.1006/jdeq.1998.3612. MR1705387

[17] José M. Arrieta, A. N. Carvalho, and A. Rodríguez-Bernal, *Attractors of parabolic problems with nonlinear boundary conditions. Uniform bounds*, Comm. Partial Differential Equations **25** (2000), no. 1-2, 1–37, DOI 10.1080/03605300008821506. MR1737541

[18] ———, *Upper semicontinuity for attractors of parabolic problems with localized large diffusion and nonlinear boundary conditions*, J. Differential Equations **168** (2000), no. 1, 33–59, Special issue in celebration of Jack K. Hale's 70th birthday, Part 1 (Atlanta, GA/Lisbon, 1998). MR1801342

[19] José M. Arrieta and Esperanza Santamaría, *$C^{1,\theta}$-estimates on the distance of inertial manifolds*, Collect. Math. **69** (2018), no. 3, 315–336, DOI 10.1007/s13348-018-0227-9. MR3842209

[20] A. V. Babin and M. I. Vishik, *Existence and estimates of the dimensions of attractors of quasilinear parabolic equations and of the Navier-Stokes system* (Russian), Uspekhi Mat. Nauk **37** (1982), no. 3(225), 173–174. MR659432

[21] A. V. Babin and M. I. Vishik, *Regular attractors of semigroups and evolution equations*, J. Math. Pures Appl. (9) **62** (1983), no. 4, 441–491 (1984). MR735932

[22] ———, *Attractor in evolutionary equations*, vol. 25, Studies in Mathemathics and its Applications, 1992.

[23] J. M. Ball, *On the asymptotic behavior of generalized processes, with applications to nonlinear evolution equations.ations*, J. Differential Equations **27** (1978), no. 2, 224–265, DOI 10.1016/0022-0396(78)90032-3. MR461576

[24] J. M. Ball, *Continuity properties and global attractors of generalized semiflows and the Navier-Stokes equations*, J. Nonlinear Sci. **7** (1997), no. 5, 475–502, DOI 10.1007/s003329900037. MR1462276

[25] Peter W. Bates, Kening Lu, and Chongchun Zeng, *Existence and persistence of invariant manifolds for semiflows in Banach space*, Mem. Amer. Math. Soc. **135** (1998), no. 645, viii+129, DOI 10.1090/memo/0645. MR1445489

[26] F. D. M. Bezerra, A. N. Carvalho, J. W. Cholewa, and M. J. D. Nascimento, *Parabolic approximation of damped wave equations via fractional powers: fast growing nonlinearities and continuity of the dynamics*, J. Math. Anal. Appl. **450** (2017), no. 1, 377–405, DOI 10.1016/j.jmaa.2017.01.024. MR3606173

[27] S. B. Bodnaruk and V. L. Kulik, *On the dependence of bounded invariant manifolds of autonomous systems of differential equations on parameters* (Ukrainian, with English and Ukrainian summaries), Ukraïn. Mat. Zh. **48** (1996), no. 6, 747–752, DOI 10.1007/BF02384170; English transl., Ukrainian Math. J. **48** (1996), no. 6, 838–845 (1997). MR1418152

[28] M. C. Bortolan, T. Caraballo, A. N. Carvalho, and J. A. Langa, *An estimate on the fractal dimension of attractors of gradient-like dynamical systems*, Nonlinear Anal. **75** (2012), no. 14, 5702–5722, DOI 10.1016/j.na.2012.05.018. MR2942948

[29] M. C. Bortolan, T. Caraballo, A. N. Carvalho, and J. A. Langa, *Skew product semiflows and Morse decomposition*, J. Differential Equations **255** (2013), no. 8, 2436–2462, DOI 10.1016/j.jde.2013.06.023. MR3082469

[30] M. C. Bortolan, A. N. Carvalho, and J. A. Langa, *Structure of attractors for skew product semiflows*, J. Differential Equations **257** (2014), no. 2, 490–522, DOI 10.1016/j.jde.2014.04.008. MR3200379

[31] M.C. Bortolan, A.N. Carvalho, J.A. Langa, and Geneviève Raugel, *Non-autonomous perturbations of morse-smale semigroups: stability of the phase diagram*, Preprint.

[32] Mahdi Boukrouche, G. Łukaszewicz, and J. Real, *On pullback attractors for a class of two-dimensional turbulent shear flows*, Internat. J. Engrg. Sci. **44** (2006), no. 13-14, 830–844, DOI 10.1016/j.ijengsci.2006.05.012. MR2255762

[33] R. C. D. S. Broche, A.N. Carvalho, and J. Valero, *A non-autonomous scalar one-dimensional dissipative parabolic problem: The description of the dynamics*, Preprint.

[34] P. Brunovský and P. Poláčik, *The Morse-Smale structure of a generic reaction-diffusion equation in higher space dimension*, J. Differential Equations **135** (1997), no. 1, 129–181, DOI 10.1006/jdeq.1996.3234. MR1434918

[35] P. Brunovský and P. Poláčik, *The Morse-Smale structure of a generic reaction-diffusion equation in higher space dimension*, J. Differential Equations **135** (1997), no. 1, 129–181, DOI 10.1006/jdeq.1996.3234. MR1434918

[36] P. Brunovsky and Geneviève Raugel, *Genericity of the Morse-Smale property for damped wave equations*, J. Dynam. Differential Equations **15** (2003), no. 2-3, 571–658, DOI

10.1023/B:JODY.0000009749.10737.9d. Special issue dedicated to Victor A. Pliss on the occasion of his 70th birthday. MR2046732

[37] S. M. Bruschi and A. N. Carvalho, *Upper semicontinuity of attractors for the discretization of strongly damped wave equations*, Mat. Contemp. **32** (2007), 39–62. MR2428426

[38] S. M. Bruschi, A. N. Carvalho, J. W. Cholewa, and Tomasz Dlotko, *Uniform exponential dichotomy and continuity of attractors for singularly perturbed damped wave equations*, J. Dynam. Differential Equations **18** (2006), no. 3, 767–814, DOI 10.1007/s10884-006-9023-4. MR2264043

[39] T. Caraballo, A. N. Carvalho, Henrique B. da Costa, and J. A. Langa, *Equi-attraction and continuity of attractors for skew-product semiflows*, Discrete Contin. Dyn. Syst. Ser. B **21** (2016), no. 9, 2949–2967, DOI 10.3934/dcdsb.2016081. MR3567795

[40] T. Caraballo, J. A. Langa, Felipe Rivero, and A. N. Carvalho, *A gradient-like nonautonomous evolution process*, Internat. J. Bifur. Chaos Appl. Sci. Engrg. **20** (2010), no. 9, 2751–2760, DOI 10.1142/S0218127410027337. MR2738731

[41] T. Caraballo, A. N. Carvalho, J. A. Langa, and Felipe Rivero, *Existence of pullback attractors for pullback asymptotically compact processes*, Nonlinear Anal. **72** (2010), no. 3-4, 1967–1976, DOI 10.1016/j.na.2009.09.037. MR2577594

[42] T. Caraballo, A. N. Carvalho, J. A. Langa, and Felipe Rivero, *A non-autonomous strongly damped wave equation: existence and continuity of the pullback attractor*, Nonlinear Anal. **74** (2011), no. 6, 2272–2283, DOI 10.1016/j.na.2010.11.032. MR2781757

[43] T. Caraballo and Xiaoying Han, *Applied nonautonomous and random dynamical systems*, SpringerBriefs in Mathematics, Springer, Cham, 2016. Applied dynamical systems. MR3586630

[44] T. Caraballo, Juan C. Jara, J. A. Langa, and Zhenxin Liu, *Morse decomposition of attractors for non-autonomous dynamical systems*, Adv. Nonlinear Stud. **13** (2013), no. 2, 309–329, DOI 10.1515/ans-2013-0204. MR3076793

[45] T. Caraballo, Peter E. Kloeden, and P. Marín-Rubio, *Weak pullback attractors of setvalued processes*, J. Math. Anal. Appl. **288** (2003), no. 2, 692–707, DOI 10.1016/j.jmaa.2003.09.039. MR2020190

[46] T. Caraballo, Peter E. Kloeden, and J. Real, *Pullback and forward attractors for a damped wave equation with delays*, Stoch. Dyn. **4** (2004), no. 3, 405–423, DOI 10.1142/S0219493704001139. MR2086936

[47] T. Caraballo and J. A. Langa, *On the upper semicontinuity of cocycle attractors for non-autonomous and random dynamical systems*, Dyn. Contin. Discrete Impuls. Syst. Ser. A Math. Anal. **10** (2003), no. 4, 491–513. MR1978585

[48] T. Caraballo, J. A. Langa, and Zhenxin Liu, *Gradient infinite-dimensional random dynamical systems*, SIAM J. Appl. Dyn. Syst. **11** (2012), no. 4, 1817–1847, DOI 10.1137/120862752. MR3032851

[49] T. Caraballo, J. A. Langa, V. S. Melnik, and J. Valero, *Pullback attractors of nonautonomous and stochastic multivalued dynamical systems*, Set-Valued Anal. **11** (2003), no. 2, 153–201, DOI 10.1023/A:1022902802385. MR1966698

[50] T. Caraballo, J. A. Langa, and R. Obaya, *Pullback, forward and chaotic dynamics in 1D non-autonomous linear-dissipative equations*, Nonlinearity **30** (2017), no. 1, 274–299, DOI 10.1088/1361-6544/30/1/274. MR3604611

[51] T. Caraballo, G. Łukaszewicz, and J. Real, *Pullback attractors for asymptotically compact non-autonomous dynamical systems*, Nonlinear Anal. **64** (2006), no. 3, 484–498, DOI 10.1016/j.na.2005.03.111. MR2191992

[52] T. Caraballo and J. Real, *Attractors for 2D-Navier-Stokes models with delays*, J. Differential Equations **205** (2004), no. 2, 271–297, DOI 10.1016/j.jde.2004.04.012. MR2091818

[53] T. Caraballo and Stefanie Sonner, *Random pullback exponential attractors: general existence results for random dynamical systems in Banach spaces*, Discrete Contin. Dyn. Syst. **37** (2017), no. 12, 6383–6403, DOI 10.3934/dcds.2017277. MR3690309

[54] Vera Lúcia Carbone, A. N. Carvalho, and Karina Schiabel-Silva, *Continuity of attractors for parabolic problems with localized large diffusion*, Nonlinear Anal. **68** (2008), no. 3, 515–535, DOI 10.1016/j.na.2006.11.017. MR2372362

[55] Vera Lúcia Carbone, A. N. Carvalho, and Karina Schiabel-Silva, *Continuity of the dynamics in a localized large diffusion problem with nonlinear boundary conditions*, J. Math. Anal. Appl. **356** (2009), no. 1, 69–85, DOI 10.1016/j.jmaa.2009.02.037. MR2524216

[56] C. A. Cardoso, J. A. Langa, and R. Obaya, *Characterization of cocycle attractors for nonautonomous reaction-diffusion equations*, Internat. J. Bifur. Chaos Appl. Sci. Engrg. **26** (2016), no. 8, 1650135, 20, DOI 10.1142/S0218127416501352. MR3533671

[57] A. N. Carvalho, *Spatial homogeneity in damped hyperbolic equations*, Dynam. Systems Appl. **1** (1992), no. 3, 221–249. MR1182646

[58] A. N. Carvalho, *Infinite-dimensional dynamics described by ordinary differential equations*, J. Differential Equations **116** (1995), no. 2, 338–404, DOI 10.1006/jdeq.1995.1039. MR1318579

[59] A. N. de Carvalho, *Parabolic problems with nonlinear boundary conditions in cell tissues*, Resenhas **3** (1997), no. 1, 123–138. Workshop on Differential Equations and Nonlinear Analysis (Águas de Lindóia, 1996). MR1474306

[60] A. N. Carvalho and J. W. Cholewa, *Attractors for strongly damped wave equations with critical nonlinearities*, Pacific J. Math. **207** (2002), no. 2, 287–310, DOI 10.2140/pjm.2002.207.287. MR1972247

[61] A. N. Carvalho and J. W. Cholewa, *Local well posedness for strongly damped wave equations with critical nonlinearities*, Bull. Austral. Math. Soc. **66** (2002), no. 3, 443–463, DOI 10.1017/S0004972700040296. MR1939206

[62] A. N. Carvalho and J. W. Cholewa, *Continuation and asymptotics of solutions to semilinear parabolic equations with critical nonlinearities*, J. Math. Anal. Appl. **310** (2005), no. 2, 557–578, DOI 10.1016/j.jmaa.2005.02.024. MR2022944

[63] ———, *Strongly damped wave equations in* $W_0^{1,p}(\Omega) \times L^p(\Omega)$, Discrete Contin. Dyn. Syst. **Dynamical systems and differential equations. Proceedings of the 6th AIMS International Conference** (2007), 230–239. MR2409217

[64] A. N. Carvalho and J. W. Cholewa, *Regularity of solutions on the global attractor for a semilinear damped wave equation*, J. Math. Anal. Appl. **337** (2008), no. 2, 932–948, DOI 10.1016/j.jmaa.2007.04.051. MR2386343

[65] A. N. Carvalho and J. W. Cholewa, *Exponential global attractors for semigroups in metric spaces with applications to differential equations*, Ergodic Theory Dynam. Systems **31** (2011), no. 6, 1641–1667, DOI 10.1017/S0143385710000702. MR2851670

[66] A. N. Carvalho, J. W. Cholewa, and Tomasz Dlotko, *Strongly damped wave problems: bootstrapping and regularity of solutions*, J. Differential Equations **244** (2008), no. 9, 2310–2333, DOI 10.1016/j.jde.2008.02.011. MR2413843

[67] A. N. Carvalho, J. W. Cholewa, and Tomasz Dlotko, *Damped wave equations with fast growing dissipative nonlinearities*, Discrete Contin. Dyn. Syst. **24** (2009), no. 4, 1147–1165, DOI 10.3934/dcds.2009.24.1147. MR2505696

[68] A. N. Carvalho, J. W. Cholewa, and Tomasz Dłotko, *Equi-exponential attraction and rate of convergence of attractors with application to a perturbed damped wave equation*, Proc. Roy. Soc. Edinburgh Sect. A **144** (2014), no. 1, 13–51, DOI 10.1017/S0308210511001235. MR3164534

[69] A. N. Carvalho, J. W. Cholewa, G. Lozada-Cruz, and M. R. T. Primo, *Reduction of infinite dimensional systems to finite dimensions: compact convergence approach*, SIAM J. Math. Anal. **45** (2013), no. 2, 600–638, DOI 10.1137/10080734X. MR3035474

[70] A. N. Carvalho and J. A. Cuminato, *Reaction-diffusion problems in cell tissues*, J. Dynam. Differential Equations **9** (1997), no. 1, 93–131, DOI 10.1007/BF02219054. MR1451746

[71] A. N. Carvalho, H. M. Rodrigues, and Tomasz Dłotko, *Upper semicontinuity of attractors and synchronization*, J. Math. Anal. Appl. **220** (1998), no. 1, 13–41, DOI 10.1006/jmaa.1997.5774. MR1612059

[72] A. N. Carvalho and Jack K. Hale, *Large diffusion with dispersion*, Nonlinear Anal. **17** (1991), no. 12, 1139–1151, DOI 10.1016/0362-546X(91)90233-Q. MR1137899

[73] A. N. Carvalho and J. A. Langa, *Non-autonomous perturbation of autonomous semilinear differential equations: continuity of local stable and unstable manifolds*, J. Differential Equations **233** (2007), no. 2, 622–653, DOI 10.1016/j.jde.2006.08.009. MR2292521

[74] A. N. Carvalho and J. A. Langa, *An extension of the concept of gradient semigroups which is stable under perturbation*, J. Differential Equations **246** (2009), no. 7, 2646–2668, DOI 10.1016/j.jde.2009.01.007. MR2503016

[75] A. N. Carvalho, J. A. Langa, and J. C. Robinson, *Lower semicontinuity of attractors for non-autonomous dynamical systems*, Ergodic Theory Dynam. Systems **29** (2009), no. 6, 1765–1780, DOI 10.1017/S0143385708000850. MR2563091

[76] A. N. Carvalho, J. A. Langa, and J. C. Robinson, *On the continuity of pullback attractors for evolution processes*, Nonlinear Anal. **71** (2009), no. 5-6, 1812–1824, DOI 10.1016/j.na.2009.01.016. MR2524394

[77] A. N. Carvalho, J. A. Langa, and J. C. Robinson, *Structure and bifurcation of pullback attractors in a non-autonomous Chafee-Infante equation*, Proc. Amer. Math. Soc. **140** (2012), no. 7, 2357–2373, DOI 10.1090/S0002-9939-2011-11071-2. MR2898698

[78] A. N. Carvalho, J. A. Langa, and J. C. Robinson, *Attractors for infinite-dimensional non-autonomous dynamical systems*, Applied Mathematical Sciences, vol. 182, Springer, New York, 2013. MR2976449

[79] A. N. Carvalho, J. A. Langa, and J. C. Robinson, *Non-autonomous dynamical systems*, Discrete Contin. Dyn. Syst. Ser. B **20** (2015), no. 3, 703–747, DOI 10.3934/dcdsb.2015.20.703. MR3331676

[80] A. N. Carvalho, J. A. Langa, J. C. Robinson, and Antonio Suárez, *Characterization of non-autonomous attractors of a perturbed infinite-dimensional gradient system*, J. Differential Equations **236** (2007), no. 2, 570–603, DOI 10.1016/j.jde.2007.01.017. MR2322025

[81] A. N. Carvalho, J. A. Langa, J. C. Robinson, and Antonio Suárez, *Characterization of non-autonomous attractors of a perturbed infinite-dimensional gradient system*, J. Differential Equations **236** (2007), no. 2, 570–603, DOI 10.1016/j.jde.2007.01.017. MR2322025

[82] A. N. Carvalho and Luiz Augusto F. Oliveira, *Delay-partial differential equations with some large diffusion*, Nonlinear Anal. **22** (1994), no. 9, 1057–1095, DOI 10.1016/0362-546X(94)90228-3. MR1279132

[83] A. N. Carvalho and Antônio L. Pereira, *A scalar parabolic equation whose asymptotic behavior is dictated by a system of ordinary differential equations*, J. Differential Equations **112** (1994), no. 1, 81–130, DOI 10.1006/jdeq.1994.1096. MR1287553

[84] A. N. Carvalho and Sergey Piskarev, *A general approximation scheme for attractors of abstract parabolic problems*, Numer. Funct. Anal. Optim. **27** (2006), no. 7-8, 785–829, DOI 10.1080/01630560600882723. MR2262543

[85] Alexandre N. de Carvalho and Marcos Roberto Teixeira Primo, *Spatial homogeneity in parabolic problems with nonlinear boundary conditions*, Commun. Pure Appl. Anal. **3** (2004), no. 4, 637–651, DOI 10.3934/cpaa.2004.3.637. MR2106293

[86] A. N. Carvalho and Stefanie Sonner, *Pullback exponential attractors for evolution processes in Banach spaces: theoretical results*, Commun. Pure Appl. Anal. **12** (2013), no. 6, 3047–3071, DOI 10.3934/cpaa.2013.12.3047. MR3060923

[87] A. N. Carvalho and Stefanie Sonner, *Pullback exponential attractors for evolution processes in Banach spaces: properties and applications*, Commun. Pure Appl. Anal. **13** (2014), no. 3, 1141–1165, DOI 10.3934/cpaa.2014.13.1141. MR3177693

[88] N. Chafee and E. F. Infante, *A bifurcation problem for a nonlinear partial differential equation of parabolic type*, Applicable Anal. **4** (1974/75), 17–37, DOI 10.1080/00036817408839081. MR440205

[89] D. N. Cheban, *Global attractors of non-autonomous dynamical and control systems*, 2nd ed., Interdisciplinary Mathematical Sciences, vol. 18, World Scientific Publishing Co. Pte. Ltd., Hackensack, NJ, 2015. MR3329680

[90] D. N. Cheban, Peter E. Kloeden, and B. Schmalfuß, *The relationship between pullback, forward and global attractors of nonautonomous dynamical systems*, Nonlinear Dyn. Syst. Theory **2** (2002), no. 2, 125–144. MR1989935

[91] Shu Ping Chen and Roberto Triggiani, *Proof of two conjectures by G. Chen and D. L. Russell on structural damping for elastic systems*, Approximation and optimization (Havana, 1987), Lecture Notes in Math., vol. 1354, Springer, Berlin, 1988, pp. 234–256, DOI 10.1007/BFb0089601. MR996678

[92] Shu Ping Chen and Roberto Triggiani, *Proof of extensions of two conjectures on structural damping for elastic systems*, Pacific J. Math. **136** (1989), no. 1, 15–55. MR971932

[93] Shu Ping Chen and Roberto Triggiani, *Characterization of domains of fractional powers of certain operators arising in elastic systems, and applications*, J. Differential Equations **88** (1990), no. 2, 279–293, DOI 10.1016/0022-0396(90)90100-4. MR1081250

[94] Shu Ping Chen and Roberto Triggiani, *Gevrey class semigroups arising from elastic systems with gentle dissipation: the case* $0 < \alpha < \frac{1}{2}$, Proc. Amer. Math. Soc. **110** (1990), no. 2, 401–415, DOI 10.2307/2048084. MR1021208

[95] Xu-Yan Chen and Hiroshi Matano, *Convergence, asymptotic periodicity, and finite-point blow-up in one-dimensional semilinear heat equations*, J. Differential Equations **78** (1989), no. 1, 160–190, DOI 10.1016/0022-0396(89)90081-8. MR986159

[96] V. V. Chepyzhov and M. I. Vishik, *Attractors of nonautonomous dynamical systems and their dimension*, J. Math. Pures Appl. (9) **73** (1994), no. 3, 279–333. MR1273705

[97] Vladimir V. Chepyzhov, Monica Conti, and Vittorino Pata, *Totally dissipative dynamical processes and their uniform global attractors*, Commun. Pure Appl. Anal. **13** (2014), no. 5, 1989–2004, DOI 10.3934/cpaa.2014.13.1989. MR3230421

[98] Vladimir V. Chepyzhov and Mark I. Vishik, *Attractors for equations of mathematical physics*, American Mathematical Society Colloquium Publications, vol. 49, American Mathematical Society, Providence, RI, 2002. MR1868930

[99] Alexey Cheskidov and Landon Kavlie, *Degenerate pullback attractors for the 3D Navier-Stokes equations*, J. Math. Fluid Mech. **17** (2015), no. 3, 411–421, DOI 10.1007/s00021-015-0214-9. MR3383921

[100] Alexey Cheskidov and Songsong Lu, *Uniform global attractors for the nonautonomous 3D Navier-Stokes equations*, Adv. Math. **267** (2014), 277–306, DOI 10.1016/j.aim.2014.09.005. MR3269180

[101] J. W. Cholewa and Tomasz Dlotko, *Global attractors in abstract parabolic problems*, London Mathematical Society Lecture Note Series, vol. 278, Cambridge University Press, Cambridge, 2000. MR1778284

[102] J. W. Cholewa, Radoslaw Czaja, and Gianluca Mola, *Remarks on the fractal dimension of bi-space global and exponential attractors* (English, with English and Italian summaries), Boll. Unione Mat. Ital. (9) **1** (2008), no. 1, 121–145. MR2388001

[103] I.D. Chueshov, *Vvedenie v teoriyu beskonechnomernykh dissipativnykh sistem*, Universitetskie Lektsii po Sovremennoui Matematike. [University Lectures in Contemporary Mathematics], AKTA, Kharkiv, 1999. MR1788405

[104] Igor Chueshov and Irena Lasiecka, *Long-time behavior of second order evolution equations with nonlinear damping*, Mem. Amer. Math. Soc. **195** (2008), no. 912, viii+183, DOI 10.1090/memo/0912. MR2438025

[105] Charles Conley, *Isolated invariant sets and the Morse index*, CBMS Regional Conference Series in Mathematics, vol. 38, American Mathematical Society, Providence, R.I., 1978. MR511133

[106] P. Constantin and C. Foias, *Global Lyapunov exponents, Kaplan-Yorke formulas and the dimension of the attractors for 2D Navier-Stokes equations*, Comm. Pure Appl. Math. **38** (1985), no. 1, 1–27, DOI 10.1002/cpa.3160380102. MR768102

[107] Michele Coti Zelati and Piotr Kalita, *Minimality properties of set-valued processes and their pullback attractors*, SIAM J. Math. Anal. **47** (2015), no. 2, 1530–1561, DOI 10.1137/140978995. MR3337999

[108] Hans Crauel and Franco Flandoli, *Attractors for random dynamical systems*, Probab. Theory Related Fields **100** (1994), no. 3, 365–393, DOI 10.1007/BF01193705. MR1305587

[109] Hongyong Cui and J. A. Langa, *Uniform attractors for non-autonomous random dynamical systems*, J. Differential Equations **263** (2017), no. 2, 1225–1268, DOI 10.1016/j.jde.2017.03.018. MR3632219

[110] Radoslaw Czaja, *Pullback exponential attractors with admissible exponential growth in the past*, Nonlinear Anal. **104** (2014), 90–108, DOI 10.1016/j.na.2014.03.020. MR3196891

[111] Radosław Czaja, Waldyr M. Oliva, and Carlos Rocha, *On a definition of Morse-Smale evolution processes*, Discrete Contin. Dyn. Syst. **37** (2017), no. 7, 3601–3623, DOI 10.3934/dcds.2017155. MR3639435

[112] Henrique B. da Costa and José Valero, *Morse decompositions and Lyapunov functions for dynamically gradient multivalued semiflows*, Nonlinear Dynam. **84** (2016), no. 1, 19–34, DOI 10.1007/s11071-015-2193-z. MR3472729

[113] E. Brian Davies, *Linear operators and their spectra*, Cambridge Studies in Advanced Mathematics, vol. 106, Cambridge University Press, Cambridge, 2007. MR2359869

[114] Emerson A. M. de Abreu and A. N. Carvalho, *Attractors for semilinear parabolic problems with Dirichlet boundary conditions in varying domains*, Mat. Contemp. **27** (2004), 37–51. MR2130211

[115] A. Eden, C. Foias, B. Nicolaenko, and R. Temam, *Exponential attractors for dissipative evolution equations*, RAM: Research in Applied Mathematics, vol. 37, Masson, Paris; John Wiley & Sons, Ltd., Chichester, 1994. MR1335230

[116] M. Efendiev, S. Zelik, and A. Miranville, *Exponential attractors and finite-dimensional reduction for non-autonomous dynamical systems*, Proc. Roy. Soc. Edinburgh Sect. A **135** (2005), no. 4, 703–730, DOI 10.1017/S030821050000408X. MR2173336

[117] Pierre Fabrie, Cedric Galusinski, Alain Miranville, and S. Zelik, *Uniform exponential attractors for a singularly perturbed damped wave equation*, Discrete Contin. Dyn. Syst. **10** (2004), no. 1-2, 211–238, DOI 10.3934/dcds.2004.10.211. Partial differential equations and applications. MR2026192

[118] Bernold Fiedler and Carlos Rocha, *Heteroclinic orbits of semilinear parabolic equations*, J. Differential Equations **125** (1996), no. 1, 239–281, DOI 10.1006/jdeq.1996.0031. MR1376067

[119] Ciprian Foias, George R. Sell, and Roger Temam, *Inertial manifolds for nonlinear evolutionary equations*, J. Differential Equations **73** (1988), no. 2, 309–353, DOI 10.1016/0022-0396(88)90110-6. MR943945

[120] Ciprian Foias and Roger Temam, *On the Hausdorff dimension of an attractor for the two-dimensional Navier-Stokes equations*, Phys. Lett. A **93** (1983), no. 9, 451–454, DOI 10.1016/0375-9601(83)90628-X. MR697674

[121] John M. Franks, *Time dependent stable diffeomorphisms*, Invent. Math. **24** (1974), 163–172, DOI 10.1007/BF01404304. MR345136

[122] Mirelson M. Freitas, Piotr Kalita, and J. A. Langa, *Continuity of non-autonomous attractors for hyperbolic perturbation of parabolic equations*, J. Differential Equations **264** (2018), no. 3, 1886–1945, DOI 10.1016/j.jde.2017.10.007. MR3721416

[123] G. Fusco, *On the explicit construction of an ODE which has the same dynamics as a scalar parabolic PDE*, J. Differential Equations **69** (1987), no. 1, 85–110, DOI 10.1016/0022-0396(87)90104-5. MR897442

[124] Giorgio Fusco and Waldyr Muniz Oliva, *Jacobi matrices and transversality*, Proc. Roy. Soc. Edinburgh Sect. A **109** (1988), no. 3-4, 231–243, DOI 10.1017/S0308210500027748. MR963029

[125] G. Fusco and W. M. Oliva, *Transversality between invariant manifolds of periodic orbits for a class of monotone dynamical systems*, J. Dynam. Differential Equations **2** (1990), no. 1, 1–17, DOI 10.1007/BF01047768. MR1041196

[126] F. R. Gantmacher, *The theory of matrices. Vols. 1, 2*, Translated by K. A. Hirsch, Chelsea Publishing Co., New York, 1959. MR0107649

[127] J.-M. Ghidaglia and R. Temam, *Attractors for damped nonlinear hyperbolic equations*, J. Math. Pures Appl. (9) **66** (1987), no. 3, 273–319. MR913856

[128] Maurizio Grasselli and Vittorino Pata, *Uniform attractors of nonautonomous dynamical systems with memory*, Evolution equations, semigroups and functional analysis (Milano, 2000), Progr. Nonlinear Differential Equations Appl., vol. 50, Birkhäuser, Basel, 2002, pp. 155–178. MR1944162

[129] G. Guerrero, J. A. Langa, and A. Suárez, *Architecture of attractor determines dynamics on mutualistic complex networks*, Nonlinear Anal. Real World Appl. **34** (2017), 17–40, DOI 10.1016/j.nonrwa.2016.07.009. MR3567947

[130] Jack K. Hale, *Ordinary differential equations*, 2nd ed., Robert E. Krieger Publishing Co., Inc., Huntington, N.Y., 1980. MR587488

[131] Jack K. Hale, *Asymptotic behaviour and dynamics in infinite dimensions*, Nonlinear differential equations (Granada, 1984), Res. Notes in Math., vol. 132, Pitman, Boston, MA, 1985, pp. 1–42. MR908897

[132] Jack K. Hale, *Asymptotic behavior of dissipative systems*, Mathematical Surveys and Monographs, vol. 25, American Mathematical Society, Providence, RI, 1988. MR941371

[133] Jack K. Hale, Luis T. Magalhães, and Waldyr M. Oliva, *Dynamics in infinite dimensions*, 2nd ed., Applied Mathematical Sciences, vol. 47, Springer-Verlag, New York, 2002. With an appendix by Krzysztof P. Rybakowski. MR1914080

[134] Jack K. Hale and Geneviève Raugel, *Upper semicontinuity of the attractor for a singularly perturbed hyperbolic equation*, J. Differential Equations **73** (1988), no. 2, 197–214, DOI 10.1016/0022-0396(88)90104-0. MR943939

[135] Jack K. Hale and Geneviève Raugel, *Lower semicontinuity of attractors of gradient systems and applications*, Ann. Mat. Pura Appl. (4) **154** (1989), 281–326, DOI 10.1007/BF01790353. MR1043076

[136] Jack K. Hale and Geneviève Raugel, *Lower semicontinuity of the attractor for a singularly perturbed hyperbolic equation*, J. Dynam. Differential Equations **2** (1990), no. 1, 19–67, DOI 10.1007/BF01047769. MR1041197

[137] Jack K. Hale and Geneviève Raugel, *A modified Poincaré method for the persistence of periodic orbits and applications*, J. Dynam. Differential Equations **22** (2010), no. 1, 3–68, DOI 10.1007/s10884-009-9155-4. MR2607154

[138] A. Haraux, *Two remarks on hyperbolic dissipative problems* (English, with French summary), Nonlinear partial differential equations and their applications. Collège de France seminar, Vol. VII (Paris, 1983), Res. Notes in Math., vol. 122, Pitman, Boston, MA, 1985, pp. 6, 161–179. MR879461

[139] A. Haraux, *Recent results on semi-linear hyperbolic problems in bounded domains*, Partial differential equations (Rio de Janeiro, 1986), Lecture Notes in Math., vol. 1324, Springer, Berlin, 1988, pp. 118–126, DOI 10.1007/BFb0100787. MR965530

[140] A. Haraux, *Systèmes dynamiques dissipatifs et applications* (French), Recherches en Mathématiques Appliquées [Research in Applied Mathematics], vol. 17, Masson, Paris, 1991. MR1084372

[141] D.B. Henry, *Invariant manifolds near a fixed point*, Handwritten notes, IME-USP.

[142] D. B. Henry, *Geometric theory of semilinear parabolic equations*, Lecture Notes in Mathematics, vol. 840, Springer-Verlag, Berlin-New York, 1981. MR610244

[143] D. B. Henry, *Some infinite-dimensional Morse-Smale systems defined by parabolic partial differential equations*, J. Differential Equations **59** (1985), no. 2, 165–205, DOI 10.1016/0022-0396(85)90153-6. MR804887

[144] D. B. Henry, *Perturbation of the boundary in boundary-value problems of partial differential equations*, London Mathematical Society Lecture Note Series, vol. 318, Cambridge University Press, Cambridge, 2005. With editorial assistance from Jack Hale and Antônio Luiz Pereira. MR2160744

[145] Luan T. Hoang, Eric J. Olson, and J. C. Robinson, *On the continuity of global attractors*, Proc. Amer. Math. Soc. **143** (2015), no. 10, 4389–4395, DOI 10.1090/proc/12598. MR3373937

[146] Mike Hurley, *Chain recurrence, semiflows, and gradients*, J. Dynam. Differential Equations **7** (1995), no. 3, 437–456, DOI 10.1007/BF02219371. MR1348735

[147] Romain Joly and Geneviève Raugel, *Generic Morse-Smale property for the parabolic equation on the circle*, Ann. Inst. H. Poincaré Anal. Non Linéaire **27** (2010), no. 6, 1397–1440, DOI 10.1016/j.anihpc.2010.09.001. MR2738326

[148] Varga Kalantarov, Anton Savostianov, and S. Zelik, *Attractors for damped quintic wave equations in bounded domains*, Ann. Henri Poincaré **17** (2016), no. 9, 2555–2584, DOI 10.1007/s00023-016-0480-y. MR3535872

[149] Peter E. Kloeden and Thomas Lorenz, *Pullback and forward attractors of nonautonomous difference equations*, Difference equations, discrete dynamical systems and applications, Springer Proc. Math. Stat., vol. 150, Springer, Cham, 2015, pp. 37–48, DOI 10.1007/978-3-319-24747-2_3. MR3477513

[150] Peter E. Kloeden, *Pullback attractors in nonautonomous difference equations*, J. Differ. Equations Appl. **6** (2000), no. 1, 33–52, DOI 10.1080/10236190008808212. MR1752154

[151] Peter E. Kloeden and Thomas Lorenz, *Pullback attractors of reaction-diffusion inclusions with space-dependent delay*, Discrete Contin. Dyn. Syst. Ser. B **22** (2017), no. 5, 1909–1964, DOI 10.3934/dcdsb.2017114. MR3627136

[152] Peter E. Kloeden and Sergey Piskarev, *Discrete convergence and the equivalence of equi-attraction and the continuous convergence of attractors*, Int. J. Dyn. Syst. Differ. Equ. **1** (2007), no. 1, 38–43, DOI 10.1504/IJDSDE.2007.013743. MR2492244

[153] Peter E. Kloeden and Martin Rasmussen, *Nonautonomous dynamical systems*, Mathematical Surveys and Monographs, vol. 176, American Mathematical Society, Providence, RI, 2011. MR2808288

[154] Peter E. Kloeden, José Real, and Chunyou Sun, *Pullback attractors for a semilinear heat equation on time-varying domains*, J. Differential Equations **246** (2009), no. 12, 4702–4730, DOI 10.1016/j.jde.2008.11.017. MR2523300

[155] Peter E. Kloeden and Meihua Yang, *Forward attraction in nonautonomous difference equations*, J. Difference Equ. Appl. **22** (2016), no. 4, 513–525, DOI 10.1080/10236198.2015.1107550. MR3486077

[156] H.-W. Knobloch, *Invariant manifolds for ordinary differential equations*, Differential equations and mathematical physics (Birmingham, AL, 1990), Math. Sci. Engrg., vol. 186, Academic Press, Boston, MA, 1992, pp. 121–149, DOI 10.1016/S0076-5392(08)63378-0. MR1126693

[157] Kazuo Kobayasi, *An L^p theory of invariant manifolds for parabolic partial differential equations on $\mathbb{R}^d$*, J. Differential Equations **179** (2002), no. 1, 195–212, DOI 10.1006/jdeq.2001.4026. MR1883742

[158] S. G. Kryzhevich and V. A. Pliss, *On the structural stability of nonautonomous systems* (Russian, with Russian summary), Differ. Uravn. **39** (2003), no. 10, 1325 1333, 1437, DOI 10.1023/B:DIEQ.0000017913.79915.b1; English transl., Differ. Equ. **39** (2003), no. 10, 1395–1403. MR1955030

[159] Olga Ladyzhenskaya, *Attractors for semigroups and evolution equations*, Lezioni Lincee. [Lincei Lectures], Cambridge University Press, Cambridge, 1991. MR1133627

[160] J. A. Langa, G. Łukaszewicz, and J. Real, *Finite fractal dimension of pullback attractors for non-autonomous 2D Navier-Stokes equations in some unbounded domains*, Nonlinear Anal. **66** (2007), no. 3, 735–749, DOI 10.1016/j.na.2005.12.017. MR2274880

[161] J. A. Langa, Alain Miranville, and José Real, *Pullback exponential attractors*, Discrete Contin. Dyn. Syst. **26** (2010), no. 4, 1329–1357, DOI 10.3934/dcds.2010.26.1329. MR2600749

[162] J. A. Langa, J. C. Robinson, A. Rodríguez-Bernal, A. Suárez, and A. Vidal-López, *Existence and nonexistence of unbounded forwards attractor for a class of non-autonomous reaction diffusion equations*, Discrete Contin. Dyn. Syst. **18** (2007), no. 2-3, 483–497, DOI 10.3934/dcds.2007.18.483. MR2291908

[163] Desheng Li and Peter E. Kloeden, *Equi-attraction and the continuous dependence of pullback attractors on parameters*, Stoch. Dyn. **4** (2004), no. 3, 373–384, DOI 10.1142/S0219493704001061. MR2085974

[164] Desheng Li and Peter E. Kloeden, *Equi-attraction and the continuous dependence of attractors on parameters*, Glasg. Math. J. **46** (2004), no. 1, 131–141, DOI 10.1017/S0017089503001605. MR2034840

[165] G. Łukaszewicz and Witold Sadowski, *Uniform attractor for 2D magneto-micropolar fluid flow in some unbounded domains*, Z. Angew. Math. Phys. **55** (2004), no. 2, 247–257, DOI 10.1007/s00033-003-1127-7. MR2047286

[166] John Mallet-Paret, *Negatively invariant sets of compact maps and an extension of a theorem of Cartwright*, J. Differential Equations **22** (1976), no. 2, 331–348, DOI 10.1016/0022-0396(76)90032-2. MR423399

[167] John Mallet-Paret and George R. Sell, *Inertial manifolds for reaction diffusion equations in higher space dimensions*, J. Amer. Math. Soc. **1** (1988), no. 4, 805–866, DOI 10.2307/1990993. MR943276

[168] Ricardo Mañé, *On the dimension of the compact invariant sets of certain nonlinear maps*, Dynamical systems and turbulence, Warwick 1980 (Coventry, 1979/1980), Lecture Notes in Math., vol. 898, Springer, Berlin-New York, 1981, pp. 230–242. MR654892

[169] P. Marín-Rubio and Jose Real, *Pullback attractors for 2D-Navier-Stokes equations with delays in continuous and sub-linear operators*, Discrete Contin. Dyn. Syst. **26** (2010), no. 3, 989–1006, DOI 10.3934/dcds.2010.26.989. MR2600726

[170] Hiroshi Matano, *Nonincrease of the lap-number of a solution for a one-dimensional semilinear parabolic equation*, J. Fac. Sci. Univ. Tokyo Sect. IA Math. **29** (1982), no. 2, 401–441. MR672070

[171] Valery S. Melnik and José Valero, *On attractors of multivalued semi-flows and differential inclusions*, Set-Valued Anal. **6** (1998), no. 1, 83–111, DOI 10.1023/A:1008608431399. MR1631081

[172] Konstantin Mischaikow, Hal Smith, and Horst R. Thieme, *Asymptotically autonomous semiflows: chain recurrence and Lyapunov functions*, Trans. Amer. Math. Soc. **347** (1995), no. 5, 1669–1685, DOI 10.2307/2154964. MR1290727

[173] M. W. Hirsch, C. C. Pugh, and M. Shub, *Invariant manifolds*, Lecture Notes in Mathematics, Vol. 583, Springer-Verlag, Berlin-New York, 1977. MR0501173

[174] Douglas E. Norton, *The fundamental theorem of dynamical systems*, Comment. Math. Univ. Carolin. **36** (1995), no. 3, 585–597. MR1364499

[175] W. M. Oliva, *Morse-Smale semiflows, openness and A-stability*, Differential equations and dynamical systems (Lisbon, 2000), Fields Inst. Commun., vol. 31, Amer. Math. Soc., Providence, RI, 2002, pp. 285–307. MR1904521

[176] W. M. Oliva, J. C. F. De Oliveira, and J. Solà-Morales, *An infinite-dimensional Morse-Smale map*, NoDEA Nonlinear Differential Equations Appl. **1** (1994), no. 4, 365–387, DOI 10.1007/BF01194986. MR1300148

[177] George Osipenko and Eugene Ershov, *Perturbation of invariant manifolds of ordinary differential equations*, Six lectures on dynamical systems (Augsburg, 1994), World Sci. Publ., River Edge, NJ, 1996, pp. 213–265. MR1441126

[178] J. Palis, *On Morse-Smale dynamical systems*, Topology **8** (1968), 385–404, DOI 10.1016/0040-9383(69)90024-X. MR246316

[179] J. Palis and S. Smale, *Structural stability theorems*, Global Analysis (Proc. Sympos. Pure Math., Vol. XIV, Berkeley, Calif., 1968), Amer. Math. Soc., Providence, R.I., 1970, pp. 223–231. MR0267603

[180] J. Palis Jr and W. de Melo, *Introdução aos sistemas dinâmicos*, Projeto Euclides [Euclid Project], vol. 6, Instituto de Matemática Pura e Aplicada, Rio de Janeiro, 1978. MR654863

[181] Mauro Patrão, *Morse decomposition of semiflows on topological spaces*, J. Dynam. Differential Equations **19** (2007), no. 1, 181–198, DOI 10.1007/s10884-006-9033-2. MR2279951

[182] Mauro Patrão and Luiz A. B. San Martin, *Semiflows on topological spaces: chain transitivity and semigroups*, J. Dynam. Differential Equations **19** (2007), no. 1, 155–180, DOI 10.1007/s10884-006-9032-3. MR2279950

[183] A. Pazy, *Semigroups of linear operators and applications to partial differential equations*, Applied Mathematical Sciences, vol. 44, Springer-Verlag, New York, 1983. MR710486

[184] P. Poláčik, *Parabolic equations: asymptotic behavior and dynamics on invariant manifolds*, Handbook of dynamical systems, Vol. 2, North-Holland, Amsterdam, 2002, pp. 835–883, DOI 10.1016/S1874-575X(02)80037-6. MR1901067

[185] Martin Rasmussen, *All-time Morse decompositions of linear nonautonomous dynamical systems*, Proc. Amer. Math. Soc. **136** (2008), no. 3, 1045–1055, DOI 10.1090/S0002-9939-07-09071-5. MR2361880

[186] Geneviève Raugel, *Singularly perturbed hyperbolic equations revisited*, International Conference on Differential Equations, Vol. 1, 2 (Berlin, 1999), World Sci. Publ., River Edge, NJ, 2000, pp. 647–652. MR1870210

[187] Geneviève. Raugel, *Global attractors in partial differential equations*, Handbook of dynamical systems, Vol. 2, North-Holland, Amsterdam, 2002, pp. 885–982, DOI 10.1016/S1874-575X(02)80038-8. MR1901068

[188] Felipe Rivero, *Forward and pullback attraction on pullback attractors*, Bol. Soc. Esp. Mat. Apl. SeMA **51** (2010), 155–161, DOI 10.1007/bf03322567. MR2675975

[189] J. C. Robinson, *Infinite-dimensional dynamical systems: An introduction to dissipative parabolic PDEs and the theory of global attractors*, Cambridge Texts in Applied Mathematics, Cambridge University Press, Cambridge, 2001. MR1881888

[190] A. Rodríguez-Bernal, *Localized spatial homogenization and large diffusion*, SIAM J. Math. Anal. **29** (1998), no. 6, 1361–1380, DOI 10.1137/S003614109731864X. MR1638046

[191] A. Rodríguez-Bernal and A. Vidal-López, *Existence, uniqueness and attractivity properties of positive complete trajectories for non-autonomous reaction-diffusion problems*, Discrete Contin. Dyn. Syst. **18** (2007), no. 2-3, 537–567, DOI 10.3934/dcds.2007.18.537. MR2291911

[192] Krzysztof P. Rybakowski, *The homotopy index and partial differential equations*, Universitext, Springer-Verlag, Berlin, 1987. MR910097

[193] George R. Sell, *Nonautonomous differential equations and topological dynamics. I. The basic theory*, Trans. Amer. Math. Soc. **127** (1967), 241–262, DOI 10.2307/1994645. MR212313

[194] George R. Sell and Yuncheng You, *Dynamics of evolutionary equations*, Applied Mathematical Sciences, vol. 143, Springer-Verlag, New York, 2002. MR1873467

[195] Stephen Smale, *Morse inequalities for a dynamical system*, Bull. Amer. Math. Soc. **66** (1960), 43–49, DOI 10.1090/S0002-9904-1960-10386-2. MR117745

[196] Haitao Song and Hongqing Wu, *Pullback attractors of nonautonomous reaction-diffusion equations*, J. Math. Anal. Appl. **325** (2007), no. 2, 1200–1215, DOI 10.1016/j.jmaa.2006.02.041. MR2270079

[197] Bruno Sportisse, *A review of current issues in air pollution modeling and simulation*, Comput. Geosci. **11** (2007), no. 2, 159–181, DOI 10.1007/s10596-006-9036-4. MR2327965

[198] Chunyou Sun, Daomin Cao, and Jinqiao Duan, *Uniform attractors for nonautonomous wave equations with nonlinear damping*, SIAM J. Appl. Dyn. Syst. **6** (2007), no. 2, 293–318, DOI 10.1137/060663805. MR2318656

[199] Chunyou Sun and Yanbo Yuan, *L^p-type pullback attractors for a semilinear heat equation on time-varying domains*, Proc. Roy. Soc. Edinburgh Sect. A **145** (2015), no. 5, 1029–1052, DOI 10.1017/S0308210515000177. MR3406460

[200] R. Temam, *Infinite-dimensional dynamical systems in mechanics and physics*, 2nd ed., Applied Mathematical Sciences, vol. 68, Springer-Verlag, New York, 1997. MR1441312

[201] M. I. Vishik, *Asymptotic behaviour of solutions of evolutionary equations*, Lezioni Lincee. [Lincei Lectures], Cambridge University Press, Cambridge, 1992. MR1215488

[202] M. I. Vishik, S. V. Zelik, and V. V. Chepyzhov, *Regular attractors and their nonautonomous perturbations* (Russian, with Russian summary), Mat. Sb. **204** (2013), no. 1, 3–46, DOI 10.1070/SM2013v204n01ABEH004290; English transl., Sb. Math. **204** (2013), no. 1-2, 1–42. MR3060075

[203] Bixiang Wang, *Sufficient and necessary criteria for existence of pullback attractors for non-compact random dynamical systems*, J. Differential Equations **253** (2012), no. 5, 1544–1583, DOI 10.1016/j.jde.2012.05.015. MR2927390

[204] Yonghai Wang, Pengrui Li, and Yuming Qin, *Upper semicontinuity of uniform attractors for nonclassical diffusion equations*, Bound. Value Probl., posted on 2017, Paper No. 84, 11, DOI 10.1186/s13661-017-0816-7. MR3659996

[205] Yejuan Wang and Shengfan Zhou, *Kernel sections and uniform attractors of multi-valued semiprocesses*, J. Differential Equations **232** (2007), no. 2, 573–622, DOI 10.1016/j.jde.2006.07.005. MR2286392

[206] S. Zelik, *Strong uniform attractors for non-autonomous dissipative PDEs with non translation-compact external forces*, Discrete Contin. Dyn. Syst. Ser. B **20** (2015), no. 3, 781–810, DOI 10.3934/dcdsb.2015.20.781. MR3331678

[207] Caidi Zhao and Shengfan Zhou, *Pullback attractors for a non-autonomous incompressible non-Newtonian fluid*, J. Differential Equations **238** (2007), no. 2, 394–425, DOI 10.1016/j.jde.2007.04.001. MR2341431

List of Figures

Index

Selected Published Titles in This Series

246 **Matheus C. Bortolan, Alexandre N. Carvalho, and José A. Langa,** Attractors Under Autonomous and Non-autonomous Perturbations, 2020
245 **Aiping Wang and Anton Zettl,** Ordinary Differential Operators, 2019
244 **Nabile Boussaïd and Andrew Comech,** Nonlinear Dirac Equation, 2019
243 **José M. Isidro,** Jordan Triple Systems in Complex and Functional Analysis, 2019
242 **Bhargav Bhatt, Ana Caraiani, Kiran S. Kedlaya, Peter Scholze, and Jared Weinstein,** Perfectoid Spaces, 2019
241 **Dana P. Williams,** A Tool Kit for Groupoid C^*-Algebras, 2019
240 **Antonio Fernández López,** Jordan Structures in Lie Algebras, 2019
239 **Nicola Arcozzi, Richard Rochberg, Eric T. Sawyer, and Brett D. Wick,** The Dirichlet Space and Related Function Spaces, 2019
238 **Michael Tsfasman, Serge Vlăduţ, and Dmitry Nogin,** Algebraic Geometry Codes: Advanced Chapters, 2019
237 **Dusa McDuff, Mohammad Tehrani, Kenji Fukaya, and Dominic Joyce,** Virtual Fundamental Cycles in Symplectic Topology, 2019
236 **Bernard Host and Bryna Kra,** Nilpotent Structures in Ergodic Theory, 2018
235 **Habib Ammari, Brian Fitzpatrick, Hyeonbae Kang, Matias Ruiz, Sanghyeon Yu, and Hai Zhang,** Mathematical and Computational Methods in Photonics and Phononics, 2018
234 **Vladimir I. Bogachev,** Weak Convergence of Measures, 2018
233 **N. V. Krylov,** Sobolev and Viscosity Solutions for Fully Nonlinear Elliptic and Parabolic Equations, 2018
232 **Dmitry Khavinson and Erik Lundberg,** Linear Holomorphic Partial Differential Equations and Classical Potential Theory, 2018
231 **Eberhard Kaniuth and Anthony To-Ming Lau,** Fourier and Fourier-Stieltjes Algebras on Locally Compact Groups, 2018
230 **Stephen D. Smith,** Applying the Classification of Finite Simple Groups, 2018
229 **Alexander Molev,** Sugawara Operators for Classical Lie Algebras, 2018
228 **Zhenbo Qin,** Hilbert Schemes of Points and Infinite Dimensional Lie Algebras, 2018
227 **Roberto Frigerio,** Bounded Cohomology of Discrete Groups, 2017
226 **Marcelo Aguiar and Swapneel Mahajan,** Topics in Hyperplane Arrangements, 2017
225 **Mario Bonk and Daniel Meyer,** Expanding Thurston Maps, 2017
224 **Ruy Exel,** Partial Dynamical Systems, Fell Bundles and Applications, 2017
223 **Guillaume Aubrun and Stanisław J. Szarek,** Alice and Bob Meet Banach, 2017
222 **Alexandru Buium,** Foundations of Arithmetic Differential Geometry, 2017
221 **Dennis Gaitsgory and Nick Rozenblyum,** A Study in Derived Algebraic Geometry, 2017
220 **A. Shen, V. A. Uspensky, and N. Vereshchagin,** Kolmogorov Complexity and Algorithmic Randomness, 2017
219 **Richard Evan Schwartz,** The Projective Heat Map, 2017
218 **Tushar Das, David Simmons, and Mariusz Urbański,** Geometry and Dynamics in Gromov Hyperbolic Metric Spaces, 2017
217 **Benoit Fresse,** Homotopy of Operads and Grothendieck–Teichmüller Groups, 2017
216 **Frederick W. Gehring, Gaven J. Martin, and Bruce P. Palka,** An Introduction to the Theory of Higher-Dimensional Quasiconformal Mappings, 2017
215 **Robert Bieri and Ralph Strebel,** On Groups of PL-homeomorphisms of the Real Line, 2016

For a complete list of titles in this series, visit the
AMS Bookstore at **www.ams.org/bookstore/survseries/**.